Deep Learning and Signal-Processing Methods for Multisensor Data Fusion

Arlene John · Barry Cardiff · Deepu John

Deep Learning and Signal-Processing Methods for Multisensor Data Fusion

Applications to Ambulatory Health Monitoring

Springer

Arlene John
Biomedical Signals and Systems, Faculty of Electrical Engineering, Mathematics, and Computer Science
University of Twente
Enschede, The Netherlands

Barry Cardiff
School of Electrical and Electronic Engineering
University College Dublin
Dublin, Ireland

Deepu John
School of Electrical and Electronic Engineering
University College Dublin
Dublin, Ireland

ISBN 978-3-031-96723-8 ISBN 978-3-031-96724-5 (eBook)
https://doi.org/10.1007/978-3-031-96724-5

This Springer imprint is published by the registered company Springer Nature Switzerland AG
The registered company address is: Gewerbestrasse 11, 6330 Cham, Switzerland

No idea is conceived in our mind independent of our five senses.

—Albert Einstein

Preface

The ability of the human brain to make informed decisions is inherently dependent on the fusion of information from multiple senses. This natural form of data integration has inspired technological advancements in many fields, particularly in health monitoring. Wearable devices have revolutionized healthcare, offering real-time insights into physiological parameters such as heart rate, respiration rate, and overall cardiovascular health. However, the accuracy and reliability of these insights hinge on the ability to process and integrate data from multiple sensors effectively.

This book, *Deep Learning and Signal-Processing Methods for Multisensor Data Fusion: Applications to Ambulatory Health Monitoring*, emerges from years of research into multi-sensor data fusion and its application in real-world healthcare scenarios. It builds upon foundational principles in signal processing, machine learning, and biomedical engineering to develop robust methodologies that enhance the reliability and efficiency of health monitoring systems. By addressing key challenges such as ambulatory signal quality, computational efficiency, and generalizability, we aim to contribute to a more comprehensive and unified framework for data fusion.

The origins of this book trace back to the Ph.D. research of one of the authors and our collaborative research efforts in the fields of electrical engineering, biomedical signal processing, and artificial intelligence. The book is based on various articles that were previously published by us. Through this compendium, we seek to bridge the gap between theory and application, providing researchers, engineers, and healthcare professionals with a structured approach to developing advanced multi-sensor fusion algorithms.

The book is structured to provide a logical progression from fundamental concepts to advanced techniques. It begins with an introduction to data fusion in health monitoring, followed by an in-depth exploration of fusion methodologies, signal quality indicators, and machine learning models. Subsequent chapters delve into the practical applications of these techniques, including heart rate estimation, heartbeat detection, and the detection of conditions such as sleep apnea and atrial fibrillation. The concluding sections focus on the generalizability of fusion algorithms, offering insights that extend beyond healthcare to other domains of multi-sensor data integration. Each chapter has been structured to stand alone, allowing readers to dive into any topic of interest without necessarily following a strict order. Whether you choose to read a single chapter or jump between sections, you will find valuable ideas and perspectives that do not rely on prior chapters for understanding. That said, for a deeper and more cohesive experience, we encourage you to read the book in sequence. Each chapter builds upon the themes and concepts introduced earlier, creating a natural flow that enhances comprehension and engagement. By following this progression, you may uncover connections and nuances that enrich the overall journey.

This work is intended for researchers, engineers, and professionals working in biomedical engineering, artificial intelligence, and wearable technology development. It is also suitable for graduate students and academics seeking a deeper understanding of multi-sensor fusion and its implications in health monitoring. We hope this book serves as a valuable resource in advancing both theoretical research and practical implementations in the field.

We extend our gratitude to our colleagues, mentors, and institutions that have supported our research endeavors. We also acknowledge the contributions of the broader scientific community, whose work has laid the foundation for our exploration of multi-sensor data fusion. We extend our gratitude to our families, our institutions—University of Twente and University College Dublin, and the various funding organizations that supported the research.

With this book, we aspire to inspire further innovation in wearable healthcare technology and contribute to the ongoing evolution of data-driven decision-making in medicine.

To our readers, we hope you enjoy reading this book as much as we enjoyed writing it, and may it inspire new ideas within the ever-growing field of multisensor data fusion in healthcare.

Rijswijk, The Netherlands — Arlene John
February 2025

Competing Interests The authors have no competing interests to declare that are relevant to the content of this manuscript.

Declarations

Acknowledgement of Sources of Funding
1. University College Dublin,
2. NSERC Canada Discovery Grant,
3. CHIST-ERA program [JEDAI project],
4. Irish Research Council under the new foundations scheme,
5. Microelectronics Circuit Centre Ireland [MCCI-2018-03, MCCI-2020-07], and
6. 4TU High tech for a sustainable future RECENTRE program.

Contents

1 Introduction 1
1.1 Wearable Health Monitoring 2
1.2 Lines of Inquiry 3
1.3 Structure of the Book 5
1.3.1 Significance of this Work 7
References 7

2 A Survey of Fusion Frameworks and Algorithms for Physiological Monitoring 9
2.1 Introduction 10
2.2 Methodology 11
2.3 Data Fusion Frameworks in Literature 11
2.3.1 Fusion Classification Based on the Relationship Between Input Data Sources (Durrant-Whyte Model) 16
2.3.2 Fusion Classification Based on the Fusion Abstraction Level (Luo and Kay Classification) 16
2.3.3 Input/Output-Based Characterization (Dasarathy Model) 19
2.4 Data Fusion Applications in Literature 21
2.4.1 Applications in Military and Defense 21
2.4.2 Applications in Robotics 22
2.4.3 Applications in Healthcare 23
2.4.4 Other Applications 23
2.5 Fusion Approaches for Physiological Monitoring 24
2.5.1 The Premise of Signal Quality Indicators 25
2.5.2 Data Fusion for Heartbeat Detection 26
2.5.3 Data Fusion for Heart Rate Estimation 31
2.5.4 Data Fusion for Respiratory Rate Estimation 32
2.5.5 Data Fusion for Sleep Apnea Detection 36
2.5.6 Data Fusion for Arrhythmia Detection 38
2.5.7 Data Fusion for Atrial Fibrillation Detection 42

2.6 Discussion ... 46
2.6.1 Challenges ... 47
2.7 Conclusions ... 48
References ... 49

3 Signal Quality Assessment Classifiers for Wearable IoT Sensors ... 57
3.1 Introduction ... 57
3.2 Method ... 59
3.2.1 Dataset ... 59
3.2.2 Method Outline ... 60
3.3 Non-fiducial Signal Quality Indicators ... 61
3.3.1 Kurtosis SQI (kSQI) ... 61
3.3.2 Skewness SQI (sSQI) ... 61
3.4 Machine Learning (ML) for Signal Quality Assessment of ECG Signals ... 63
3.4.1 *k*-Nearest Neighbor Classifier ... 64
3.4.2 Decision Trees ... 64
3.4.3 Support Vector Machines ... 65
3.4.4 Naïve Bayes Classifier ... 65
3.4.5 Linear Discriminant Analysis ... 66
3.4.6 Feedforward Neural Network ... 68
3.4.7 Ensembles and Classifier Fusion ... 68
3.5 Results ... 70
3.5.1 Performance Analysis ... 70
3.5.2 Complexity Calculation ... 75
3.6 Discussions ... 79
3.7 Conclusions ... 79
References ... 80

4 A Multi-sensor Fusion Technique for Heart Rate Estimation ... 83
4.1 Introduction ... 83
4.1.1 Related Works ... 84
4.1.2 Math Notation ... 85
4.1.3 Dataset ... 86
4.2 Single-Channel QRS Complex Detection and RR Sequence Calculation ... 87
4.3 Signal Quality Indicators (SQIs) ... 88
4.3.1 Kurtosis SQI (SQI_{kurt}) ... 88
4.3.2 Frequency SQI (SQI_{freq}) ... 88
4.3.3 Total SQI ... 89
4.3.4 The SQI per RR Interval Stream ... 89
4.4 Bayesian Probability Calculation ... 92
4.5 Fusion Algorithms ... 93
4.5.1 Selection of a Single Most Probable Value ... 93
4.5.2 Calculation of Most Probable Value by Combination ... 94

4.5.3 Calculation of Most Probable Value Through Elimination and Combination ... 96
4.5.4 Bayes-Eliminated Fusion Algorithms ... 97
4.6 Performance Comparison ... 98
4.6.1 Mean Absolute Error ... 98
4.6.2 Bland Altman Plots ... 108
4.7 Results ... 112
4.8 Discussions ... 115
4.9 Conclusions ... 116
References ... 116

5 A Multi-sensor and Multimodal Data Fusion Technique for Heartbeat Detection ... 119
5.1 Introduction ... 119
5.1.1 Related Works ... 120
5.2 Methodology ... 122
5.2.1 Datasets and Pre-processing ... 122
5.2.2 Discrete Stationary Wavelet Transform ... 123
5.2.3 Signal Quality Indicators ... 123
5.2.4 Data Fusion ... 136
5.2.5 Testing ... 138
5.3 Results and Discussions ... 140
5.3.1 Performance Comparison ... 140
5.4 Discussions ... 150
5.5 Conclusions ... 151
References ... 152

6 A Multimodal and Multiresolution Fusion Using Convolutional Neural Networks ... 155
6.1 Introduction ... 155
6.2 1D-CNN-Based Fusion for Multi-sensor Data ... 157
6.2.1 Data-Driven Fusion Approach ... 157
6.2.2 Signal-Based Selective Dropout ... 159
6.3 Application to Sleep Apnea Detection ... 160
6.3.1 Dataset and Models ... 162
6.3.2 ECG Model ... 163
6.3.3 SpO2 Model ... 166
6.4 Results ... 168
6.4.1 Performance Evaluation with Noisy Data ... 169
6.4.2 Generalization to Multiple Sensors ... 170
6.4.3 Fusion Algorithm with Abdominal Signal ... 172
6.4.4 Complexity Analysis ... 177
6.5 Discussions ... 182
6.6 Conclusions ... 184
References ... 186

7 A Multi-level and Multimodal Data Fusion Algorithm Using Neural Networks ... 189
7.1 Introduction ... 189
7.2 Atrial Fibrillation Detection and State-of-the-Art ... 191
7.3 1D-CNN-Based Multi-level Fusion for Multi-sensor Data ... 192
7.3.1 Multi-level Fusion Architecture ... 193
7.3.2 Multi-level Data-Driven Fusion Approach with Signal Quality Indicators ... 194
7.4 Application to Atrial Fibrillation Detection ... 197
7.4.1 Dataset ... 197
7.4.2 The ECG and PPG Signal Models ... 198
7.4.3 Signal Quality Indicators for ECG and PPG Signals ... 199
7.4.4 Fusion Network Architecture ... 201
7.5 Results ... 203
7.5.1 Loss Functions ... 204
7.5.2 Performance Evaluation with Noisy Data ... 205
7.5.3 Performance Evaluation Without SQI Inputs for Both Clean and Noisy Inputs ... 207
7.6 Discussions ... 209
7.7 Conclusions ... 209
References ... 210

8 Discussions and Conclusions ... 213
8.1 Summary ... 213
8.2 Discussions ... 215
8.3 Limitations and Critical Assessment ... 216
8.4 Recommendations for Future Works ... 217
8.5 Considerations for Implementation ... 218
References ... 219

Appendix A: Low Complexity Signal Quality Indicators ... 221

Appendix B: Low-Complexity Sleep-Apnea Detection Methods ... 227

Appendix C: Generalized Multi-level Fusion Architecture ... 235

Index ... 239

Acronyms

1D-CNN	One-dimensional convolutional neural network
ABP	Arterial blood pressure
AF	Atrial fibrillation
AI	Artificial intelligence
BCG	Ballistocardiogram
BP	Blood pressure
CinC	Computing in Cardiology
CNN	Convolutional neural network
cSQI	Curve SQI
CVD	Cardiovascular Diseases
DAI-DAO	Data In-Data Out
DAI-DEO	Data In-Decision Out
DAI-FEO	Data In-Feature Out
DCNN	Deep convolutional neural network
DEI-DEO	Decision In-Decision Out
DT	Decision Tree
ECG	Electrocardiogram
EEG	Electroencephalogram
EMG	Electromyogram
EOG	Electrooculogram
F SQI	Fiducial feature-based SQI
FEI-DEO	Feature In-Decision Out
FEI-FEO	Feature In-Feature Out
fSQI	Frequency SQI
GNSS	Global navigation satellite system
HR	Heart rate
HRV	Heart rate variability
IMU	Inertial measurement unit
IoT	Internet of Things
iPPG	Imaging photoplethsymogram
k-NN	*k*-Nearest Neighbor

kSQI	Kurtosis SQI
Laser	Light amplification by stimulated emission of radiation
LDA	Linear Discriminant Analysis
Lidar	Light detection and ranging
LSTM	Long short-term memory
LZ	Lempel-Ziv
MAE	Mean Absolute Error
ML	Machine learning
NB	Naïve Bayes
NF SQI	Non-fiducial feature-based SQI
NN	Neural Network
PPG	Photoplethsymogram
Radar	Radio detection and ranging
RGB	Red, green, blue
RMSE	Root Mean Square Error
RNN	Recurrent Neural Network
SaO2	Arterial oxygen saturation
SLAM	Simultaneous localization and mapping
SNR	Signal-to-noise ratio
SoC	System-on-chip
Sonar	Sound navigation and ranging
SpO2	Peripheral oxygen saturation
SQI	Signal quality indicator
sSQI	Skewness SQI
SV	Stroke volume
SVEB	Supraventricular ectopic beat
SVM	Support vector machine
VEB	Ventricular ectopic beat

List of Figures

Fig. 1.1 Flow chart of a multi-sensor fusion system 2

Fig. 1.2 Outline of this book 6

Fig. 2.1 The JDL fusion framework. Reused from [27], with permission from Elmenreich 13

Fig. 2.2 The Durrant-Whyte classification of fusion algorithms is based on the relationship between the data sources. Sensors $S1$ and $S2$ observe different objects, and therefore their fusion is complementary fusion. Sensors $S2$ and $S3$ observe the same object, and therefore their fusion is competitive/redundant. Sensors $S4$ and $S5$ observe the same object, but the information provided by the two independent sensors is fused to derive information that would not be available from single sensors. *Note*: Modified from [25]. Copyright Castanedo (CC BY 3.0) 17

Fig. 2.3 Luo and Kay's classification method categorizes fusion algorithms based on the level of information abstraction at which fusion occurs, dividing them into data-level fusion, feature-level fusion, and decision-level fusion. *Note*: Minor modifications from [32]. License obtained for reuse 17

Fig. 2.4 Dasarathy's classification of fusion based on the input-output abstraction levels. Reused from [27], with permission from Elmenreich 21

Fig. 2.5 Summarized flow diagram of voting-based decision fusion algorithms for heartbeat detection as discussed in [35, 72, 73] 27

Fig. 2.6 The blind deconvolution-based source-filter approach to fusion for heartbeat detection. *Note*: Figure reused from [74], with permission from Antink 27

Fig. 2.7 Summarized flow diagram of SQI/quality assessment-based lead switching for heartbeat detection as discussed in [75–79] . . . 28
Fig. 2.8 Summarized flow diagram of CNN-based heartbeat detection algorithms as discussed in [29, 80]. A variation of this flow is proposed in [81] . . . 29
Fig. 2.9 Summarized flow diagram of heart rate estimation through fusion as discussed in [42, 82–88] . . . 32
Fig. 2.10 Summarized flow diagram of the machine learning-based fusion methods for sleep apnea detection as discussed in [105, 117, 118, 122] . . . 36
Fig. 2.11 Summarized flow diagram of 2D-CNN-based fusion methods for sleep apnea detection (used in [116]) . . . 37
Fig. 2.12 Summarized flow diagram of CNN/ LSTM-based fusion methods for sleep apnea detection as discussed in [113, 119, 123–125] . . . 38
Fig. 2.13 Summarized flow diagram of the rule-based fusion algorithms for arrhythmia detection as discussed in [128–134]. Similar algorithm flow, but without SQIs, is used in [135–138] . . . 40
Fig. 2.14 Summarized flow diagram of the machine learning-based fusion algorithms for arrhythmia detection as discussed in [139–145] . . . 41
Fig. 2.15 Summarized flow diagram of the deep learning-based fusion algorithms for arrhythmia detection as discussed in [149–152] . . . 42
Fig. 2.16 The Markov model-based classifier for atrial fibrillation detection discussed in [31]. *Note*: Modified from [31], with permission from Bonomi . . . 45
Fig. 2.17 Summarized flow diagram of the machine learning-based fusion algorithms for atrial fibrillation detection as discussed in [30, 157, 158] . . . 45
Fig. 3.1 ECG signal with fiducial characteristic points—the P, Q, R, S, and T waves . . . 59
Fig. 3.2 Flow diagram of the proposed methodology with kSQI and sSQI indicating the signal quality indicators used in this study . . . 60
Fig. 3.3 Block diagram of the kSQI and sSQI calculation stage. The green boxes indicate inputs and the pink boxes indicate final outputs. LUTs stand for look-up tables. The bold connectors indicate vectors, and *N* indicates the number of samples per window. The details of kSQI and sSQI are discussed in detail in Sects. 3.3.1 and 3.3.2 . . . 62
Fig. 3.4 A record of unacceptable signal quality and acceptable signal quality with the corresponding feature vectors . . . 63

Fig. 3.5 Schematic of prediction using k-NN classifier 64
Fig. 3.6 Indicative schematic of prediction using DT 65
Fig. 3.7 Schematic diagram of prediction process using SVMs 66
Fig. 3.8 Flow diagram of the prediction process using NB classifier 67
Fig. 3.9 Flow diagram of the prediction process using LDA 67
Fig. 3.10 Schematic of prediction using the NN 69
Fig. 3.11 Comparison bar plot of accuracy, sensitivity, and specificity of an unweighted k-NN classifier for $k = 1, 3, 5, 7$ 71
Fig. 3.12 Comparison bar plot of accuracy, sensitivity, and specificity of a weighted k-NN classifier for $k = 3, 5, 7$ 72
Fig. 3.13 Box plot of accuracy values during testing for all learners over 10-fold cross-validation 74
Fig. 3.14 Accuracy versus training data size 75
Fig. 4.1 Flow diagram of the proposed fusion methodology for heart rate estimation 86
Fig. 4.2 Figure detailing how the RR sequence is extracted from the beat annotations 87
Fig. 4.3 The total SQI versus SNR plot for when the ECG signal is corrupted by **a** electrode motion noise, **b** muscle artifacts, and **c** baseline wander noise 90
Fig. 4.4 SQI Variations with noise in channel 1 91
Fig. 4.5 The plot of MAE versus SNR when channel 1 is corrupted by muscle artifacts and channel 2 is clean 98
Fig. 4.6 The plot of MAE versus SNR when channel 2 is corrupted by muscle artifacts and channel 1 is clean 106
Fig. 4.7 The plot of MAE versus SNR when both channels 1 and 2 are corrupted by muscle artifacts 106
Fig. 4.8 Bland Altman plot showing the agreement between the RR interval information obtained from **a** Channel 1, **c** Channel 2, and **e** α-trim mean filtering when channel 1 is corrupted by electrode motion noise at –20 dB and channel 2 is clean, and **b** Channel 1, **d** Bayes α-trim mean filtering, and **f** SWA with Bayes probabilities when Channel 1 is corrupted by electrode motion noise at 20 dB and when channel 2 is clean, with the reference RR interval for all records. The limits of agreements of the differences are shown through P_{99} and P_1 in the figure to indicate where 98% of the estimates lie. The values recorded are in beats per minute (The y-axis limits of the plots have been adjusted for ease of comparison and, therefore, some outliers are not visible in the figures) 109

Fig. 4.9 Bland Altman plot showing the agreement between the RR interval information obtained from **a** Channel 2, **c** Channel 1, and **e** α-trim mean filtering when channel 2 is corrupted by baseline wander noise at –20 dB and channel 1 is clean, and **b** Channel 2, **d** Bayes α-trim mean filtering, and **f** SWA with Bayes probabilities when Channel 2 is corrupted by baseline wander noise at 20 dB and when channel 2 is clean, with the reference RR interval for all records. The limits of agreements of the differences are shown through P_{99} and P_1 in the figure to indicate where 98% of the estimates lie. The values recorded are in beats per minute (The y-axis limits of the plots have been adjusted for ease of comparison and, therefore, some outliers are not visible in the figures) 110
Fig. 4.10 Bland Altman plot showing the agreement between the RR interval information obtained from **a** Channel 1, **c** Channel 2, and **e** Kalman fusion when both channel 1 and channel 2 are corrupted by muscle artifacts at –20 dB, and **b** Channel 1, **d** Channel 2, and **f** SWA with Bayes probabilities when both channel 1 and channel 2 is corrupted by muscle artifacts at 20 dB, with the reference RR interval for all records. The limits of agreements of the differences are shown through P_{99} and P_1 in the figure to indicate where 98% of the estimates lie. The values recorded are in beats per minute (The y-axis limits of the plots have been adjusted for ease of comparison and, therefore, some outliers are not visible in the figures) 111
Fig. 5.1 Flow diagram of the proposed fusion algorithm. The sum of the $d2$ and $d3$ bands obtained from the discrete stationary wavelet transforms of the ECG and PPG signals fused in the ratio of the linearized SQIs 122
Fig. 5.2 Flow diagram for SQI-based weight calculation for fusion 124
Fig. 5.3 Block diagram for implementing cSQI calculation (instantaneous c values). The boxes with green outlines show the inputs 127
Fig. 5.4 Overlay of selected cycles for template generation of the ECG signal 128
Fig. 5.5 Figure detailing how for a given ECG signal window centered at a specific point and at two different SNRs, the template is rearranged to align with the point of maximum correlation. It also shows how the cSQI changes with varying levels of noise 129

Fig. 5.6 cSQI plots at each sample point of an ECG signal when it is corrupted by **a** 50 Hz power-line interference, **b** baseline wander, **c** muscle artifacts, and **d** electrode motion at –10 dB. The vertical dotted lines delimit regions of clean signal and regions where noise is added to the clean signal 131
Fig. 5.7 cSQI versus SNR plot for when the ECG signal is corrupted by **a** 50 Hz power-line interference, **b** baseline wander, **c** muscle artifacts, and **d** electrode motion 132
Fig. 5.8 Overlay of selected cycles for template generation of the PPG signal 133
Fig. 5.9 Figure detailing how for a given PPG signal window centered at a particular point at 2 different SNRs, the template rearranges such that it is adjusted according to the point of maximum correlation and how the cSQI varies depending on noise 134
Fig. 5.10 cSQI plots at each sample point of a PPG signal when it is corrupted by **a** 50 Hz power-line interference, **b** baseline wander, **c** muscle artifacts, and **d** electrode motion at –10 dB. The vertical dotted lines delimit regions of clean signal and regions where noise is added to the clean signal 135
Fig. 5.11 cSQI versus SNR plot for when the PPG signal is corrupted by **a** 50 Hz power-line interference, **b** baseline wander, **c** muscle artifacts, and **d** electrode motion 136
Fig. 5.12 cSQI characteristics for ECG and PPG signals, with Fig. 3a showing the raw calculated cSQIs and Fig. 3b showing the linearly mapped cSQIs 137
Fig. 5.13 ECG signal and PPG signal corrupted by additive Gaussian noise, resulting in SNR of -10 dB from 1 to 3 s, and 4.4 to 7.4 s respectively (for illustrative purposes), SQI sequences, wavelet $d2 + d3$ transform sequences, fused sequence with heartbeat locations from the fusion algorithm, and the ECG signal with heartbeat locations from GQRS algorithm 139
Fig. 5.14 Performance comparison results of the data fusion algorithm against single-channel ECG GQRS, SQRS, and WQRS algorithm when the ECG signal is corrupted by the addition of **a** high-frequency noise, **b** low-frequency noise, **c** 50 Hz power-line interference, **d** baseline wander, **e** electrode motion, **f** muscle artifacts, **g** combination noise 1, and **h** combination noise 2 at SNRs from -50 to 50 dB 141
Fig. 5.15 Performance comparison results of the data fusion algorithm against single-channel PPG beat detection algorithm when the PPG signal is corrupted by the addition of Gaussian Noise at SNRs from -50 to 50 dB 143

Fig. 5.16 Fusion performance when both ECG and PPG signals are corrupted by Gaussian noise at different SNRs 144
Fig. 5.17 Bland Altman plot showing the agreement between the RR interval information obtained from the **a** Fusion, **b** GQRS, **c** SQRS, and **d** WQRS algorithms and the reference RR interval for all records when the ECG signal is corrupted by muscle artifacts such that the signal SNR is −50 dB. The limits of agreements of the differences are shown through P_{99} and P_1, which are also shown in the figure to indicate where 98% of the estimates lie. The values recorded are in beats-per-minute (The y-axis limits of the plots have been adjusted for ease of comparison and therefore some outliers are not visible in the figures) 149
Fig. 6.1 Proposed 1D-CNN-based fusion technique for multi-sensor multiresolution data fusion 158
Fig. 6.2 The ECG signal-based 1D-CNN model proposed for sleep apnea detection ($M_1$1) 163
Fig. 6.3 The SpO2 signal-based 1D-CNN model proposed for sleep apnea detection ($M_1$2) 166
Fig. 6.4 Independent 1D-CNN fusion model for sleep apnea detection from abdominal movements ($M_1$3) 171
Fig. 6.5 Performance of the single signal models and fusion models employing selective dropout for sleep apnea detection 173
Fig. 6.6 Comparison of the fusion models with and without selective dropout for sleep apnea detection 174
Fig. 6.7 Performance of the single signal models and fusion models with selective dropout in the presence of noise for sleep apnea detection 176
Fig. 6.8 Comparison of the fusion models with and without selective dropout when the dataset contains noisy signal samples for sleep apnea detection 178
Fig. 6.9 Performance of the pruned models for sleep apnea detection with sparsity varying from 50 to 80% for the **a** ECG signal model, **b** SpO2 signal model, and **c** abdomen movement signal model 181
Fig. 6.10 Performance of the pruned models for sleep apnea detection with sparsity varying from 50 to 80% when the **a** ECG and SpO2 signal are fused, **b** SpO2 and abdomen movement signal are fused, **c** ECG and abdomen movement signal are fused, and **d** SpO2, ECG, and abdomen movement signal are fused 182
Fig. 7.1 Multi-level fusion architecture as proposed in [7] 194
Fig. 7.2 Proposed multi-level fusion architecture of two sensors with SQIs as priors 197
Fig. 7.3 ECG and PPG signals with their corresponding cSQIs 200

Fig. 7.4 Bar chart illustrating the performance of the single-signal ECG and PPG networks, along with the fusion network for AF detection, trained using two different loss functions (central loss and average loss) under clean signal conditions. The legend: ECG–ECG signal-only model, PPG–PPG signal-only model, Fusion (Central loss)–Fusion model trained with central loss, Fusion (Average loss)–fusion model trained with average loss 204
Fig. 7.5 Bar chart illustrating the performance of the single-signal ECG and PPG networks, along with the fusion network for AF detection, trained using two different loss functions (central loss and average loss) in the noisy signal scenario. The legend: ECG–ECG signal-only model, PPG–PPG signal-only model, Fusion (Central loss)–Fusion model trained with central loss, Fusion (Average loss)–fusion model trained with average loss 206
Fig. A.1 4×4 multiplier using two 4×2 multipliers [2] 222
Fig. A.2 Boxplot of accuracy values during testing over the 10 folds for signal quality assessment 223
Fig. B.1 Performance of the pruned ECG model on the test set when sparsity is increased from 50 to 80%, for sleep apnea detection 228
Fig. B.2 Performance of the pruned SpO2 model on the test set when sparsity is increased from 50 to 80%, for sleep apnea detection 231
Fig. C.1 Multi-level fusion architecture with signal quality indicators for fusion multi-sensor inputs 236

List of Tables

Table 2.1	Timeline of the development of fusion frameworks	12
Table 2.2	Summary of fusion algorithms for heartbeat detection	30
Table 2.3	Summary of fusion algorithms for heart rate estimation	33
Table 2.4	Summary of fusion algorithms for respiratory rate estimation	35
Table 2.5	Summary of fusion algorithms for sleep apnea detection	39
Table 2.6	Summary of fusion algorithms for arrhythmia detection	43
Table 2.7	Summary of fusion algorithms for atrial fibrillation detection	46
Table 3.1	Performance of the classification algorithms in identifying unacceptable signal records	72
Table 3.2	Computational complexity and energy consumption of the SQIs and normalization	76
Table 3.3	Computational complexity and energy consumption of the classifiers during prediction	76
Table 3.4	Comparison of the best-performing classifier with state-of-the-art ECG record quality assessment algorithms	78
Table 4.1	Steps involved for RR interval estimation for each algorithm	99
Table 4.2	Mean absolute errors (MAEs) in seconds indicating the performance of fusion algorithms when lead 1 is corrupted by noise and lead 2 is clean	100
Table 4.3	Mean absolute errors in seconds indicating the performance of fusion algorithms when lead 2 is corrupted by noise and lead 1 is clean	102
Table 4.4	Mean absolute errors in seconds indicating the performance of fusion algorithms when both lead 1 and lead 2 are corrupted by noise	104
Table 4.5	Strategies for fusion methods for different scenarios for robust heart rate estimation	113

Table 4.6 SQI-based algorithm selection to obtain reliable RR interval estimate ... 114
Table 4.7 Comparison with state-of-the-art RR interval estimation methods ... 115
Table 5.1 Performance of the fusion algorithm, GQRS, SQRS, and WQRS algorithms for heartbeat detection when ECG signal is corrupted by different kinds of noise at −50 and 50 dB SNRs ... 145
Table 5.2 Performance of the fusion algorithm and single-channel PPG heartbeat detection algorithm when the PPG signal is corrupted by Gaussian of noise at −50 dB and 50 dB SNRs ... 147
Table 5.3 Comparison of the proposed method with other fusion methods for heartbeat detection in terms of gross sensitivity and positive predictive value ... 150
Table 6.1 ECG signal model parameters for sleep apnea detection ... 164
Table 6.2 Comparison of the performance of the proposed CNN model for apnea detection with state-of-the-art ECG record-based sleep apnea detection algorithms ... 165
Table 6.3 SpO2 signal model parameters for sleep apnea detection ... 167
Table 6.4 Comparison of the performance of the proposed CNN model for apnea detection with state-of-the-art SpO2 record-based sleep apnea detection algorithms ... 168
Table 6.5 The model parameters for the abdomen movement signal-based model for sleep apnea detection ... 171
Table 6.6 Comparison of state-of-the-art abdominal movement signals based sleep apnea detection algorithms with the performance of the proposed 1D-CNN-based models for sleep apnea detection ... 172
Table 6.7 Performance of the single signal models for sleep apnea detection ... 172
Table 6.8 Performance of the fusion models with and without selective dropout for sleep apnea detection ... 175
Table 6.9 Performance of the single signal models when the dataset contains noisy samples for sleep apnea detection ... 176
Table 6.10 Performance of the fusion models with and without selective dropout when the dataset contains noisy samples for sleep apnea detection ... 179
Table 6.11 Computational complexities of the sleep apnea detection models in terms of the number of multiplications and additions required, and the estimated energy consumption ... 180

Table 6.12 Computational complexity in terms of multiplication and addition operations and the energy consumption of the sparsified sleep apnea detection models during prediction 183
Table 6.13 Comparison of state-of-the-art fusion-based sleep apnea detection algorithms with the performance of the proposed 1D-CNN-based fusion models for apnea detection 184
Table 7.1 ECG and PPG signal network architecture and parameters for atrial fibrillation detection 199
Table 7.2 ECG and PPG signal SQI network architecture and parameters 201
Table 7.3 Central network architecture and parameters of the multi-level fusion model for atrial fibrillation detection 202
Table 7.4 Performance of the single-signal ECG and PPG networks and the fusion network for atrial fibrillation detection using two different loss functions (central loss and average loss) under clean signal conditions 205
Table 7.5 Performance of the single-signal ECG and PPG networks and the fusion network using two different loss functions (central loss and average loss) for atrial fibrillation detection in the noisy signal scenario 206
Table 7.6 Performances of the fusion architecture for AF detection, with and without SQI inputs for both clean and noisy signal inputs 207
Table 7.7 Comparison of the proposed 1D-CNN-based signal-quality-aware multi-level fusion model for detecting atrial fibrillation against state-of-the-art fusion-based atrial fibrillation detection algorithms 208
Table A.1 Performance of *k*-NN in identifying noisy records with approximate multipliers 223
Table A.2 Multipliers required at each stage for quality assessment of one record 224
Table A.3 Comparison of the approximate classifier with state-of-the-art ECG record quality assessment algorithms 224
Table A.4 Complexity of template generation algorithm (Algorithm 1) for template of length 2M+1 with N cycles 225
Table A.5 Complexity of cSQI calculation during initialization or when calculating for the first sample, c[0] 225
Table A.6 Complexity of cSQI calculation for every subsequent sample after initialization 225

Table B.1 Average performance of the models M_1, $M_{1,\text{pruned}}1$, and $M_{1,\text{binarized}}1$ for sleep apnea detection, in terms of accuracy, sensitivity, and specificity 228
Table B.2 Computational complexity and energy consumption of the sleep apnea detection models during prediction 230
Table B.3 Average Performance of the three models M_12, $M_{1,\text{pruned}}2$, and $M_{1,\text{binarized}}2$ for sleep apnea detection, in terms of accuracy, sensitivity, and specificity 232
Table B.4 Computational Complexity and Energy consumption of the sleep apnea detection models during prediction 233

Chapter 1
Introduction

Keywords Data fusion · Wearable health monitoring · Heartbeat detection · Heart rate estimation · Sleep apnea detection · Atrial fibrillation detection

The subject of this book is multi-sensor data fusion for wearable health monitoring. "The process of combining observations from a number of different sensors to provide a robust and complete description of an environment or process of interest" is referred to as multi-sensor data fusion [1]. This definition implies that the obtained sensor data has multiple measurements that need to be combined, where the multiple measurements are produced by different sensors, or the measurements are produced by the same sensors, but obtained at different times. It is quite instructive to compare the general concept of multi-sensor data fusion to the manner in which humans and animals use a combination of multiple senses and experiences to improve their chances of survival. In performing sensor fusion, the aim is to improve the quality of the information, so that the fusion system can achieve improved accuracy and robust inferences over that which can be obtained using a single sensor source. A summarized flow chart of a multi-sensor fusion system is shown in Fig. 1.1.

The basic problem of multi-sensor data fusion is determining the best procedure for combining the multi-sensor data inputs. The fusion operation can happen at any stage during the sensor information-processing stream as shown in Fig. 1.1, which is discussed in the subsequent chapters. Although data fusion is a very commonly used solution to enhance the performance of an algorithm or task, no generalized frameworks or methods of usage or methodology have been established for data fusion of physiological signals. This book started with the goal of developing a generalizable fusion framework within the context of biomedical time-series data obtained from wearable devices.

A. John et al., *Deep Learning and Signal-Processing Methods for Multisensor Data Fusion*, https://doi.org/10.1007/978-3-031-96724-5_1

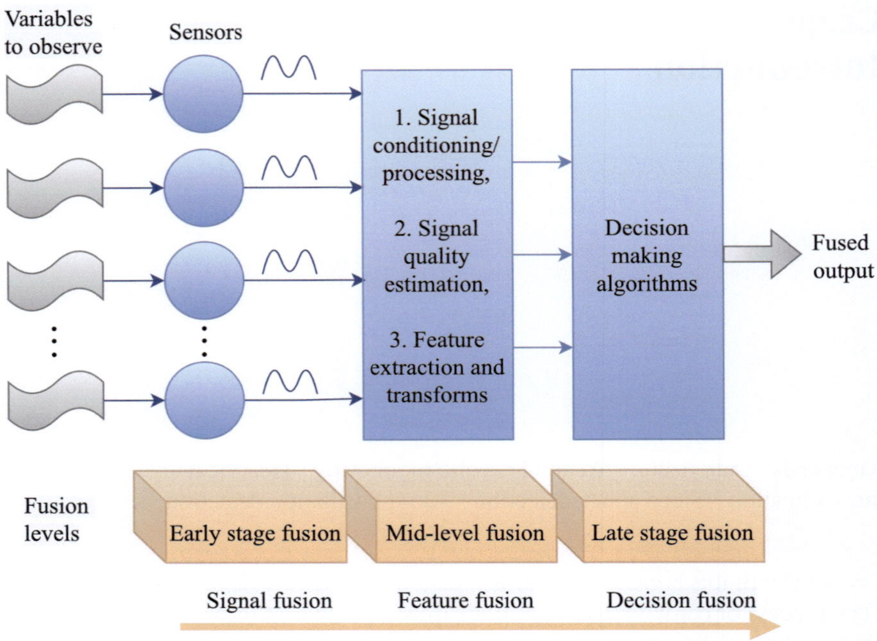

Fig. 1.1 Flow chart of a multi-sensor fusion system

1.1 Wearable Health Monitoring

Healthcare expenditure has been drastically increasing worldwide over the past several decades [2]. The rapid advancements in sensor technologies, Internet of Things (IoT), and artificial intelligence (AI) have enabled continuous, real-time tracking of physiological signals outside traditional clinical settings. Continuous and pro-active monitoring of vital health signs outside a lab environment using wearable sensors is widely seen as the most attractive option for health analysis these days [3, 4]. However, there are several challenges involved in making wearable sensors a reality. One of the major challenges is the low and unpredictable quality of the signals acquired due to various types of noises corrupting the signals, such as motion artifacts, lack of robustness due to electrode motion, etc., in wearable devices. Corrupted signals can lead to drastic variations in measured or derived parameters, leading to false alarms in ambulatory health monitoring devices, which would further lead to alarm fatigue and slow down response times to adverse events [5]. In other cases, a single lead can fail, leading to a loss of information. Also, often due to the miniaturized construction of sensors, electrode contacts are placed very close to each other instead of the standard measurement locations. This results in low signal amplitude which adds to the signal quality issue and results in poor feature extraction and data interpretation [6]. Therefore, the spurious measurements from sensors must be identified and eliminated as their incorporation in the interpretation task may lead to inaccurate estimation.

Data fusion algorithms enhance the performance of a particular task by combining information from multiple sensor sources. The principal motivation for multi-sensor data fusion is to improve the quality of the information output in a process through increased spatial and temporal coverage, increased robustness to sensor and algorithmic failures, better noise suppression, and increased estimation accuracy [7–9]. However, in the case of wearable sensor signal processing, the elimination of spurious data and utilizing information from the noise-free sensor sources for final inferences is the most appealing application of data fusion. Elimination of spurious data can be carried out through various methods in fusion frameworks, but in this book, the focus is on developing signal-quality-aware methods to aid the fusion process. The book aims to derive from the principles and theory discussed in the literature for multi-sensor fusion [1, 7–11] and to extend it to one-dimensional temporal signals for the robust derivation of primary health parameters like heartbeat locations and heart rate, and for the detection of health-related events such as sleep apnea and atrial fibrillation. This book aims to sequentially build up to the development of a generalized data fusion framework for ambulatory physiological signal monitoring and benchmark against available datasets.

The field of wearable health monitoring presents several challenges, including signal variability, motion artifacts, environmental interferences, and the need for energy-efficient computational techniques. Traditional single-sensor approaches often struggle with noise corruption, leading to unreliable readings and false alarms. To overcome these limitations, multi-sensor fusion strategies are required to combine multi-sensor data effectively and derive more accurate physiological interpretations. This book aims to provide an in-depth exploration of multi-sensor data fusion techniques in the context of wearable health monitoring, with a focus on developing a generalizable fusion framework capable of handling diverse physiological signals and improving real-time inference accuracy. It builds upon the foundational principles of signal processing and machine learning while incorporating novel deep learning frameworks that enhance data reliability. Although the body of work focuses on the development of data fusion algorithms for wearable health monitoring, the specific applications used to demonstrate the effectiveness of the proposed frameworks vary. Therefore, a brief description of how the book is framed based on the topics addressed follows.

1.2 Lines of Inquiry

1. *Can the use of signal quality indicators help reduce the possibility of failure in fusion systems in the presence of spurious data?*
   Fusion of the data obtained from various sensors is employed to provide a meaningful representation of the physiological data of the subjects. However, the data obtained from wearable devices while subjects move through their day-to-day life

is often corrupted by noise [12]. From the literature, it is often observed that as the number of sensors increases, the performance of the fusion system improves, but when one of the sensors provides spurious data, the fusion process leads to worse performance than when only one of the clean sensor sources is used for inference [7]. Although there are works based on Bayesian approaches to automatically identify the inconsistency in sensor measurements [13], wherein the effect of increasing the variance of the posterior distribution when measurement from one of the sensors is inconsistent with respect to the other, such methods are only useful within a statistical fusion framework. In this book, signal quality assessment methods that can provide a signal quality indicator for a signal segment as well as for every sample in 1D time-series data, such as electrocardiogram (ECG) [14] and photoplethysmogram (PPG) [15], based on limited prior information is proposed. The signal quality indicators (SQIs) are developed so that they can be used as inputs to any fusion framework that can accept SQIs for information fusion. The developed SQIs are used to aid in the fusion frameworks developed in this book.

2. *How can a fusion framework be developed for fusing signals with similar signal waveforms?*
   The simplest type of fusion framework takes multi-sensor inputs of uniform waveform shapes or structures as inputs. For example, two ECG sensors can be referred to as two sensors with inputs of similar data or modality, whereas data from an ECG sensor and an accelerometer are signals of different modalities. The advantage of fusing sensor sources of the same modality is that there is no need to transform the signals into a common representational format prior to fusion [7]. Various works on unimodal signal fusion have been discussed in the literature, however, no conclusive methodology or paradigm is proposed based on the input signal quality. Here, in this book, various algorithms that are suited for signal-level as well as the feature-level fusion of multi-sensor unimodal signals with the signal quality indicators estimated to aid in the fusion process are explored. The performances of various fusion algorithms at varying signal quality levels are analyzed to develop a fusion strategy based on the signal quality level. Here, the algorithms were developed in the context of heart rate estimation from multi-channel ECG recordings.

3. *How can a fusion framework be developed for signals of different modalities?*
   Developing from the results of the fusion of multi-sensor unimodal data, a fusion framework for multimodal signals was also developed in this book. Instantaneous (or sample-by-sample) fusion of signals for temporal inferences of high resolution has not been explored in the literature previously. To achieve this, the multimodal signals need to be transformed into a common representational format without the loss of temporal information prior to fusion to capture the events of interest. A generalized signal quality indicator that can be used for capturing the instantaneous signal quality of signals of different modalities was developed to aid in the fusion process. Here, the developed fusion algorithm was applied to heartbeat detection through the fusion of ECG and PPG signals.

4. *How can the issue of different sampling frequencies of the input signals be handled in fusion systems without additional computational costs involved in signal resampling or signal padding*?
Often, the different sensors used in the context of wearable health monitoring may have different sampling frequencies, and therefore extraction of temporal information from fused data often requires signal padding, resampling, etc. [16]. Deep learning methods that are data-driven function approximators have been found to be useful in extracting the requisite features for event detection or pattern recognition problems. Therefore, a data-driven fusion architecture that utilizes a one-dimensional convolutional neural network (1D-CNN) was developed to fuse signals with different sampling resolutions without resorting to padding or resampling for temporal inferences. The developed fusion architecture was applied to the sleep apnea detection problem by fusing ECG, peripheral oxygen saturation (SpO2), and abdominal movement data, where the SpO2 signal was sampled at a lower frequency compared to the other signals.

5. *How can a fusion framework be developed that automatically addresses the issue of the information abstraction level in fusion systems?*
Data fusion algorithms are most commonly classified based on the information abstraction level: signal level, feature level, or decision level [17]. However, oftentimes, the designer decides based on empirical evidence and prior knowledge at what information abstraction-level fusion needs to be carried out. Here, a data-driven approach for the fusion architecture using a convolutional neural network (CNN) is proposed to automatically learn the best information abstraction level at which fusion needs to be carried out. The SQIs are also fed into the network to aid in the fusion process in the presence of noisy data. The proposed multi-level fusion framework is applied to the atrial fibrillation detection problem where ECG and PPG signals are fused to detect atrial fibrillation events.

1.3 Structure of the Book

The book is organized as shown in Fig. 1.2. Chapter 2 expands on the context discussed in this chapter and provides an overview of existing literature on data fusion frameworks and fusion characterizations, and its applications in biomedical signal processing.

Chapter 3 discusses machine learning-based methods for signal quality assessment of signal segments, with the developed features and methods applied to ECG signal segments.

Chapter 4 focuses on developing fusion algorithms for signals obtained from multi-sensor sources acquiring data of similar modalities. The developed methods are applied to the two-channel ECG records for heart rate detection.

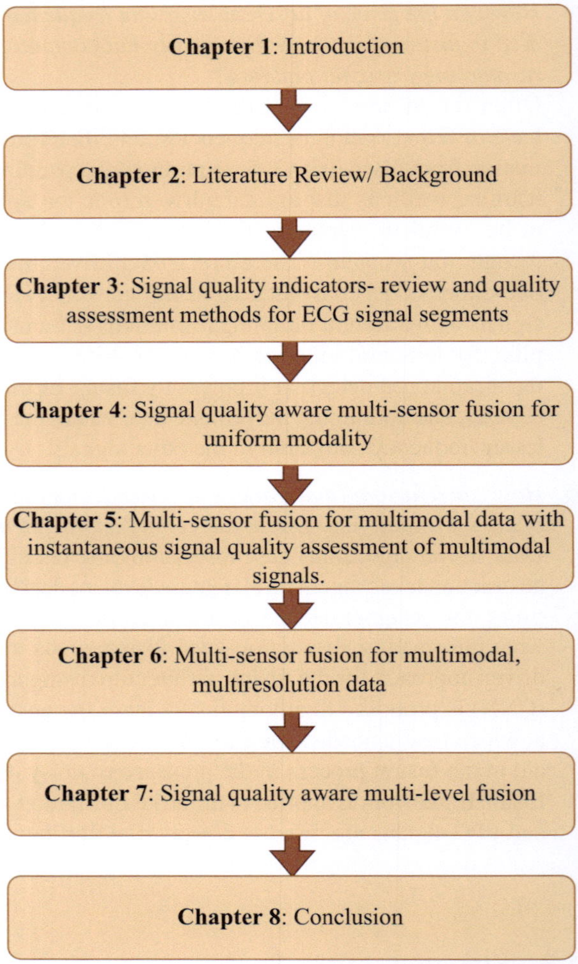

Fig. 1.2 Outline of this book

Chapter 5 further develops on the methods discussed in Chap. 4 to extend fusion algorithms to multimodal data. The developed framework is applied to the problem of heartbeat detection from ECG and PPG signals.

Chapter 6 tackles the issue of the fusion of signals sampled at different sampling frequencies for temporal inferences without having to resort to signal padding/resampling to correct for resolution differences. The proposed framework is applied to the sleep apnea detection problem by fusing ECG, SpO2, and abdominal movement data.

Chapter 7 focuses on developing a fusion framework that automatically decides the optimal information abstraction level at which fusion needs to be carried out. The proposed architecture is verified on the atrial fibrillation detection problem through the fusion of ECG and PPG signals with the aid of the signal quality indicators.

Chapter 8 summarizes the most important findings and conclusions from Chaps. 3, 4, 5, 6, and 7 by answering the lines of inquiry mentioned in this chapter. Potential future research directions are detailed to build on the conclusions and limitations that will be drawn from this book.

1.3.1 Significance of this Work

The growing reliance on wearable technology in healthcare requires robust data fusion methods to enhance diagnostic accuracy and reliability. This book provides a comprehensive roadmap for researchers, engineers, and healthcare professionals interested in integrating multi-sensor fusion techniques into real-world applications. By bridging the gap between signal processing, machine learning, and biomedical engineering, this work contributes to the advancement of intelligent health monitoring systems that can support early disease detection and personalized medicine.

References

1. H. Durrant-Whyte, T.C. Henderson, *Multisensor Data Fusion* (Springer International Publishing, Cham, 2016), pp. 867–896
2. OECD, Health spending (indicator) (2019). https://doi.org/10.1787/8643de7e-en
3. R. Pickus, Continuous monitoring of patients provides versatile, improved care case studies (Sept 2021)
4. T.O Malley, The need for continuous monitoring in today's healthcare system (Nov 2020)
5. M.-C. Chambrin, Alarms in the intensive care unit: how can the number of false alarms be reduced? Crit. Care **5**(4), 184–188 (2001)
6. J. Gierałtowski, K. Ciuchciński, I. Grzegorczyk, K. Stępień, M. Soliński, P. Podziemski, R S slope detection algorithm for extraction of heart rate from noisy, multimodal recordings. Physiol. Meas. **36**(8), 1743–1761 (2015)
7. H.B. Mitchell, *Data Fusion: Concepts and Ideas* (Springer Science & Business Media, 2012)
8. B.V. Dasarathy, Sensor fusion potential exploitation-innovative architectures and illustrative applications. Proc. IEEE **85**, 24–38 (1997)
9. D.L. Hall, J. Llinas, An introduction to multisensor data fusion. Proc. IEEE **85**, 6–23 (1997)
10. W. Elmenreich, *An introduction to Sensor Fusion* (Technische Universität Wien, Institut für Technische Informatik, Technical Report, May 2001)
11. R.R. Brooks, S.S. Iyengar, *Multi-Sensor Fusion: Fundamentals and Applications with Software* (Prentice-Hall Inc, USA, 1998)
12. I. Silva, G.B. Moody, L. Celi, Improving the quality of ECGs collected using mobile phones: the PhysioNet/computing in cardiology challenge 2011, in *2011 Computing in Cardiology* (Sept 2011), pp. 273–276
13. M. Kumar, D.P. Garg, R.A. Zachery, A method for judicious fusion of inconsistent multiple sensor data. IEEE Sens. J. **7**(5), 723–733 (2007)
14. U. Satija, B. Ramkumar, M.S. Manikandan, A review of signal processing techniques for electrocardiogram signal quality assessment. IEEE Rev. Biomed. Eng. **11**, 36–52 (2018)

15. C. Orphanidou, *Signal Quality Assessment in Physiological Monitoring: State of the Art and Practical Considerations* (Springer International Publishing, 2018)
16. A. Jafari, A. Ganesan, C.S.K. Thalisetty, V. Sivasubramanian, T. Oates, T. Mohsenin, SensorNet: A scalable and low-power deep convolutional neural network for multimodal data classification. IEEE Trans. Circuits Syst. I Regul. Pap. **66**(1), 274–287 (2019)
17. F. Castanedo, A review of data fusion techniques. Sci. World J. (2013)

Chapter 2
A Survey of Fusion Frameworks and Algorithms for Physiological Monitoring

Abstract As the need for precise, continuous, and non-invasive health monitoring increases, multi-sensor data fusion has become a critical component of contemporary healthcare systems. This chapter examines the development and present state of data fusion frameworks, providing an in-depth overview of the terminology, classification systems, and algorithmic approaches that define the field. Beginning with foundational concepts and key classifications, the chapter examines the most widely adopted fusion technique, including Kalman filters, Bayesian inference models, and machine learning approaches, and assesses their roles in enhancing the reliability and precision of health monitoring systems. Drawing from fusion applications in defense, autonomous driving, robotics, and image analysis, the discussion contextualizes how advancements in data fusion have informed healthcare solutions. A particular focus is placed on biomedical applications such as heartbeat detection, respiration rate estimation, and the detection of sleep apnea, arrhythmias, and atrial fibrillation. The chapter also addresses critical challenges, such as data heterogeneity and sensor reliability, which emphasizes the need for intelligent fusion algorithms. Ultimately, this chapter provides both a technical foundation and a forward-looking perspective on the role of multi-sensor data fusion in healthcare.

Keywords Sensor fusion · Health monitoring · Fusion classification · Signal quality indicators · Kalman fusion · Bayesian filtering · Heartbeat detection · Respiration rate estimation · Sleep apnea detection · Arrhythmia detection · Atrial fibrillation detection

Data fusion is a widely used solution to handle issues like missing data, noisy information, or low quality signals. However, there are no definitive methods or methodologies established for fusing physiological signals to enhance the quality of health monitoring or inference. This chapter aims to condense the diverse and disparate field of data fusion to identify relevant fusion methodologies applicable to monitoring and inference, specifically for one-dimensional time-series data obtained from health monitoring sensors, such as electrocardiograms (ECG), photoplethysmogram (PPG), accelerometer data etc. However, with this chapter, the expectation is that the fusion methodologies identified could also be extended to broader applications in

A. John et al., *Deep Learning and Signal-Processing Methods for Multisensor Data Fusion*, https://doi.org/10.1007/978-3-031-96724-5_2

healthcare. This chapter provides an overview of the development of fusion frameworks in the literature and common terminology used in fusion literature. The chapter introduces the fusion classification standards and methods that are most relevant from a sensing and estimation perspective. Applications of the reviewed fusion frameworks in applications such as defense, robotics, and healthcare are also discussed. Finally, the chapter converges on applications of fusion algorithms in biomedical inference tasks like heartbeat detection, sleep apnea detection, respiration rate estimation, arrhythmia detection, and atrial fibrillation detection.

2.1 Introduction

Non-invasive ambulatory physiological signal monitoring research focuses on monitoring primary physiological signals like heart rate, respiration rate, peripheral capillary oxygen saturation as well as derived physiological parameters. However, there are several challenges involved in making wearable sensors a reality. One of the major challenges associated with making wearable sensors robust is the low quality of the signals acquired due to motion artifacts, power demands, silicon area, and loss of information due to lead failure. Data fusion has emerged as a solution that can achieve improved accuracy and specific inferences over that which can be obtained using a single sensor source. Although data fusion is a very commonly used solution, no conclusive methods of usage or methodology have been established for data fusion of physiological signals for improving the quality of monitoring or inference. The development of a standard data fusion framework would provide holistic and context-aware information about an individual's health. This could provide more effective diagnosis and prognosis, eliminate cumbersome lab test sessions, and improve the overall quality of life. The theory of data fusion is also highly relevant to fields such as autonomous vehicles, environmental monitoring, structural health monitoring of civil infrastructure, and robotics.

This chapter is organized as follows: Section 2.2 describes the procedure used to extract relevant literature. Section 2.3 discusses the fusion methodologies and frameworks developed in literature in chronological order. Further details on the most suitable fusion framework classification methods from a sensing and estimation perspective are also discussed in this section. Section 2.4 discusses the fusion methods used in the field of defense, robotics, and healthcare. A review of fusion algorithms applied to biomedical inference tasks is discussed in Sect. 2.5 and the premise of signal quality indicators and the need for signal quality indicators in fusion systems for wearable devices are discussed in Sect. 2.5.1. The concluding remarks on the gaps addressed in this book are provided in Sect. 2.6.

2.2 Methodology

In this chapter, we summarize details of existing fusion frameworks in disparate fields to converge on a suitable framework classification strategy for the multi-sensor fusion works carried out in the field of biomedical signal processing. This review was carried out using a systematic search across multiple databases, including Scopus, Web of Science, PubMed, IEEEXplore, and Google Scholar. The keywords used included "data fusion," "multisensor (multi-sensor) data fusion," "fusion methodology," and "fusion frameworks." A snowballing search on the included articles was carried out to ensure no works were missed. The included studies are discussed in this chapter and are used to develop or compile a preference list of fusion frameworks that are useful in classifying multi-sensor fusion algorithms developed for health monitoring for biomedical 1-dimensional time-series signals. After the inclusion of the articles on fusion frameworks, a similar search was carried out for the monitoring applications included in this review, such as multi-sensor data fusion for heartbeat detection, heart rate estimation, respiration rate estimation, sleep apnea detection, arrhythmia detection, and atrial fibrillation detection. The inclusion criterion is that the works should include the fusion of multiple sensors for the development of a multi-sensor fusion algorithm for these applications. Fusion of different features obtained from a single sensor source is excluded.

2.3 Data Fusion Frameworks in Literature

This section provides a brief description of data fusion architectures discussed in the literature. Fusion methods are often categorized based on the application domain, fusion objective, sensor type, sensor suite configuration, etc. A number of data fusion frameworks have been developed over the years to aid in the identification of the most appropriate fusion methodologies for varying applications. Table 2.1 depicts the development timeline of various fusion architectures and data fusion reviews. In this section, we elaborate on these fusion methodologies and present findings chronologically, helping the reader trace potential relationships and influences within the literature. The Joint Directors of Laboratories (JDL) data fusion framework, the oldest and most widely used framework, was developed to support advancements in military applications [1]. JDL introduces processes, functions, and different techniques with four key processes: object refinement, situation refinement, threat refinement, and process refinement (Fig. 2.1). While this model has been applied in large-scale military projects, it appears to be highly specific to this domain.

A data fusion framework based on a behavioral knowledge framework was proposed by Pau in [2]. The initial stages of the framework involve feature vector extraction and vector alignment to obtain a fused feature vector. A set of behavioral rules that formalize the world model is applied to the fused representation of the feature vectors. In [3], multiple smart sensors are serviced by a knowledge-based sensor

Table 2.1 Timeline of the development of fusion frameworks

1987	JDL [1].
1988	Pau [2]. Schoess and Castore [3]. Durrant-Whyte [4].
1989	Luo and Kay [5]. Thomopoulos [6].
1993	Clement *et al.* [7].
1997	Dasarathy [8].
1998	Harris *et al.* [9]. Pohl and Genderen [10].
2000	Bedworth and O'Brien [11].
2001	Keithley [12].
2003	Laskey and Mahoney [13]. Chang *et al.* [14]. Carvalho *et al.* [15] .
2004	Brooks *et al.* [16]. Ahmed and Pottie [17]. Blasch *et al.* [18].
2005	Esteban *et al.* [19].
2008	Cohen and Edan [20].
2009	Bowman and Steinberg *et al.* [21]. Llinas [22]. Dong *et al.* [23].
2010	Chen [24].
2013	Castanedo [25].
2017	Fung [26].

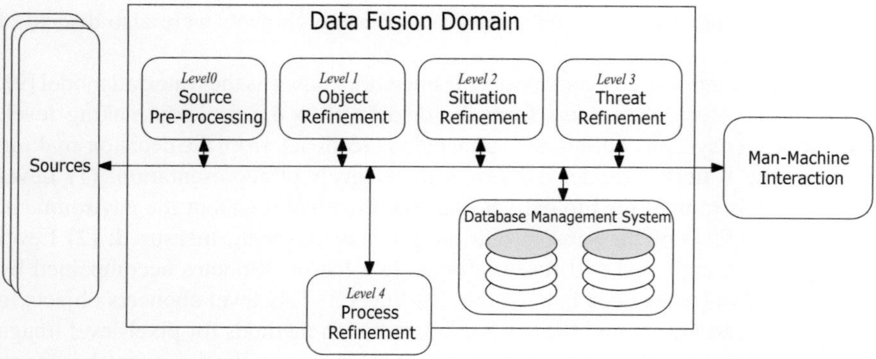

Fig. 2.1 The JDL fusion framework. Reused from [27], with permission from Elmenreich

supervisor that uses the confidence level assigned to each sensor to process data from the sensors as an integrated group, and these multiple sensor groups are then combined to create a reconfigurable and fault-tolerant sensor fusion architecture. A data fusion classification framework was proposed by Durrant-Whyte, which was based on the relationship between data sources or sensors: complementary, redundant, and cooperative [4].

Luo and Kay proposed a fusion classification based on the information abstraction level at which fusion happens: signal level, pixel level, characteristic/feature level, and symbol/decision level [5]. A comparable fusion classification approach was introduced in [6], based on the presence of mathematical or statistical models. This method includes (1) signal-level fusion, where data correlation is achieved through learning in the absence of a mathematical model describing the measured phenomenon; (2) evidence-level fusion, which combines data at various inference levels using a statistical model tailored to the user's requirements, such as decision-making or hypothesis testing; and (3) dynamics-level fusion, where data fusion is performed utilizing an existing mathematical model.

Clement et al. proposed a multi-specialist sensor fusion architecture wherein possibility theory is used to model knowledge of sensors and the uncertainty and imprecision of models in the context of image fusion in remote sensing [7]. Three kinds of specialists are present in the architecture: generic specialists (scene and conflict), semantic object specialists, and low-level specialists, and a blackboard structure with centralized control is used for fusion.

Dasarathy [8] proposed a fusion classification architecture based on the input and output information abstraction level: data in-data out, data in-feature out, feature in-feature out, feature in-decision out, decision in-decision out. Dasarathy also discusses the concept of self-improving multi-sensor fusion system architectures, wherein the central (fusion system) and local (individual sensor subsystems) decision-makers mutually enhance the other's performance by providing reinforced learning. Fusion system architectures for environments where local decision-makers can only make limited decisions, covering only a subset of possible choices, were also discussed in [8].

Harris et al. introduced a hierarchical framework known as the waterfall model [9]. In this architecture, data flows from the data level to the decision-making level, with the sensor system continuously updated via feedback from the decision-making module. The waterfall model comprises three levels of representation: (1) Level 1: Raw data is transformed to extract essential information about the environment, leveraging models of the sensors and the phenomena being measured; (2) Level 2: Features are extracted and fused, producing a list of estimates accompanied by probabilities and associated beliefs; and (3) Level 3: This level connects objects to events. Pohl and Genderen [10] reviewed data fusion methods for pixel-level image fusion and classified the various fusion algorithms based on the fusion methodology into (1) band selection methods, (2) color-related techniques, (3) statistical/numerical methods, and (4) other combined approaches.

Bedworth and O'Brien [11] proposed the Omnibus model, which was an approach toward creating a generalized fusion framework. This process model is a hybrid of Dasarathy's framework of I/O-based categorization [8] and the Waterfall model [9], along with a feedback loop based on decisions and actions, inspired by military strategist John Boyd's work *A discourse on winning and losing* [28].

Keithley introduced a fusion evaluation methodology centered on information needs, which inherently addresses questions of "quality" [12]. This approach employs a canonical structure that integrates quality constraints of the information needs with corresponding actions, forming a "knowledge matrix." This matrix enables the mapping of varied information needs onto a unified framework for evaluation. The overall evaluation method is inherently multi-intelligence, with fusion serving as just one component of the broader process.

Laskey and Mahoney explore the application of probabilistic network-based knowledge and data fusion within the domain of medical diagnosis [13]. A method to characterize the relationship between sensed information inputs and the quality of fused information output was proposed by Chang et al. to evaluate and understand the fusion system performance [14]. A data fusion architecture utilizing the Unified Modeling Language (UML) was developed, incorporating a taxonomy based on the definitions of raw data, variables, or tasks that can be applied to network-based dynamic distributed systems [15]. The key advantage of this architecture lies in its adaptability, as it can be reconfigured to align with the characteristics of the measured environment and the availability of sensing units or data sources. This flexibility facilitates the creation of a generalized data fusion algorithm.

A brief overview of soft computing methodologies employed in decision-making for challenging optimization problems, particularly in the context of information fusion, was presented in [16]. These methodologies include linear programming, fuzzy logic, and artificial neural networks. The approaches discussed are considered a class of heuristics aimed at minimizing a specified objective function to derive a solution. Ahmed and Pottie proposed a probabilistic, information-processing approach to data fusion in multi-sensor networks, for which a Bayesian approach was provided and justified using information measures [17]. Blasch et al. discuss the development of metrics to analyze fusion performance in dynamic situation analysis [18], both for user assessment and algorithm comparison. The proposed metrics for the evaluation of fusion algorithms included accuracy (error), timeliness (delay), cost (dollars), confidence (probability), and throughput (amount).

A system-based approach for fusion architecture definition was proposed in [19], featuring a framework structured around three key steps in system analysis: identification, estimation, and validation.

Cohen and Edan propose a sensor fusion framework that employs a selection algorithm to identify the most reliable logical sensors in real time and determine the most appropriate algorithm for fusing data from these sensors on a robotic platform [20]. The framework incorporates measures to assess sensor performance online. The fusion architecture prioritizes using the simplest sensor fusion algorithm alongside the most reliable sensors.

Bowman and Steinberg proposed a process for developing data fusion systems by offering guidelines for selecting fusion algorithms tailored to specific applications [21]. They conceptualized data fusion systems engineering as a resource management process, where available techniques and design resources are utilized to achieve an objective criterion. Resource management was framed as a process of mapping the problem space to the solution space. Llinas discusses a four-step test and assessment methodology to identify the criteria, measures, and metrics to evaluate a multi-sensor fusion process performance [22]. A review of multi-sensor satellite image fusion systems was provided in [23], with the review focusing on various fusion frameworks and improvements and developments in existent fusion algorithms.

Chen explores the concept of characterizing the relationship between the sensed information inputs available to the fusion system and the quality of the fused information output, aiming to enhance the evaluation and understanding of the fusion system's performance [24], similar to the discussion provided in [14]. The proposed fusion performance model was built upon Bayesian theory, combined with other simulation and analytical methods.

Castanedo reviews various fusion classification schemes and common fusion algorithms [25]. These methods and algorithms are classified into three categories: (1) data association, (2) state estimation, and (3) decision fusion.

Fung reviews the most common fusion algorithms discussed in the literature, and it was identified that most fusion architectures are variations of (1) Kalman filtering, (2) machine learning, (3) Bayesian inference techniques, (4) sequential Monte Carlo (particle filtering), (5) Dempster-Shafer theory of evidencing, and (6) artificial neural networks [26].

From the history of data fusion studies, it can be observed that various fusion frameworks and classification strategies were developed, but fusion systems are heavily influenced by the application they are used for, and no single generalized fusion framework can capture the nature of all fusion algorithms used in scientific research. It can also be observed that over time, studies essentially built on existing fusion algorithms, leading to a few algorithms that ended up being most commonly used. For example, fusion frameworks and architectures in the field of military systems and robotics place importance on architecture, control flow, and feedback systems.

Here we further discuss fusion algorithm development in the field of biomedical signal processing for one-dimensional time-series signals for health monitoring, and therefore the fusion classification techniques used in this chapter will utilize fusion classification theories that are suitable for one-dimensional signal-processing tasks. Therefore, in this chapter and also further in this book, fusion algorithms are categorized based on the Durrant-Whyte classification, Luo and Kay classification, and Dasarathy's models, which are discussed in detail next.

2.3.1 Fusion Classification Based on the Relationship Between Input Data Sources (Durrant-Whyte Model)

A data fusion classification framework was proposed by Durrant-Whyte [4], which was based on the relationship between input data sources and classified as follows:

1. Complementary fusion: In complementary fusion, the information from the input sources offers different perspectives on the scene, allowing them to be combined to generate a more comprehensive understanding of the target. For example, ECG and blood pressure (BP) signals were fused to identify heartbeat locations in [29].
2. Redundant/competitive fusion: In redundant fusion, two or more sensors gather information about the same target, which can be combined to enhance the confidence in the target's estimation. For example, eight channels of PPG signals are fused to detect atrial fibrillation events in [30].
3. Cooperative fusion: Cooperative fusion occurs when the provided information of the same source is integrated to create new, more complex insights than that can be obtained with a single sensor. For example, in [31], accelerometer data is used to remove noisy segments of ECG signals to aid in the detection of atrial fibrillation events.

The Durrant-Whyte classification is detailed in Fig. 2.2.

2.3.2 Fusion Classification Based on the Fusion Abstraction Level (Luo and Kay Classification)

Fusion algorithms were classified based on the information abstraction level at which fusion happens by Luo and Kay [5]. The classification is shown in Fig. 2.3.

Data-Level Fusion

Data-level fusion (also referred to as signal-level fusion) focuses on fusing raw signals obtained from the sensors. This method ensures that no information is lost

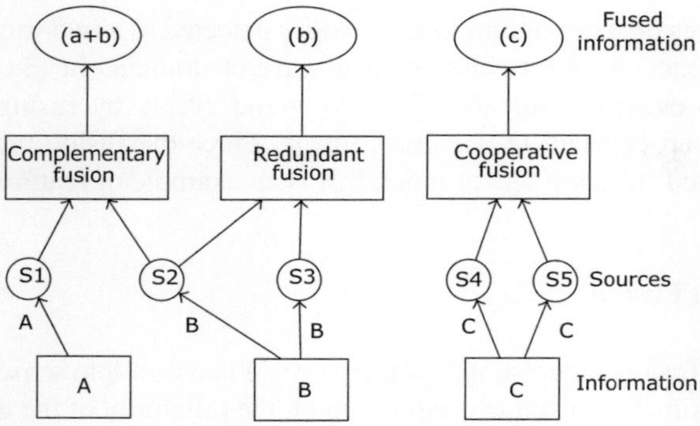

Fig. 2.2 The Durrant-Whyte classification of fusion algorithms is based on the relationship between the data sources. Sensors $S1$ and $S2$ observe different objects, and therefore their fusion is complementary fusion. Sensors $S2$ and $S3$ observe the same object, and therefore their fusion is competitive/redundant. Sensors $S4$ and $S5$ observe the same object, but the information provided by the two independent sensors is fused to derive information that would not be available from single sensors. *Note*: Modified from [25]. Copyright Castanedo (CC BY 3.0)

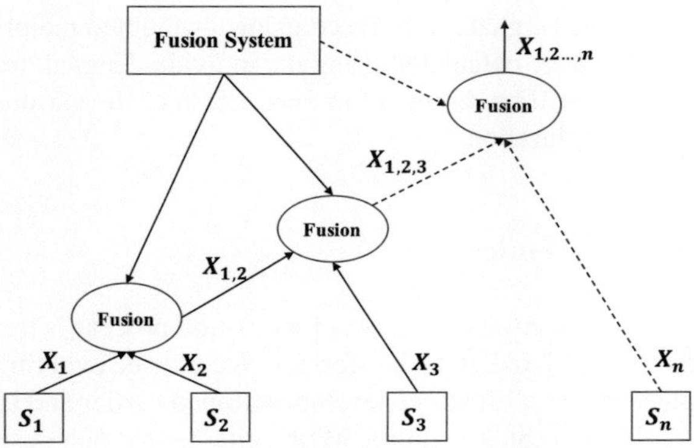

Fig. 2.3 Luo and Kay's classification method categorizes fusion algorithms based on the level of information abstraction at which fusion occurs, dividing them into data-level fusion, feature-level fusion, and decision-level fusion. *Note*: Minor modifications from [32]. License obtained for reuse

during processing or feature extraction. However, this method can give faulty results if the signals are corrupted, which can then be corrected/adjusted later if fusion is introduced at the feature level. An example of this is image stitching for object tracking, where the input images from multiple cameras are stitched together to provide a wider field of view [33]. Signal-level fusion also needs to address the issue of spatial or temporal alignment (synchronization) to prevent mismatch during fusion.

Feature-Level Fusion

Feature-level fusion involves the fusion of features extracted from the sensor sources. These features can be time-frequency images of 1D signals such as short-term Fourier transform images or spectrogram images, edges detected in a two-dimensional (2D) image, time-frequency information in the wavelet domain. In [34], CNNs were used for audio event classification into 50 sound events by fusing spectrogram, cochleogram, and constant Q-transform image. Since the fusion process involved features extracted from the sensor input, this is an example of feature-level fusion.

Decision-Level Fusion

Decision-level fusion occurs at the output stage. When multiple sensor and processing units are estimating a single desired output, the reliability of the detected output can be enhanced through fusion at the decision level, using methods such as majority voting, weighted voting, or other more advanced techniques. For example, the fusion of the R-peak (QRS-peak in ECG signals) positions identified from multiple sources was used to improve the reliability of heart rate (HR) calculation in [35]. Here, R-peak positions detected from ECG signals; R-peak positions detected from ECG artifacts in electromyogram (EMG), electrooculogram (EOG), and electroencephalogram (EEG) signals; and R-peak positions detected from arterial blood pressure (ABP) signals were fused using a weighted majority voting approach [35]. The weights were calculated using the individual signal quality (signal quality detection methods will be discussed in Sect. 2.5.1) of the various signal inputs in windows of 5-second duration.

Temporal Fusion

An often overlooked aspect of the fusion process is temporal fusion, which refers to the integration of data or information collected over time. This type of fusion can take place at any of the three levels mentioned earlier and is considered orthogonal to the level-based categorization. Many other terms such as spatial and spectral fusion have been occasionally used in literature to characterize a few specific fusion methods. Temporal fusion leverages the time-based relationships between multiple sensors. An example of temporal fusion at the feature level is the use of particle filtering for extracting HR by fusing ECG, PPG, and accelerometer data [36]. HR was extracted from ECG using time-series peak detection methods and using a time-frequency spectrogram in case of the PPG signal. Here, the most dominant frequencies close to human HRs in the accelerometer data were used to correct the errors due to motion artifacts in the proposed particle filtering-based fusion method.

2.3.3 *Input/Output-Based Characterization (Dasarathy Model)*

The three-level (data-feature-decision) hierarchy of fusion discussed above can be expanded into five fusion process I/O-dependent modes, which was proposed by Dasarathy [8]:

1. Data in-data out (DAI-DAO) fusion.
2. Data in-feature out (DAI-FEO) fusion.
3. Feature in-Feature out (FEI-FEO) fusion.
4. Feature in-Decision out (FEI-DEO) fusion.
5. Decision in-Decision out (DEI-DEO) fusion.

An example of each of the I/O modes of fusion is discussed below.

DAI-DAO Fusion

Fusion of ECG signals with ABP and EEG signals for enhanced reliability of heartbeat detection using multimodal data association methods was proposed in [37]. Rule-based association models used an ABP signal to find the position of the R-peak in an ECG signal when the ECG lead was corrupted by noise. Similarly, to determine the location of the R-peak, an adaptive linear filter was used to extract the ECG signal corrupting EEG signals, when the ECG lead and ABP lead were simultaneously corrupted. The signal corruption levels were identified using signal quality indices. This is an example of DAI-DAO fusion, with temporal fusion done based on when each signal is clean or noisy.

DAI-FEO Fusion

Respiratory rate detection using accelerometers and non-structured rule-based fusion of accelerometer information was discussed in [38]. The accelerometers placed on either side of the chest wall were used to measure the thoracic volume displacement during breathing. Here, the accelerometer data (displacements and rotations in three axes) was used as input to the fusion system, with thoracic volume displacement being the output. This is an example of DAI-FEO fusion.

FEI-FEO Fusion

Extraction of respiration rate intervals by fusing the signals derived from ECG and PPG signals is one of the common applications of fusion in health monitoring. Amplitude modulation, frequency modulation, and baseline modulation of an ECG signal, and amplitude modulation, frequency modulation, and baseline modulation of a PPG signal were fused based on a simple weighted addition in [39]. The weights for simple weighted addition were calculated using the individual signal quality of the derived signals in windows of varying durations. This is an example of FEI-FEO fusion with the modulated signals being feature inputs and the weighted sums being the outputs.

FEI-DEO Fusion

In [40], a fuzzified rule-based method for the fuzzy probability distribution of left ventricular failure, right ventricular failure, and pulmonary oedema through the fusion of multimodal cardiovascular signals, namely, HR, respiration rate, peripheral oxygen saturation, and mean, systolic, and diastolic arterial blood pressures was proposed. The fuzzy probabilities of left ventricular failure, right ventricular failure, and pulmonary oedema were then used to calculate a patient deterioration rate index. This is an example of FEI-DEO fusion where the features extracted (HR, respiration rate, peripheral oxygen saturation and mean, systolic, and diastolic arterial blood pressures) from the raw sensor inputs (ECG, respiratory volume, PPG, ABP) were fed into the fuzzified rule-based model for fusion. The outputs from the fusion model were the probabilities of left ventricular failure, right ventricular failure, and pulmonary oedema, which were the requisite outputs or decisions. Similarly, a novel hierarchical heartbeat classification system was constructed to accurately classify ventricular ectopic beat (VEB) and supraventricular ectopic beat (SVEB) as discussed in [41]. This was done by using random projections of two leads around the beat locations and feeding the random projections as features to an ensemble of support vector machines (SVMs) to detect VEB. Then, the ratio of the HR was compared to a predetermined threshold to detect SVEB. Here fusion was carried out by extracting features from two leads and feeding them to a machine learning (ML) algorithm for beat classification. This is also an example of FEI-DEO fusion.

DEI-DEO Fusion

Bayesian fusion using the R-peak-to-R-peak (RR) intervals (to calculate HR) obtained from three capacitive ECG sensors and three PPG sensors was proposed in [42]. Instead of using a signal-quality-based fusion, Bayesian fusion based on prior probabilities and likelihoods of R-R frequencies obtained per sensor is used. The difference between the estimated HR calculated using this method and the gold standard was less than ± 2 beats-per-minute (bpm). This is an example of DEI-DEO fusion as the RR intervals were taken in as inputs for fusion to make a "decision" on the most accurate RR interval value.

Dasarathy's I/O mode-based classification of fusion systems is detailed in Fig. 2.4.

Although the term fusion is also used in literature for the integration of multiple features extracted from the same sensor source, in this book, fusion refers to the fusion of signals, features, or decisions from multiple input sensor sources. In the next section, literature on multi-sensor fusion in the most common fusion application

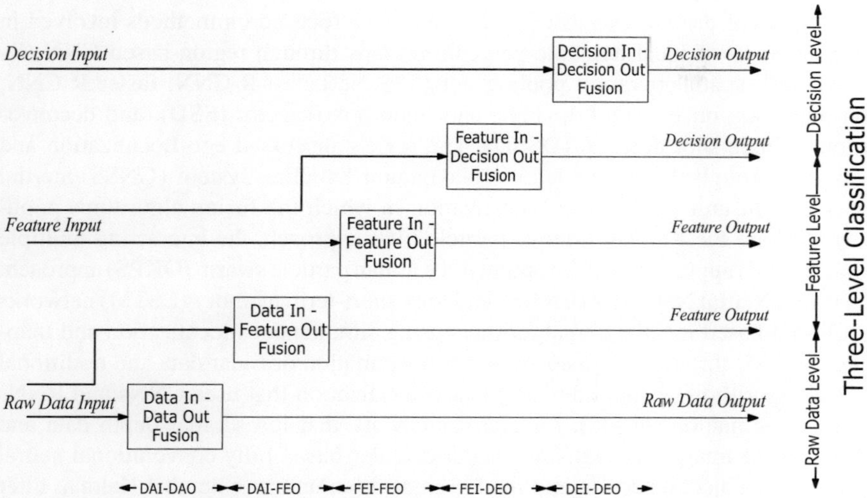

Fig. 2.4 Dasarathy's classification of fusion based on the input-output abstraction levels. Reused from [27], with permission from Elmenreich

domains, namely, robotics, military, and healthcare is reviewed. A detailed review of fusion for physiological parameter monitoring in wearable devices follows after the next section.

2.4 Data Fusion Applications in Literature

Although there are a large number of publications on fusion systems for applications in various domains, they are scattered. A review of data fusion literature carried out from 2014 to 2019 focused mainly on fusion applications in robotics, military, and healthcare [43]. These works are discussed in detail in the next few subsections.

2.4.1 Applications in Military and Defense

In the military domain, major applications include three-dimensional (3D) object detection in autonomous driving [44], perception and localization in autonomous driving [45, 46], moving object detection and tracking [47], and obstacle detection [48]. In [44], light detection and ranging (LiDAR) point clouds and red-green-blue (RGB) camera images are fused through the use of deep learning networks and convolutional neural networks (CNNs) for 3D object detection in autonomous driving. In [46], a review of various fusion methods for perception and localization in autonomous driving was discussed. The review focused on methods involved in the generation of local dynamic perception maps through region-based CNN (R-CNN) [49], spatial pyramid pooling net (SPP-Net), fast-R-CNN, faster R-CNN, you only look once (YOLO), single-shot multibox detector (SSD), and deconvolutional single-shot detector (DSSD). The review discussed ego-Localization and mapping through the use of Global Navigation Satellite System (GNSS)/inertial measurement unit (IMU)-based localization in which the fusion algorithms popularly used are the Kalman filter or particle filter approach, the interacting multiple model (IMM) approach, and the optimized Kalman particle swarm (OKPS) approach, Recurrent Neural Networks (RNNs), and long short-term memory (LSTM) networks and visual-based localization algorithms using simultaneous localization and mapping (SLAM) through the fusion of stereo information or Lidar data and traditional SLAM algorithms, visual odometry-based localization that uses CNNs and RNNs for pose estimation. In [45], a fusion framework that fuses Lidar depth data and RGB camera images through an encoder-decoder-based fully convolutional neural network for object detection and road segmentation and an extended Kalman filter nonlinear state estimator method for moving object tracking was proposed. In [47], an evidential framework (which is a generalized Bayesian framework of subjective probability) was proposed to solve the detection and tracking of moving objects problem through the fusion of radio detection and ranging (radar), Lidar, and camera images. In [48], a two-stage multi-task fusion for obstacle detection is proposed, which fuses sparse radar and camera images using CNNs, which are then used to detect big obstacles. The bottleneck features of this CNN are further fused with the raw camera image and passed through convolution layers to detect small obstacles.

2.4.2 Applications in Robotics

In the field of robotics, the most common applications for which fusion is used include localization [50], self-positioning [51], navigation [52], and human-robot interaction [53]. Fusion of a three-axis accelerometer and a three-axis gyroscope with the use of an extended Kalman filter (EKF) to estimate the position and orientation of a mobile robot was proposed in [50]. The task of self-positioning (such as walking on carpet and robot crouch)/self-activity recognition is an important aspect of the robot interacting with its environment. In [51], self-activity recognition is carried out through the use of LSTMs to model temporal information conveyed in multiple sensory streams such as the servomotor data, IMU, and pressure sensors. The features learned by the LSTMs are fused in the fully connected layer in the proposed approach. A review on underwater robot navigation was carried out in [52], where the fusion of sensors, such as light amplification by stimulated emission of radiation (Laser), sound navigation and ranging (Sonar), and underwater cameras, is discussed and fusion algorithms can be categorized into the following types: (1) filtering and estimation-based methods, such as Kalman filtering, filtering to eliminate erroneous data, and averaging; (2) mapping-oriented techniques, including SLAM, Bayesian inference, and the extended Kalman filter for map generation; (3) behavior-based architectures, which involve two layers of sensor fusion: module-layer sensor fusion, where sensor data is combined at the signal, pixel, or feature level, and global-layer sensor fusion, where sensor information is indirectly fused by merging specific behaviors at the symbol level (e.g., obstacle avoidance, target tracking); and (4) machine learning-based systems, utilizing methods such as support vector machines and neural networks. A review on wearable sensor fusion for human-robot interaction was carried out in [53], which discussed various rule-based fusion methods to fuse EMG, ECG, and IMUs for various applications such as gait phase recognition in lower limb robots, eye tracking, prediction of arm trajectories, and tremor suppression.

2.4.3 Applications in Healthcare

In healthcare, applications include monitoring and wearable technologies [54], wireless body sensor networks [55], medical diagnostics [56], and medical imaging [57]. A review on wearable sensor fusion methods for human activity recognition through the fusion of ECG, PPG, HR, wearable camera, inertial sensors, etc, using ML and deep learning-based methods was discussed along with popular fusion algorithms, such as Kalman filtering and Dempster-Shafer theory [54]. In [55], a review on sensor fusion in body sensor networks, which is closely related to wearable sensor fusion, was carried out, which focused on body sensor networks for human activity recognition, emotion recognition, and general health monitoring. Fusion of surface EMG recordings through feature extraction and classification using support vector machines to detect lower limb movement tasks was proposed in [56]. A two-stage multimodal fusion framework was proposed in [57] to obtain a single image for efficient clinical diagnosis: a cascaded combination of stationary wavelet transform and non-sub-sampled Contourlet transform for fusing images acquired using magnetic resonance imaging and computed tomography scan.

2.4.4 Other Applications

Fusion algorithms are also used in various other image fusion applications. A review on pixel-level image fusion for various applications was discussed in [58], where image fusion algorithms were classified into simple weighted averaging fusion, multi-scale fusion which includes the pyramid scheme and wavelet fusion, point-based rules, and area-based rules with application to digital camera applications, medical imaging, surveillance and navigation, and remote sensing. In a reverse approach, tomography is introduced as a sensing task with hard-field tomography defined as a fusion process and various features of fusion algorithms were attributed to tomography images in [59].

Another application wherein sensor fusion plays an important role is in machine health monitoring. A review on the fusion algorithms used to fuse accelerometer data for engine fault detection in flights was carried out in [60], which discussed various fusion algorithms using blind deconvolution, adaptive noise cancellation, active band selection/rejection, and wavelet transform techniques to remove unwanted signal components and decompose damage event-related signals along with a discussion on approaches for normalization, resampling, and phase compensation techniques needed to align data within the same reference frame. A deep neural network approach based on stacked autoencoders was constructed to enhance the feature learning and deep feature fusion to fuse data from accelerometers, acoustic emission sensors, etc. for fault diagnosis in rotating machinery in [61].

Another interesting fusion application is in plant water stress assessment through the fusion of thermal images, aerial images, environmental data, soil and moisture data, satellite data, and plant image data using various neural networks and ML-based methods as reviewed in [62]. Multimodal fusion of thermal images and gas sensor array sequence data for gas detection using LSTM and CNNs was proposed in [63].

The articles reviewed in this section capture the scattered nature of data fusion literature, and it can be observed that the frameworks developed for fusion are heavily dependent on the application. However, some common trends can be observed, and a few repeatedly used methods are as follows:

1. Kalman filtering for state estimation once sensor fusion problems are converted into estimation problems.
2. Multiscale fusion of multimodal data in image fusion.
3. Simple weighted averaging or sample-wise/pixel-wise operations in unimodal data fusion.
4. ML-based methods for feature fusion which also includes deep learning and CNNs with raw data as input.

The next section examines various fusion methodologies employed in human physiology monitoring. It evaluates whether the previously established classification effectively categorizes fusion algorithms for physiological monitoring tasks and explores whether common trends in fusion algorithms emerge across different applications in this field.

2.5 Fusion Approaches for Physiological Monitoring

Traditional physiological signal monitoring with bedside monitors is ideal for accurate real-time inferences. The advantage of this method is that the stationary patient and sterile environment provides signals of high diagnostic quality. On the other hand, wearable devices which are more suitable for long-term ambulatory health monitoring and enhance patient mobility are highly prone to ambulatory noises of unpredictable nature. Moreover, in physiological signal monitoring, underlying physiological characteristics can be determined or inferred from a wide variety of signals.

For example, heartbeats can be detected from multiple ECG signals placed on the body, from PPG sensors, ballistocardiogram (BCG) sensors, and more. Even in the context of non-wearable monitoring, video-based methods for HR estimation have been discussed in literature [64]. Therefore, multiple sensors can track the physiological characteristic of interest, and can be used in wireless/wearable devices to obtain more accurate inferences through data fusion. Moreover, fusing information from various sensors can prevent cases of inaccurate inferences due to ambulatory noise. Hence, multi-sensor data fusion is the most coherent course of action for physiological parameter tracking or estimation in wearable devices. Watrzek et al. [42] clearly illustrate the advantages of fusion compared to single sensor (non-fusion) methods. In [42], various multi-sensor fusion algorithms were explored such as the median of the sensor readings, the best sensor reading based on a previously calculated quality index, and a Bayesian approach to fusion for HR estimation. In this work, it was observed that the fusion algorithms exhibit lower errors over no fusion, especially when paired with artifact detection. They also showed that artifact detection or fusion of signals free of artifacts is important to achieve maximum gains from fusion. The next section deals with the premise of signal quality indicators (SQIs) in the context of fusing signals free of artifacts to prevent catastrophic fusion.

2.5.1 The Premise of Signal Quality Indicators

Fusion of the data obtained from various sensors is employed to provide a meaningful representation of the physiological data of the subjects. The acquired data is used to assess the user's fitness or health condition, and in certain cases, clinicians utilize this data for more detailed clinical analysis. The data obtained from wearable devices while subjects move through their day-to-day life is often corrupted by noise. Moreover, the placement locations of the sensors on the subjects' bodies may not be ideal for measurement.

From the literature, it is often observed that as the number of sensors increases, the performance of the fusion system improves. This conclusion, however, is flawed, as there are instances when the performance of a data fusion system, F, is actually inferior to that of the individual sensors. This phenomenon is known as catastrophic fusion and clearly should be avoided at all times [65]. In ambulatory health monitoring, the cause of catastrophic fusion can be attributed to spurious signals or noisy data obtained from the sensors, as model generation and testing are often carried out in ideal or sterile conditions. Therefore, the quality of the acquired signal plays a crucial role in the accuracy of the signal analysis, particularly for wearable physiological monitoring devices that offer real-time feedback to the user. This is where sensor fusion becomes necessary to enhance inferences, with the automatic estimation of noisy or clean segments (signal quality) in continuously monitored data being crucial for health and fitness monitoring devices. This is the role of SQIs in fusion for ambulatory monitoring.

The signal quality indicators represent the level of sensor uncertainty in the sensor models and are useful in preventing the phenomenon of catastrophic fusion. The use of signal quality indicators is similar to online sensor performance quantification as discussed by Cohen and Edan in [20]. It can be observed from data fusion literature that the usage of SQIs was formalized and clearly defined for fusion algorithms in health monitoring, with various SQI algorithms developed for this purpose [66]. In the next few subsections, a review of data fusion literature for specific applications in health monitoring, such as heartbeat detection, heart rate estimation, respiration rate estimation, sleep apnea detection, arrhythmia detection, and atrial fibrillation detection, is carried out. For this review, specific keywords connected to the application in conjunction with the keywords discussed in the methods section were used to carry out a systematic search across multiple databases.

2.5.2 Data Fusion for Heartbeat Detection

In the literature, most methods for heartbeat detection focus on identifying the QRS-peak in ECG signals to calculate HR. This method is preferred because the QRS-peak is typically easy to detect, as it accounts for 90% of the signal energy in a beat cycle. Although most heartbeat detection algorithms focus on using ECG signals for inference, fusion with other indicative signals, such as PPG and EEG, is also used. The Physionet/CinC 2014 challenge aimed to improve heartbeat detection using multimodal data. The challenge dataset included signals like ECG, stroke volume (SV), BP, EMG, etc., [67]. An example of signal-level fusion for heartbeat detection is presented in [68], where a novel fusion algorithm utilizing 12-lead ECG signals is introduced, based on the concept of a local weighted linear prediction algorithm. Fusion is performed to obtain a more accurate estimate of the ECG signal from 12 noisy channels taken from the CinC 2011 challenge database, which is subsequently used for HR estimation. Few works in the literature focus on voting-based methods wherein the heartbeat location is selected based on the agreement between the peaks obtained from the different sensors. In these works, the beat locations are obtained using existing peak detection algorithms like the Pan-Tompkins algorithm [69], SQRS [70], WQRS [71], variations of these algorithms, etc. This method is used in [72] where 12 leads of ECG signal are used to finalize the beat location, and in [73], where ECG, PPG, BP, SV signal, etc. are used for beat location and weighted voting. For the voting stage, the ECG, BP, and PPG signals were assigned higher weights due to their performance on the PhysioNet/CinC challenge 2014 training set. These works are examples of decision-level fusion using a voting approach. A flow diagram summarizing the fusion algorithms for heartbeat detection based on voting-based methods is shown in Fig. 2.5.

A modeling approach where the heart is modeled as a single-input multi-output system with the outputs being ECG, PPG, and BCG signals obtained from the sensors was proposed in [74]. In this approach, fusion is carried out through a blind deconvolution to estimate the heartbeat locations (Fig. 2.6).

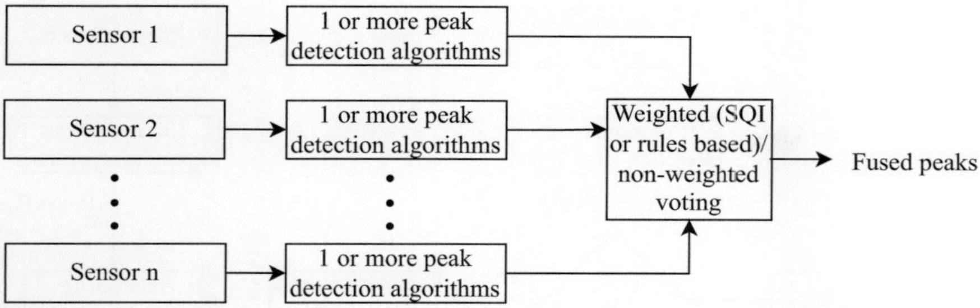

Fig. 2.5 Summarized flow diagram of voting-based decision fusion algorithms for heartbeat detection as discussed in [35, 72, 73]

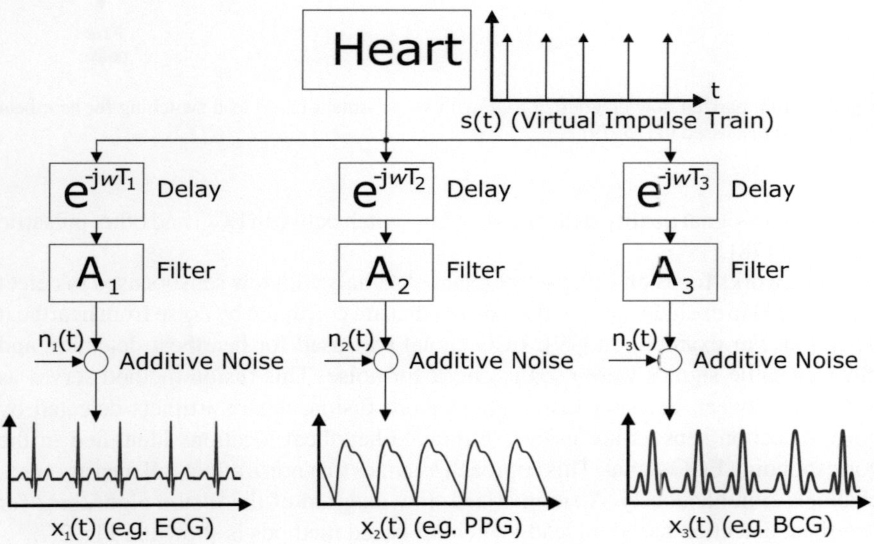

Fig. 2.6 The blind deconvolution-based source-filter approach to fusion for heartbeat detection. *Note*: Figure reused from [74], with permission from Antink

An enhanced approach to fusion in wearable devices involves utilizing signal quality indices, thereby mitigating the risk of catastrophic fusion. Many works focus on signal quality index-based lead switching using ECG signals and other pulsatile signals like PPG, BP signal, etc. This method is used in [75–77] to select and switch between heartbeat locations obtained from ECG and blood pressure signals, wherein the beat locations are obtained using existing peak detection algorithms. This is an example of decision-level fusion where an SQI-based selection on the final position of heartbeats is carried out. In [35], a novel majority voting fusion algorithm based on beat SQI was proposed for robust HR estimation using both cardiovascular and non-cardiovascular signals. A modified slope sum function and the Teager-Kaiser energy were employed for beat detection from ECG and other non-cardiovascular

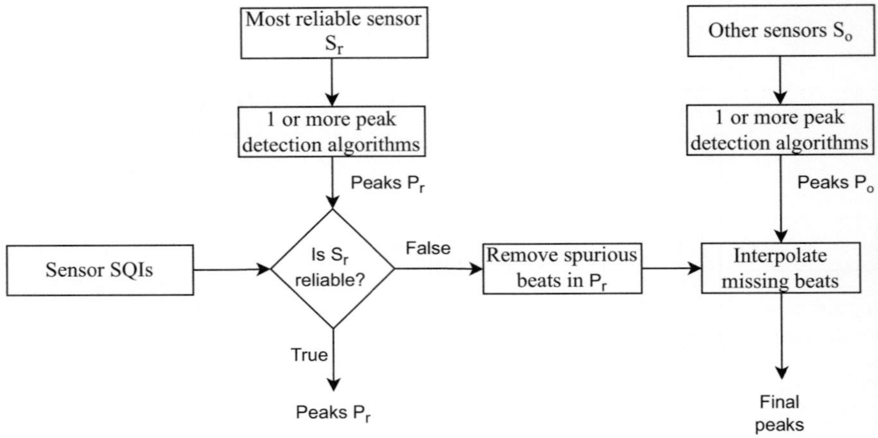

Fig. 2.7 Summarized flow diagram of SQI/quality assessment-based lead switching for heartbeat detection as discussed in [75–79]

signals, with signal quality indicators used to switch between ECG and other pulsatile signals in [78].

Some works focus on using multi-channel signals with few sensors used to detect artifacts and to exclude the signal segments that are corrupted by noise from heartbeat detection. For example, in [79], ECG signal was used for heartbeat detection and other pulsatile signals were used to check for noise. This fusion method serves as a hybrid between decision fusion and feature fusion, where artifacts detected by noise-detection sensors are used to eliminate heartbeat locations identified in the corresponding ECG signal. This approach assumes that noise affects all measurement channels simultaneously. A summarized flow diagram of the fusion algorithms for heartbeat detection based on lead-switching-based methods is shown in Fig. 2.7.

With advances in ML, CNNs have become popular in processing time-series data. This has led to various works that focus on using CNNs for beat detection. This is carried out by transforming an event detection problem (heartbeat detection) into a classification problem by windowing sensor signal segments and classifying whether or not the windowed segment contained a beat approximately in the central location of the signal window. This methodology is important in fusion as any number of sensor inputs can be fed into independent convolutional layers for independent filtering of the signal segments and then fusing the transformed features in the flatten stage prior to the fully connected network for classification. This approach is used in [29] for the fusion of ECG and BP data, and in [80] for the fusion of ECG and BP with application to capacitive ECG signals obtained from drivers during driving. These two works are examples of feature-level fusion algorithms as the final features after the convolutional layers are fused prior to decision-making. These works can also be considered as examples of FEI-DEO fusion. Another method of fusing image-based PPG (iPPG), PPG, BCG, and ECG signals was proposed in [81].

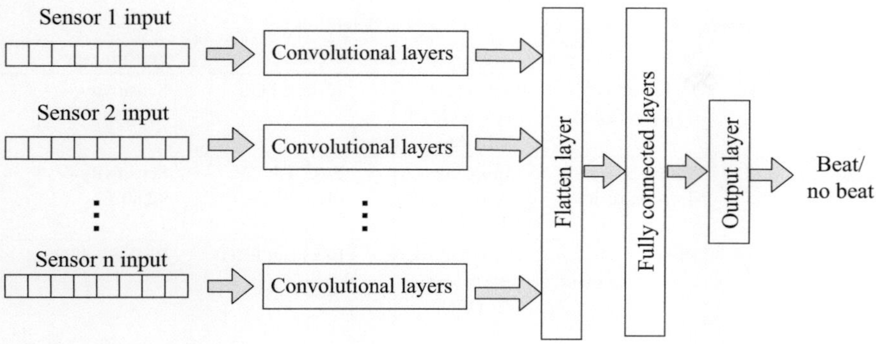

Fig. 2.8 Summarized flow diagram of CNN-based heartbeat detection algorithms as discussed in [29, 80]. A variation of this flow is proposed in [81]

This approach employs a hybrid fusion strategy using a voting mechanism across three different CNN networks. These networks utilize: (1) early signal fusion, (2) sensor-based late fusion (or feature fusion) with two filters per layer for each sensor, and (3) sensor-based late fusion (or feature fusion) with one filter per layer for each sensor. Here, signal-level and feature-level fusions are carried out in estimating the heartbeat position in the three networks individually, and a decision fusion is carried out in the end through a voting approach. This is an example of decision-level fusion. A summarized flow diagram of the CNN-based fusion algorithms for heartbeat detection is shown in Fig. 2.8.

A summary of the fusion algorithms discussed for heartbeat detection is provided in Table 2.2. The performances of the fusion algorithms are also provided for comparison purposes. However, it should be noted that performance metrics vary from article to article. Therefore, a brief explanation of the various performance metrics used for this application is provided next:

1. Accuracy—The proportion of correct predictions (both true positives and true negatives) among the total number of cases examined. In the case of heart beat detection, a true positive is a correctly identified heart beat location, where an estimated location is deemed accurate if it is no further than 150 ms from the corresponding ground truth location, as per the recommendation of the American National Standard for ambulatory ECG analyzers (ANSI/AAMI EC38-1998).
2. Sensitivity—The proportion of correct predictions of the positives among the total number of positive cases examined. Sensitivity is also termed as recall. In the case of heartbeat detection, a true positive is a correctly identified heartbeat location and positives are the total number of heartbeats.
3. Positive predictive value—The proportion of correct predictions of the positives among the total number of positives detected. Positive predictive value is also known as precision.
4. Root mean square error (RMSE) represents the square root of the second sample moment of the differences between predicted values and observed values. In the

Table 2.2 Summary of fusion algorithms for heartbeat detection

Article	Method	Fusion architecture	Signals	Performance
Zhao et al. [72]	Voting-based decision fusion	Redundant, decision level, DEI-DEO	12-lead ECG	Sensitivity—93.75%
Galeotti et al. [73]	Voting-based decision fusion	Complementary, decision level, DEI-DEO	ECG, BP, SV, EEG, EOG	Sensitivity—92.80%
Antink et al. [74]	Blind deconvolution of single-input multi-output systems	Complementary, feature level, FEI-DEO	BCG and ECG	RMSE—42.9 ms
Johnson et al. [75]	SQI-based lead switching	Complementary, decision level, DEI-DEO[b]	ECG and ABP	Score[a]—87.88%
Vollmer [76]	SQI-based lead switching	Complementary, decision level, DEI-DEO[b]	ECG and BP	Sensitivity—99.89%
Mollakazemi et al. [77]	SQI-based lead switching	Complementary, decision-level, DEI-DEO[b]	ECG and BP	Sensitivity—95.87%
Rankawat et al. [78]	SQI-based lead switching and Taeger-Kaiser energy based R-peak fusion	Complementary, decision-level, DEI-DEO[b]	EMG, EOG, and EEG	Sensitivity—99.81%
Ding et al. [79]	SQI-based lead switching	Complementary, decision-level, DEI-DEO[b]	ECG, BP, EEG, EOG, EMG	Accuracy—81.19%
Rankawat and Dubey [35]	SQI weights-based voting for decision fusion	Complementary, decision-level, DEI-DEO[b]	ECG, EMG, EOG, and EEG	Score[a]—94.93%
Chandra et al. [29]	CNN	Complementary, signal level, FEI-DEO	ECG and BP	Score[a]—94.00%
Antink et al. [80]	CNN	Redundant, signal level, FEI-DEO	3 lead ECG	Sensitivity—88.00%
Warnecke et al. [81]	Hybrid CNN	Complementary, signal, feature, and decision level, hybrid	iPPG, PPG, BCG , ECG	Average of Sensitivity and Positive Predictive Value—94.38%

[a]Score in this table refers to the custom performance metric used in the Physionet 2014 challenge [67]

[b]It can also be noted that the SQI-based lead switching and voting methods are also an example of FEI-DEO fusion as the SQI is a feature based on which the final decision is made

case of heartbeat detection, the quality of the prediction/ estimation is quantified based on the comparison of the estimated beat intervals to the ground truth beat intervals.

2.5.3 Data Fusion for Heart Rate Estimation

Building on the heartbeat detection discussed in the previous section, most health monitoring devices primarily focus on estimating heart rate. Fusion algorithms for heart rate estimation, where the beat locations are obtained from individual sensors and the inter-beat interval (or R-peak-to-R-peak interval (RR) in ECG signals) is calculated to generate a continuous-valued annotation along time for each sensor and then fused are quite common in literature. Popular algorithms involve using various algorithms to detect heart rate from a single sensor and carrying out fusion. For instance, a probabilistic model was proposed in [82] to synthesize the HR derived from multiple QRS detection algorithms applied to ECG signals. Another example of a similar method is an unsupervised Bayesian framework that synthesizes heart rate (HR) through the fusion of data from multiple heartbeat location annotators [83]. Similar approaches can be used for multi-sensor fusion for heart rate estimation.

Other works focus on utilizing the range of HR variations from multiple sensors to help estimate the true HR. A method based on particle filtering to fuse HRs obtained from ECG and PPG signals was proposed in [84], Bayesian fusion to fuse HRs obtained from ECG, ABP, and PPG signals was proposed in [85], and to fuse HRs obtained from capacitive ECG sensors and optical pulse signals were proposed in [42]. In [86], two methodologies for fusing cardiac vibration signals to calculate heart rate (HR), obtained from force sensors mounted on a bed, were proposed. The first method involved analyzing the cepstrum derived from the average spectra of the individual channels, while the second method utilized Bayesian fusion with three beat interval estimators applied to each channel. The fusion of HR obtained from ECG and PPG signals is posed as "the shortest path detection problem" on an acyclic graph to obtain a more accurate HR in [87]. A summarized flow diagram of the RR interval-based fusion algorithms for RR interval estimation is shown in Fig. 2.9.

In [88], a signal-quality-aware weighted averaging of the RR intervals obtained from ECG and BP signals was proposed, where the weights are calculated based on the signal quality, Kalman filter residuals from RR estimation, and inverse quality of the other signal. This is an example of decision fusion, as the RR intervals are the final output of the fusion algorithm (Fig. 2.9). Another work focuses on using multi-channel signals with few sensors used to detect artifacts, and to exclude the signal segments that are corrupted by noise from RR interval calculation [89]. Here, BCG signals were used for RR interval calculation and signals from other peripheral BCG signals were used for noise estimation.

A summary of the fusion algorithms discussed for HR estimation is provided in Table 2.3. The performances of the fusion algorithms are also provided for comparison purposes. However, it should be noted that performance metrics vary from

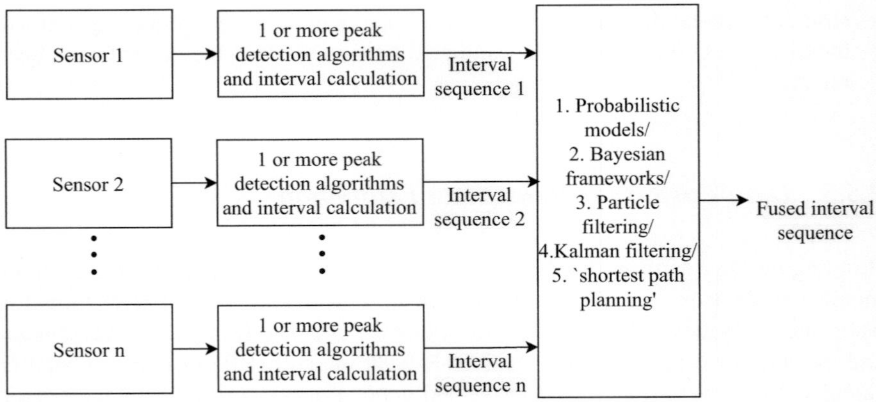

Fig. 2.9 Summarized flow diagram of heart rate estimation through fusion as discussed in [42, 82–88]

article to article, and a few metrics that were not discussed previously are used in the context of HR estimation. Therefore, a brief explanation of the various metrics used is provided next:

1. Mean absolute error (MAE) represents an arithmetic average of the absolute errors between predicted values and observed values. In the case of HR estimation, the quality of the estimation is quantified based on the comparison of the estimated beat intervals to the ground truth beat intervals.
2. Relative error represents the ratio of the absolute error of a measurement to the ground truth.
3. Mean error represents the arithmetic average of the errors between estimated values and observed values.
4. Correlation coefficient is a measure of linear correlation between two sets of data. It is the ratio between the covariance of two variables and the product of their standard deviations. Here the correlation is calculated between the estimated HR and the ground truth HR.

2.5.4 Data Fusion for Respiratory Rate Estimation

Another important vital sign along with heart rate is the respiration rate (RR). Autonomic respiratory rate detection is convenient for patients with critical illness, as RR variation can signal deterioration in hospital wards [90], as well as the risk of patient hospitalization during remote patient monitoring [91]. Respiratory modulation signals—with modulatory effects of respiration on other biomarkers—can be extracted from multiple biomedical signals, such as PPG, ECG, ABP, and peripheral arterial

Table 2.3 Summary of fusion algorithms for heart rate estimation

Article	Method	Fusion architecture	Signals	Performance
Xie and Minn [82]	Probabilistic model	Redundant, decision level, DEI-DEO	ECG	RMSE—13.96 bpm
Di et al. [83]	Bayesian fusion	Redundant, decision level, DEI-DEO	ECG	RMSE—14.37 bpm
De Morais Borges and Brusamarello [85]	Bayesian fusion	Complementary, decision level, DEI-DEO	ECG, ABP, and PPG	Relative error—23%
Wartzek et al. [42]	Bayesian fusion	Complementary, decision level, DEI-DEO	Capacitive ECG and optical pulse signals	Mean error—0.007
Brüser et al. [86]	Bayesian fusion	Redundant, decision level, DEI-DEO	BCG sensors	Relative error—2.2%
Aygun et al. [87]	Direct acyclic graph based shortest path search problem	Complementary, signal level, DAI-DAO	ECG and PPG	correlation coefficient, r = 0.89
Nathan and Jafari [84]	Particle filtering	Complementary, decision level, DEI-DEO	ECG and PPG	MAE—2 bpm
Huang et al. [89]	Signal-quality-aware linear regression and Kalman filtering	Redundant, decision level, DEI-DEO[a]	BCG sensors	MAE—31 ms

[a]It can also be noted that the SQI-based linear regression or algorithm selection are also examples of FEI-DEO fusion as the SQI is a feature based on which the final decision is made

tonometry (PAT) signals [92–94]. There are three main types of respiratory modulations on these signals: baseline modulation (BW), amplitude modulation (AM), and frequency modulation (FM) [95]. However, the performance of the RR estimation with the respiratory signal extracted using one modulation and a single-channel signal can be easily affected by noise, leading to poor RR estimation. Due to such a limitation, data fusion techniques can be applied to multi-channel signals to improve the accuracy and robustness of RR estimation. Since various modulations can be obtained from a single sensor or multiple sensors, an important term that is commonly used in respiration rate estimation is the respiration quality index (RQI). RQIs are used to determine which of the extracted modulations are of the highest quality [96]. They are used in fusion literature in a manner similar to that of SQIs.

For RR estimation with a single sensor input, e.g., PPG signal, the performance can be improved by fusing the RR values obtained from the three respiratory-induced modulations (BW, AM, and FM). A fusion algorithm termed Smart Fusion employs this method, for which, if the standard deviations (SD) from the three RR values is lower than four breaths-per-minute (brpm), the mean of the three values is used as the final RR result [97]. Otherwise, the corresponding segment is labeled with a low RR estimation quality [97]. It should be noted that in [97], Smart Fusion was employed for fusing modulation signals obtained from a single sensor source, but the popularity of the algorithm in subsequent multi-sensor fusion literature merits mention here. Smart Fusion was adapted for the multi-sensor fusion scenario for RR values extracted using pulse transit time (PTT), the heart rate interval (HRI), and pulse rate interval (PRI) methods [93]. In [94], a Kalman filter is used to fuse the RR values extracted from ECG and PAT signals. It presents an accurate robust RR estimation framework using the Signal Purity Index (SPI) as the SQI which is based on the Hjorth descriptor to control the Kalman filter noise covariance estimate for each channel and the SQI could be adapted to fuse any number of channels' RR estimates. Besides RR values, fusion methods could also be applied to multiple modulation signals or features extracted from these signals [92]. ECG-derived respiration (EDR), PPG-derived respiration (PDR), and BP-derived respiration (BDR) signals are filtered using a Kalman filter and fused through state vector fusion here.

Another commonly used fusion method is a weighted sum for feature fusion or RR interval fusion. The weight could be calculated through multiple methods such as RQI or based on sensor reliability. In [98], the weight is calculated based on the kurtosis of the power spectral density (PSD) of respiratory-related signals. A kurtosis-based weighted sum is utilized to fuse the PSD of PPG and the interval between two valleys of PPG waveforms (PPI). A similar weighted sum is applied to the two PSDs of the RR interval and the RS amplitude signals extracted from ECG. Then, one of these two channels (PPG and ECG) is selected based on an SQI presented by Orphanidou et al. [99]. In [39], three RQIs are measured based on the fast Fourier transform, autoregression, and autocorrelation methods for all the six modulations (AM, FM, BW) of ECG and PPG signals. Linear regression is used to fuse RQIs into a single RQI for each modulation. With the fused RQI, the RR estimates based on the modulation signals are merged into a single RR estimate. In [100], the clean heart rate variability (HRV) or Pulse Rate Variability (PRV) and AM signals extracted from noisy ECG and PPG signals are obtained based on an Ensemble Empirical Mode Decomposition (EEMD) method. Then, a quality-based fusion approach, i.e., a selection between the RR_{E-HRV} and RR_{E-AM} dependent on which of the two has the smallest variability in the duration of intervals between successive zero-crossings, is applied on the clean EDR signals to get the RR_{E-FUS}. The same fusion method is used to generate RR_{P-FUS} and the final fused RR, RR_{EP-FUS}.

Machine learning-based fusion methods are also popularly used in the domain of RR estimation. In [101], a neural network is used to fuse three input signals, i.e., pulse wave transit time, respiratory sinus arrhythmia (RSA) extracted from the ECG signal, and RSA from the PPG signal. All input points are normalized to a range of (0, 1), and the neural network is trained to predict an output of 1 when the inputs follow an inspiration. A deep learning (DL) architecture named U-Net [102] is applied to separate the demodulated respiratory (DR) signals and the non-respiratory

Table 2.4 Summary of fusion algorithms for respiratory rate estimation

Article	Method	Fusion architecture	Signals	Performance
Ding et al. [93]	Smart fusion	Complementary, decision level, DEI-DEO	ECG, PPG, and ABP	RMSE—1.76 brpm
Nemati et al. [94]	SQI-based weights and Kalman filter	Complementary, decision level, DEI-DEO	ECG, PPG, ABP, and PAT	RMSE—2.72 brpm
He et al. [98]	SQI-based lead switching and a kurtosis-based fusion method	Complementary, decision level, DEI-DEO[a]	ECG and PPG	–
Orphanidou [100]	SQI-based lead switching	Complementary, decision level, DEI-DEO[a]	ECG and PPG	mean error—1.8 brpm
Ahlstrom et al. [101]	Neural network	Complementary, feature level, FEI-DEO	ECG and PPG	error rate—9%
Can et al. [103]	U-Net	Complementary, decision level, DEI-DEO	ECG and SCG	MAE—0.82 brpm
Birrenkott et al. [39]	RQI-based weights and linear regression	Complementary, decision level, DEI-DEO[a]	ECG and PPG	MAE—0.71 brpm
Baker et al. [96]	RQI-based weights and neural network	Complementary, decision level, FEI-DEO	ECG and PPG	MAE—0.638 brpm

[a]It can also be noted that the SQI- or RQI-based lead switching or linear regression is also an example of FEI-DEO fusion as the SQI is a feature based on which the final decision is made

artifacts in ECG and seismocardiogram (SCG) signals in [103]. Fusion is achieved through the summation of all DR inputs at the first convolutional layer of the U-Net architecture. Additionally, ML methods can be combined with RQI to improve the performance in RR estimation. A similar framework that combines neural networks (NNs) with a novel RQI scheme is presented in [96]. A bidirectional long short-term memory (BiLSTM) network is employed to predict the respiratory rate using RR and RQI features. RRs are calculated from modulation-extracted signals, and the RQI is measured based on the quantity of variation in the signal.

A summary of the fusion algorithms discussed for respiration rate estimation is provided in Table 2.4. The metrics used are the same as those used for heart rate estimation. Another metric that was used is the error rate, which is defined as the total error out of the total measurements. The diversity of fusion algorithms for respiration rate estimation makes it challenging to create comprehensive summary diagrams that effectively integrate the various approaches used in this application.

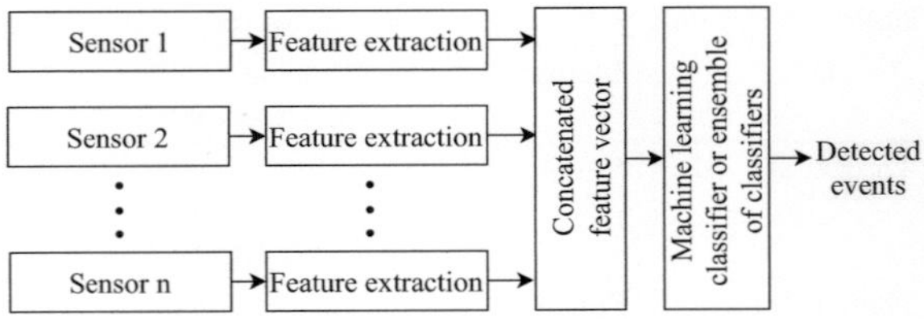

Fig. 2.10 Summarized flow diagram of the machine learning-based fusion methods for sleep apnea detection as discussed in [105, 117, 118, 122]

2.5.5 Data Fusion for Sleep Apnea Detection

Sleep apnea-hypopnea syndrome is characterized by abnormal reduction or pauses in breathing during sleep and affects 10% of middle-aged adults [104]. It can cause neurological arousal, impairing sleep quality and resulting in daytime sleepiness and fatigue [105]. Monitoring sleep apnea is crucial because it helps identify disruptions in breathing that impair sleep quality, leading to effective treatment and improvement in overall health and daytime alertness. Sleep-related disorders are typically diagnosed through overnight polysomnography under clinical supervision, a process that is both expensive and often uncomfortable for patients. As a result, developing a non-intrusive, automated method for detecting sleep apnea is crucial, particularly one that can integrate seamlessly with existing smartwatch systems that monitor sleep quality. Numerous studies have demonstrated that ECG signals [106–113], peripheral oxygen saturation (SpO2) signals [114–116], abdominal movements [117–119], and respiratory airflow [116, 120] can be used to detect sleep apnea events. The fusion of these signals for the accurate detection of sleep apnea events has also been explored in the literature.

A fusion approach using state-space-based modeling of the cardiorespiratory system for sleep apnea detection was proposed in [121], where the ECG and SpO2 signals were considered state inputs. In [120], fuzzy structural algorithms were applied to identify and characterize apnea and hypopnea episodes using respiratory airflow and SpO2 data.

ML-based models, where the features from individual signals are extracted for sleep apnea detection are a popular approach. In [105], fusion methods combining features extracted from ECG and SpO2 signals were proposed for apnea detection, utilizing various classifiers and classifier ensembles. In [122], features derived from ECG and SpO2 signals were fed as feature inputs to an SVM for apnea detection. Wavelet domain features extracted from nasal airflow, abdominal movement, and chest movement signals (representing breathing patterns) were fused with a classifier ensemble for apnea detection in [118]. Lin et al. proposed an adaptive nonharmonic model to extract features associated with thoracic and abdominal movement signals collected from wearable piezoelectric bands and used these features to classify sleep apnea events with SVMs [117]. These studies are examples of feature-level and FEI-DEO fusion. A summarized flow diagram of ML-based feature fusion methods for apnea detection is shown in Fig. 2.10.

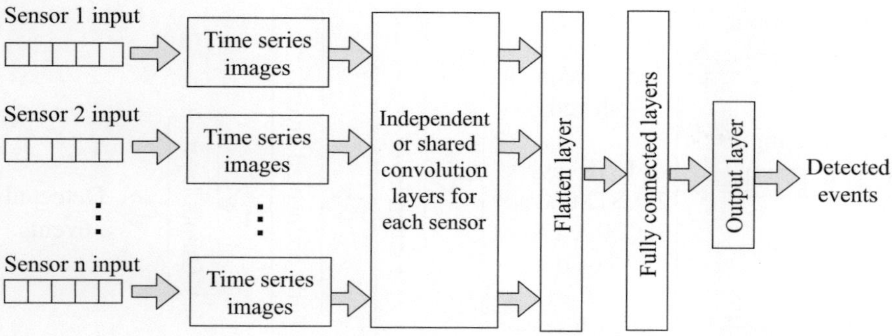

Fig. 2.11 Summarized flow diagram of 2D-CNN-based fusion methods for sleep apnea detection (used in [116])

CNNs are very popular in the research of automatic sleep apnea detection. In literature, 1D time-series data is often transformed into time-series images for apnea detection. In [116], a 2D-CNN-based method was proposed for sleep apnea detection using SpO2, oronasal airflow, ribcage, and abdominal movement data, where the signal inputs were stacked to form an image for 2D convolution (Fig. 2.11 shows a flow diagram of this fusion method). A method for fusing ECG and SpO2 signals using a combination of CNNs and LSTMs was proposed in [113]. In [119], a measurement module integrating abdominal and thoracic triaxial accelerometers, ECG and SpO2 sensors was developed, and an LSTM RNN model was proposed to classify four types of sleep breathing patterns. Nasal airflow, abdominal, and thoracic channels (NAT) were used as inputs to a CNN for sleep apnea detection in [123]. A novel CNN-based algorithm for sleep apnea detection through the fusion of HR signals, arterial oxygen saturation (SaO2), abdominal movement, and thoracic movement data was proposed in [124]. In the proposed model, the initial CNN layers are independent for the individual signals and fusion occurs in the later stages (flatten stage), and backward shortcut connections were proposed to improve the learning of the initial stages of the model. In [125], a feature fusion network model (CNN) was proposed, where each signal record (airflow respiratory signal, thoracic respiratory signal, abdominal respiratory signal) has a multi-level feature fusion network to combine features extracted in the earlier stages and later stages. The final features of each signal record are concatenated to generate a super feature vector which is finally used for apnea detection. These studies are examples of feature-level and FEI-DEO fusion, with the variation depending only on how early in the feature extraction process fusion happens. A summarized flow diagram of the CNN- or LSTM-based feature fusion methods for apnea detection is shown in Fig. 2.12.

Table 2.5 provides a summary of the fusion algorithms discussed for sleep apnea detection. It should be noted that performance metrics vary from article to article. The performance metrics include accuracy and sensitivity, where a true positive stands for a correctly identified hypopnea/apnea event. Another metric used is the area under the precision-recall curve, which shows the tradeoff between precision and recall for

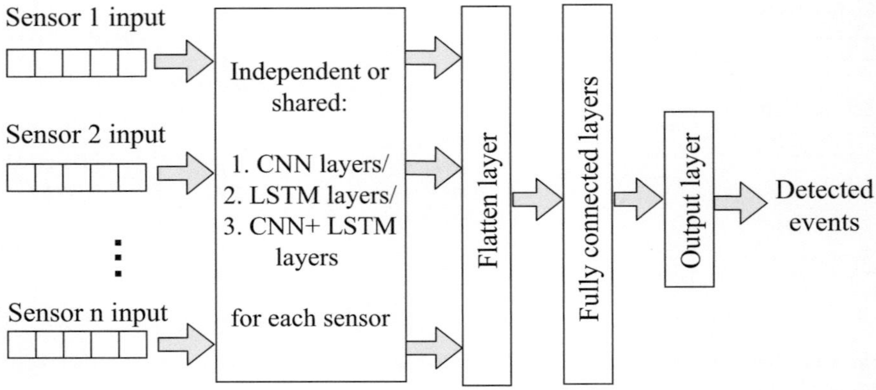

Fig. 2.12 Summarized flow diagram of CNN/ LSTM-based fusion methods for sleep apnea detection as discussed in [113, 119, 123–125]

different thresholds. A high area under the curve represents both high recall and high precision. Area under the precision-recall curve is a useful measure of success of prediction when the classes are very imbalanced, as is the case with the sleep apnea detection problem.

2.5.6 Data Fusion for Arrhythmia Detection

Arrhythmia is a condition or problem associated with irregularities in the heartbeat, including irregular rhythms. ECGs are well suited for arrhythmia detection because they provide detailed, real-time measurements of the heart's electrical activity [126]. Additionally, the results of the Physionet/CinC 2015 challenge showed that data fusion-based approaches are suited for arrhythmia detection, especially to reduce false alarms [127]. Various studies focusing on arrhythmia/abnormality detection through the fusion of ECG and other pulsatile signals are reviewed in this section.

Many fusion algorithms focused on developing rule-based approaches to detect arrhythmia events in conjunction with the use of SQIs to reduce the cases of false alarms. In [128], the Kalman filter is used to extract the HR from ECG and PPG signals based on the individual SQIs. Further to which the extracted HRs and individual SQIs were used in a rule-based fusion algorithm to classify five different arrhythmias and normal sinus rhythm. In [129], a Hilbert transform-based QRS detector was used to detect R waves from ECG signals to calculate the HR. Other open-source beat detection algorithms were used to calculate the HR from ABP and PPG signals, and the combined RR intervals obtained in conjunction with the individual signal SQIs were used to identify the type of arrhythmia. A method to trigger the arrhythmia alarm based on the agreement between the individual HRs obtained from ECG and SpO2 signals, and their corresponding SQIs based on thresholding rules was

Table 2.5 Summary of fusion algorithms for sleep apnea detection

Article	Method	Fusion architecture	Signals	Performance
Gutt et al. [121]	State-space model	Complementary, feature level, FEI-DEO	ECG and SpO2	Accuracy—82.33%
Otero et al. [120]	Fuzzy structural algorithm	Complementary, feature level, FEI-DEO	Respiratory airflow and SpO2	Sensitivity—95%
Xie and Minn [105]	ML classifiers and ensembles	Complementary, feature level, FEI-DEO	ECG and SpO2	Sensitivity—86.81%
Memis and Sert [122]	SVM	Complementary, feature level, FEI-DEO	ECG and SpO2	Accuracy—96.64%
Avci and Akbaş [118]	ML classifiers and ensembles	Complementary, feature level, FEI-DEO	Nasal airflow, abdominal movement, and chest movement signals	Accuracy—98.68%
Lin et al. [117]	SVM	Complementary, feature level, FEI-DEO	Thoracic movement and abdominal movement signals	Accuracy—81.8% ± 9.4%
Cen et al. [116]	2D-CNN	Complementary, signal-level, FEI-DEO	SpO2, oronasal airflow, ribcage, and abdominal movement signals	Accuracy—79.61%
Bernardini et al. [113]	CNN+LSTM	Complementary, signal level, FEI-DEO	ECG and SpO2	Accuracy—81.50%
Chang et al. [119]	LSTM	Complementary, signal level, FEI-DEO	thoracic movement signals, SpO2, and ECG	Accuracy—92.3%
Haidar et al. [123]	CNN	Complementary, signal level, FEI-DEO	Nasal airflow, abdominal movement, and thoracic movement signals	Accuracy—83.5%
Van Steenkiste et al. [124]	CNN	Complementary, signal level, FEI-DEO	SaO2, HR, abdominal movement, and thoracic movement signals	Area under precision-recall curve—0.74
Lv and Li [125]	CNN	Complementary, signal level, FEI-DEO	Airflow respiratory signal, thoracic movement, and abdominal movement signals	Accuracy—85.5%

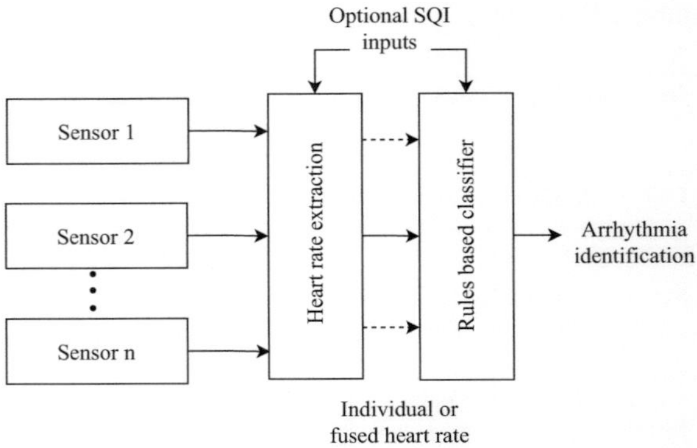

Fig. 2.13 Summarized flow diagram of the rule-based fusion algorithms for arrhythmia detection as discussed in [128–134]. Similar algorithm flow, but without SQIs, is used in [135–138]

proposed in [130]. Similar rule-based approaches for arrhythmia detection in conjunction with SQIs using ECG, PPG, and ABP signals were proposed in [131–134]. Various rule-based fusion algorithms fusing ECG and other pulsatile signals for accurate arrhythmia detection were proposed in [135–138]. A summarized flow diagram of the rule-based fusion algorithms for arrhythmia detection is shown in Fig. 2.13.

Various multi-sensor methods for arrhythmia detection rely on machine learning models. An SQI-based peak detection algorithm is proposed to detect peaks from ECG, ABP, and PPG signals to calculate the HR, from which features are extracted to feed into an SVM classifier for arrhythmia detection in [139]. A decision tree (DT) classifier was developed for arrhythmia classification using the HR features extracted from ECG, PPG, and ABP signals and the SQIs in [140]. Arrhythmia detection through a classifier combination of two ECG signal channels that use a linear discriminant classifier for arrhythmia detection individually is proposed in [141]. Five different random forest classifiers were used to classify five different types of arrhythmias, where the features given as input to the classifiers are derived from ECG, ABP, and PPG signals along with the corresponding SQIs in [142], and a similar approach was discussed in [143]. A signal similarity-based approach for interval detection from ECG and ABP/PPG signals was used to extract HR-related features for arrhythmia detection as inputs to a binary classification tree, a discriminant analysis classifier, and an SVM classifier to classify arrhythmia events and normal sinus rhythm in [144]. ECG and PPG signal features were used to classify five different types of arrhythmias, with each arrhythmia condition being classified with an individual SVM classifier in [145]. A summarized flow diagram of the ML-based fusion algorithms for arrhythmia detection is shown in Fig. 2.14.

HR features and signal quality indices from ECG, PPG, and ABP signals were used as inputs to develop a case-based reasoning model for arrhythmia detection in [146]. ECG and ABP signals are used as features for a k-Nearest Neighbor (k-NN) classifier model for arrhythmia classification in [147]. In [148], various SVM-based noise classifiers for ECG, BP, and PPG signals were used to find signal segments of good reliability, and asystole, bradycardia, and tachycardia were detected from

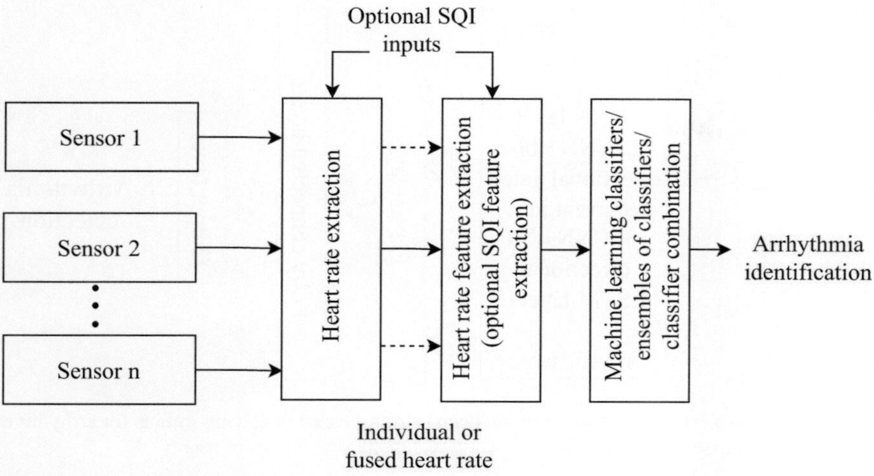

Fig. 2.14 Summarized flow diagram of the machine learning-based fusion algorithms for arrhythmia detection as discussed in [139–145]

the HRs obtained from the clean signal segments. Additionally, ventricular flutter fibrillation was detected through spectrum analysis of the ECG signal, and a heartbeat classifier based on QRS template matching was used to detect ventricular tachycardia.

CNN-based models are also commonly studied for arrhythmia detection and analysis. In [149], a CNN-based fusion algorithm using 2-lead ECG signals, where the initial layers of the CNN for both leads do not interact and get fused only in the final few layers was proposed for distinguishing between five different arrhythmia types. A novel network called MLBF-Net that uses CNNs in conjunction with bidirectional gated recurrent units to extract feature maps from 12-lead ECG signals for fusion for arrhythmia detection of five different arrhythmia types was proposed in [150]. A similar architecture is proposed in [151] for the fusion of 12-lead ECG signals with skip connection operation to fuse different levels of features extracted at different stages of the CNN, and channel-wise attention modules are used for effectively extracting the features learned at the different stages to classify five different arrhythmia types. Another algorithm that extracts local features using CNNs and then extracts temporal features by bidirectional LSTM for arrhythmia classification for 12 leads separately was proposed in [152], where an eXtreme Gradient Boosting (XGBoost) classifier is used to fuse the 12-lead models to distinguish between 5 different arrhythmia types. A summarized flow diagram of the deep learning-based fusion algorithms for arrhythmia detection is shown in Fig. 2.15.

A method for the detection of premature ventricular contraction through the mean fusion of the wavelet image of six ECG leads and feeding the Tucker-decomposed images of the inverse wavelet transform of the fused fusion image to a CNN was proposed in [153] (the flow of this method is similar to as shown in Fig. 2.11). A summary of the fusion algorithms discussed for arrhythmia detection is provided

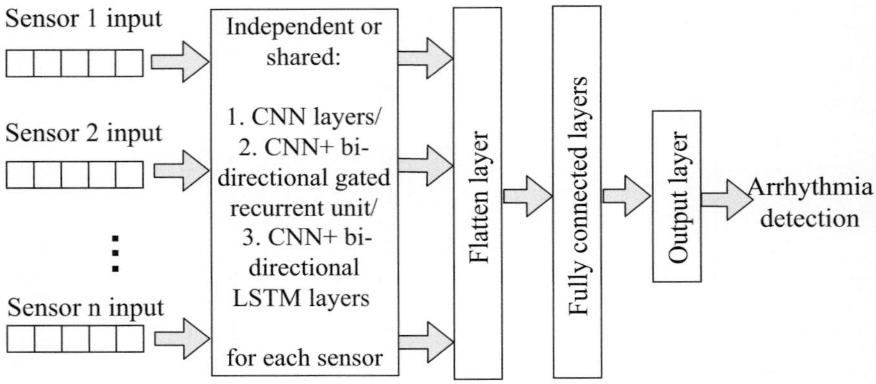

Fig. 2.15 Summarized flow diagram of the deep learning-based fusion algorithms for arrhythmia detection as discussed in [149–152]

in Table 2.6. The performances of the fusion algorithms are also provided for comparison purposes. However, it should be noted that performance metrics vary from article to article, and a few metrics not discussed previously in this chapter are used in the context of arrhythmia detection. Therefore, a brief explanation of the various metrics used for this application is provided next:

1. False alarm suppression rate—The proportion of true negatives detected from among the total number of negatives, which is also known as specificity. True negatives correspond to normal sinus rhythm in case of arrhythmia detection. False alarm suppression is the terminology used when the work focuses on reducing the number of false arrhythmia alarms.
2. F1-score—The harmonic mean of sensitivity (recall) and positive predictive value (precision).

2.5.7 Data Fusion for Atrial Fibrillation Detection

Atrial fibrillation (AF) is the most prevalent cardiac arrhythmia affecting 1–2% of the general population. It is characterized by rapid and disorganized atrial activation leading to impaired atrial function [154]. AF can be detected from single-lead short-term ECG recordings in clinical settings and in home environments [155, 156]. Literature on the use of data fusion using multiple ECG signals or a combination of ECG and other pulsatile signals for AF detection is sparse, despite the popularity of multi-sensor fusion in arrhythmia detection. In this section, we discuss a few data fusion-based AF detection algorithms.

In [31], a Markov model-based classifier that used inter-pulse interval features from pulses detected from the ECG signal was proposed to detect AF, while using the wrist accelerometer data to reject noisy pulses (Fig. 2.16).

Table 2.6 Summary of fusion algorithms for arrhythmia detection

Article	Method	Fusion architecture	Signals	Performance
He et al. [128]	SQI- and HR-based rules	Complementary, feature level, FEI-DEO	ECG and ABP	Sensitivity—65%
Sadr et al. [129]	SQI- and HR-based rules	Complementary, feature level, FEI-DEO	ECG, PPG, and ABP	Score[a]—74.03
Su et al. [130]	SQI- and HR-based rules	Complementary, feature level, FEI-DEO	ECG and SpO2	False alarm suppression—50%
Fallet et al. [131]	SQI- and HR-based rules	Complementary, feature level, FEI-DEO	ECG, PPG, and ABP	Score[a]—76.11
Couto et al. [132]	SQI- and HR-based rules	Complementary, feature level, FEI-DEO	ECG, PPG, and ABP	Score[a]—78.65
Teo et al. [133]	SQI- and HR-based rules	Complementary, feature level, FEI-DEO	ECG, PPG, and ABP	Score[a]—69.92
Liu et al. [134]	SQI- and HR-based rules	Complementary, feature level, FEI-DEO	ECG, PPG, and ABP	False alarm suppression—56.7%
Liu et al. [135]	Rules based	Complementary, feature level, FEI-DEO	ECG, PPG, and ABP	Score[a]—75.91
Plesinger et al. [136]	Rules based	Complementary, feature level, FEI-DEO	ECG, PPG, and ABP	Score[a]—81.39
Zong [137]	Rules based	Complementary, feature level, FEI-DEO	ECG, PPG, and ABP	Score[a]—70.10
Gierałtowski et al. [138]	Rules based	Complementary, feature-level, FEI-DEO	ECG, PPG, and ABP	Score[a]—57.72
Zhang et al. [139]	SVM	Complementary, feature level, FEI-DEO	ECG, PPG, and ABP	Score[a]—84.4
Caballero and Mirsky [140]	DT	Complementary, feature level, FEI-DEO	ECG, PPG, and ABP	Score[a]—65.19
de Chazal et al. [141]	Linear discriminant and voting	Redundant, feature level, FEI-DEO	Two-channel ECG	SVEB sensitivity—75.9%, VEB sensitivity—77.7%
Eerikäinen et al. [142]	ML classifiers along with SQI-based features	Complementary, feature level, FEI-DEO	ECG, PPG, and ABP	Score[a]—75.54

(continued)

Table 2.6 (continued)

Article	Method	Fusion architecture	Signals	Performance
Srivastava et al. [143]	ML classifiers along with SQI-based features	Complementary, feature level, FEI-DEO	ECG, PPG, and ABP	Accuracy—83.96%
Hoog Antink and Leonhardt [144]	DT, discriminant analysis, and SVM	Complementary, feature level, FEI-DEO	ECG and PPG/ABP	Score—75.55
Kalidas and Tamil [145]	SVM	Complementary, feature level, FEI-DEO	ECG and PPG	Sensitivity—96%
Xu et al. [146]	Case-based reasoning model	Complementary, feature level, FEI-DEO	ECG, PPG, and ABP	Sensitivity—83%
Manna and Swetapadma [147]	K-NN	Complementary, feature level, FEI-DEO	ECG and ABP	Accuracy—99.95%
Daluwatte et al. [148]	SVM	Complementary, feature-level, FEI-DEO	ECG, PPG, and ABP	Score[a]—70.2
Li et al. [149]	CNN	Redundant, signal level, FEI-DEO	2-lead ECG	SVEB sensitivity-81.1%, VEB sensitivity—97.7%
Zhang et al. [150]	MLBF-net + CNN	Redundant, signal level and feature level, DAI-FEO and FEI-DEO	12-lead ECG	F1-score—0.855
Wang et al. [151]	CNN with channel-wise attention	Redundant, signal level, FEI-DEO	12-lead ECG	F1-score—0.813
Ye and Lu [152]	CNN+ Bidirectional LSTMs	Redundant, signal level, FEI-DEO	12-lead ECG	F1-score—0.812
Hoang et al. [153]	CNN	Redundant, signal level, FEI-DEO	6-lead ECG	Sensitivity—78.60%

[a]Score in this table refers to the custom performance metric used in the Physionet 2015 challenge [127]

A few ML-based approaches for AF detection are also discussed in the literature. A hybrid classifier that leveraged an ML classifier and a statistical rule-based classifier for AF detection was proposed in [157], where the inter-beat intervals obtained from the PPG signal and related features were used for classification along with PPG SQIs as well as motion detection from the accelerometer data were used to eliminate noisy inter-beat intervals from the decision classifier. In [158], a similar approach was used to distinguish between AF, atrial flutter, and normal sinus rhythm using other waveform features extracted from the PPG signal as well as the accelerometer data using a random forest classifier. An elastic net logic was used to detect AF using a combined feature vector obtained from a CNN network (whose input was eight-channel PPG recordings) as well as features obtained from eight-channel PPG

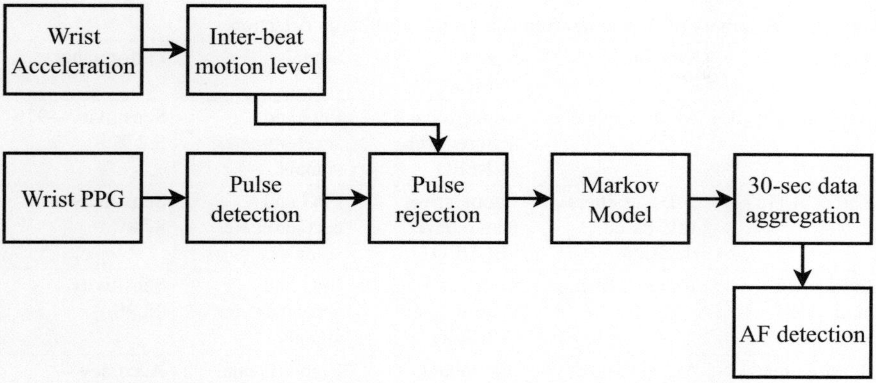

Fig. 2.16 The Markov model-based classifier for atrial fibrillation detection discussed in [31]. *Note*: Modified from [31], with permission from Bonomi

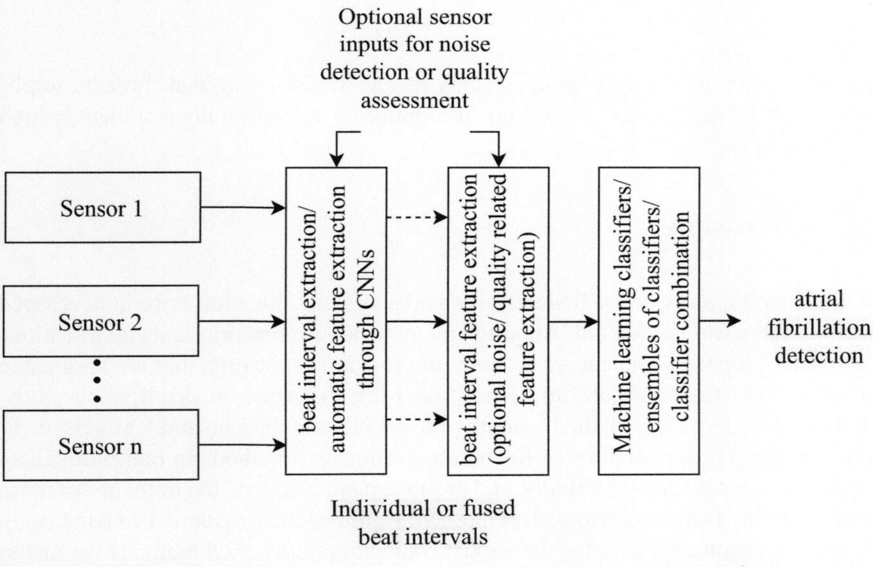

Fig. 2.17 Summarized flow diagram of the machine learning-based fusion algorithms for atrial fibrillation detection as discussed in [30, 157, 158]

recordings and its SQIs in [30]. A summarized flow diagram of the ML-based fusion algorithms for AF detection is shown in Fig. 2.17.

A summary of the fusion algorithms discussed for AF detection is provided in Table 2.7.

Now that we have thoroughly examined the key findings and insights from existing research on multi-sensor data fusion for specific applications, it is now essential to

Table 2.7 Summary of fusion algorithms for atrial fibrillation detection

Article	Method	Fusion architecture	Signals	Performance
Bonomi et al. [31]	Markov model	Cooperative, feature level, FEI-FEO	PPG and accelerometer signals	Sensitivity—97% ±2 %
Zhu et al. [157]	ML classifier+ rule-based decisions	Cooperative, feature level, FEI-DEO	PPG and accelerometer signals	Sensitivity—87.8%
Eerikäinen et al. [158]	Random forests	Cooperative, feature-level, FEI-DEO	PPG and accelerometer signals	Sensitivity—98.2%
Shashikumar et al. [30]	ML classifiers using features obtained from CNNs and SQIs	Redundant, feature level, FEI-DEO	Eight-channel PPG	Accuracy—91.8%

transition from analyzing individual study results to discussing their broader implications within the field of physiological monitoring algorithm development, which is discussed next.

2.6 Discussion

In this chapter, a review of fusion frameworks and classification criteria developed over the years was carried out. Although many fusion frameworks and classifications have been proposed over the years, only the fusion frameworks that are applicable to biomedical fusion algorithm development are discussed in detail, such as the Durrant-Whyte model [4], the Dasarathy model [8], and the Luo and Kay classification [5]. Various applications of fusion algorithms and methods in non-biomedical applications are discussed initially, and in subsequent chapters, the influence of these methods in the fusion algorithm development would become evident. From the study of various fusion applications, there exist some commonly used methods for fusion in literature:

1. Kalman filtering or variants of the Kalman filter for state estimation once sensor fusion problems are converted into estimation problems.
2. Multiscale fusion of multimodal data in image fusion.
3. Simple weighted averaging or sample-wise/pixel-wise operations in unimodal data fusion.
4. Machine learning-based methods for feature fusion, including deep learning and convolutional neural networks with raw data inputs.

A review of various fusion methodologies used for the physiological signal monitoring applications examined in this book is also carried out. From the study of application-based analysis of fusion techniques, few similarities in fusion methodologies could be observed. Therefore, from the reviewed literature on fusion for heartbeat detection, heart rate estimation, respiration rate estimation, sleep apnea detection, arrhythmia detection, and atrial fibrillation detection, it can be observed that fusion algorithms can be broadly categorized into:

1. State estimation algorithms that make use of Kalman filtering, Bayesian inference, particle filtering, and other probabilistic models.
2. Rule-based fusion algorithms: selection-based, combination-based, and elimination- and combination-based fusion.
3. SQI and rule-based algorithms.
4. Machine learning or deep learning for feature fusion.
5. Convolutional neural network-based fusion, where multi-sensor data is fed into a network and fusion is learned by the network.

2.6.1 Challenges

Although the various works discussed in this review individually address many challenges associated with multi-sensor data fusion for wearables, the primary challenges still remain, with most works providing solutions for specific applications. Ensuring high data quality is a primary challenge in multi-sensor fusion for wearable-based health monitoring, as data can often be noisy due to environmental factors or patient movement. Accurately fusing these data sources of uncertain quality requires algorithms capable of managing ambiguity and providing robust insights for healthcare applications. Uncertainty handling is critical in clinical settings, where decisions based on fused data can have significant implications for patient care. The use of SQIs has been found to tackle this issue with sufficient reliability, however, a generalized approach for all applications is not available. Additionally, health monitoring involves various data types—such as physiological signals, movement data, and others—which differ in format, precision, and sampling rate. Integrating these heterogeneous data sources requires sophisticated methods that can handle inconsistencies and align diverse data sources accurately for meaningful fusion. Some fusion works have proposed methods to address such challenges, but such considerations and approaches are not yet mainstream.

Real-time data fusion can be computationally intensive, especially when using advanced algorithms on wearable devices with limited processing power and battery life. Low latency and energy efficiency are crucial for continuous health monitoring applications, where data needs to be processed quickly without excessive power consumption. As these systems scale to integrate multiple data streams, managing and analyzing the high data volume adds to the computational demands, challenging the system's ability to provide timely health insights. Some fusion works have explored the computational cost associated with increasing the number of sensors and explored opportunities to reduce computational costs. Additionally, the use of SQIs also adds to the computational overhead and the development of energy-efficient SQI algorithms is an important research direction to enable multi-sensor data fusion.

In real-world applications, data may also be intermittently unavailable due to connectivity issues or sensor malfunctions, resulting in missing or incomplete datasets. Developing methods that can align temporally misaligned data and interpolate or manage missing values is essential to maintain the reliability of health monitoring systems.

2.7 Conclusions

From the fusion works discussed in this chapter, it can be observed that most fusion algorithms are dependent on the applications. It can also be noted that the fusion architectures discussed in Sect. 2.3 do not explicitly consider the quality of the signals used in fusion. This is a major concern in the case of biomedical wearable devices due to the ambulatory noise captured by the sensors, the miniaturized construction of the sensors, and the placement of sensors on the body away from the location of interest. Therefore, signal quality indicator-based fusion algorithms are important in the context of biomedical wearable devices, as can be observed from the discussion in Sect. 2.5. It should be noted from the existing literature that rule-based fusion algorithms dependent on SQIs for sample-wise signal fusion are not explored. This is the gap that Chap. 4 aims to address through the proposal of SQI-based selection algorithms, SQI-based combination algorithms, and SQI-based elimination and combination algorithms. This book also describes the development of an SQI-based temporal fusion framework for multimodal signals to address the gap in signal-quality-aware temporal fusion in Chap. 5.

Multi-sensor signal acquisition in wearable devices may follow a strategy of different sampling frequencies for different signals. This can be a design choice to reduce power consumption due to which the frequency of individual sensor-specific events of interest may vary. However, for developing fusion algorithms for temporal inferences, resolution matching is important. Most fusion algorithms would resort to signal interpolation or resampling to address the issue of resolution matching. However, such methods would add to the computational overhead of the fusion algorithm and the wearable device. It can also be observed that fusion works in literature often overlook the issue of multiresolution data, i.e., multi-sensor inputs sampled at different sampling frequencies (ignore or overcome the issue through resampling/ signal padding) for detection tasks with quick detection rates, and therefore in Chap. 6 the issue of multiresolution fusion in the context of CNN-based fusion for high detection rates is addressed.

From both Dasarathy's [8] and Luo's and Kay's classification [5] of fusion algorithms, it can be observed that the information abstraction level at which fusion occurs is often decided by the designer. Therefore, this method is not the ideal strategy for identifying the information abstraction level which is optimal for a given task and is found to vary based on the experiences of different designers. But with advances in deep neural network-based models, methods to develop the networks that automatically learn the best information abstraction level for fusion can be developed. This is the topic addressed in Chap. 7, where a network that performs fusion at the appropriate information abstraction level and utilizes signal quality indicators for fusion is proposed. In deep learning-based fusion, automatic identification of the fusion level using a data-driven approach enhances flexibility. Relevant information is fused early and carried forward, ensuring it remains available for later-stage fusion if needed.

References

1. F. White, Data fusion lexicon. Technical report, joint directors of laboratories, technical panel for c3 (1987)
2. L.-F. Pau, Sensor data fusion. J. Intell. Robot. Syst. 1, 103–116 (1988)
3. J. Schoess, G. Castore, A distributed sensor architecture for advanced aerospace systems, in *Sensor Fusion*, ed. by C.B. Weaver, vol. 0931 (International Society for Optics and Photonics, SPIE, 1988), pp. 74–87
4. H.F. Durrant-Whyte, Sensor models and multisensor integration. Int. J. Robot. Res. **7**(6), 97–113 (1988)
5. R. Luo, M. Kay, Multisensor integration and fusion in intelligent systems. IEEE Trans. Syst. Man Cybern. **19**(5), 901–931 (1989)
6. S.C.A. Thomopoulos, Sensor integration and data fusion, in *Sensor Fusion II: Human and Machine Strategies*. Society of Photo-Optical Instrumentation Engineers (SPIE) Conference Series, , vol. 1198 (Mar 1990), pp. 178–191
7. V. Clement, G. Giraudon, S. Houzelle, F. Sandakly, Interpretation of remotely sensed images in a context of multisensor fusion using a multispecialist architecture. IEEE Trans. Geosci. Remote Sens. **31**(4), 779–791 (1993)
8. B.V. Dasarathy, Sensor fusion potential exploitation-innovative architectures and illustrative applications. Proc. IEEE **85**, 24–38 (1997)
9. C.J. Harris, A. Bailey, T. Dodd, Multi-sensor data fusion in defence and aerospace. Aeronaut. J. (1968) **102**(1015), 229–244 (1998)
10. C. Pohl, J.L.V. Genderen, Review article multisensor image fusion in remote sensing: Concepts, methods and applications. Int. J. Remote Sens. **19**(5), 823–854 (1998)
11. M. Bedworth, J. O'Brien, The omnibus model: a new model of data fusion? IEEE Aerosp. Electron. Syst. Mag. **15**(4), 30–36 (2000)
12. H. Keithley, *An Evaluation Methodology for Fusion Processes Based on Information Needs* (CRC Press, Boca Raton, FL, 2001), pp. 527–539
13. K.B. Laskey, S.M. Mahoney, Knowledge and data fusion in probabilistic networks, in*Submitted to Machine Learning Special Issue on Knowledge and Data Fusion, A paraıtre* (2003)
14. K.C. Chang, Y. Song, M.E.L. II, Performance modeling for multisensor data fusion, in *Signal Processing, Sensor Fusion, and Target Recognition XII*, ed. by I. Kadar. International Society for Optics and Photonics, vol. 5096 (SPIE, 2003), pp. 354–363
15. H. Carvalho, W. Heinzelman, A. Murphy, C. Coelho, A general data fusion architecture, in *Proceedings of the Sixth International Conference of Information Fusion, 2003*, vol. 2 (2003), pp. 1465–1472
16. R. Brooks, S.S. Iyengar, R. Brooks, *Soft Computing Techniques* (Chapman and Hall/CRC Boca Raton, FL, 2004), pp. 321–333
17. M. Ahmed, G. Pottie, *Fusion in the Context of Information Theory* (Chapman and Hall/CRC Boca Raton, FL, 2004), pp. 419–435

18. E.P. Blasch, M. Pribilski, B. Daughtery, B. Roscoe, J. Gunsett, Fusion metrics for dynamic situation analysis, in *Signal Processing, Sensor Fusion, and Target Recognition XIII*, vol. 5429. International Society for Optics and Photonics (SPIE, 2004), pp. 428–438
19. J. Esteban, A. Starr, R. Willetts, P. Hannah, P. Bryanston-cross, A review of data fusion models and architectures: towards engineering guidelines. Neural Comput. Appl. **14**, 273–281 (2005)
20. O. Cohen, Y. Edan, A sensor fusion framework for online sensor and algorithm selection. Robot. Auton. Syst. **56**, 762–776 (2008)
21. C. Bowman, S.A.N., *Systems Engineering Approach for Implementing Data Fusion Systems* (CRC Press, Boca Raton, FL, 2009), pp. 561–596
22. J. Llinas, *Assessing the Performance of Multisensor Fusion Processes* (CRC Press, Boca Raton, FL, 2009), pp. 655–675
23. J. Dong, D. Zhuang, Y. Huang, J. Fu, Advances in multi-sensor data fusion: algorithms and applications. Sensors **9**(10), 7771–7784 (2009)
24. J. Chen, Research on fusion performance model for multisensor data fusion, in *2010 International Conference on E-Product E-Service and E-Entertainment* (2010), pp. 1–4
25. F. Castanedo, A review of data fusion techniques. Sci. World J. (2013)
26. M.L. Fung, M.Z.Q. Chen, Y.H. Chen, Sensor fusion: a review of methods and applications, in *2017 29th Chinese Control and Decision Conference (CCDC)* (2017), pp. 3853–3860
27. W. Elmenreich, An introduction to sensor fusion, *Technische Universitét Wien, Institut für Technische Informatik, Technical Report* (May 2001)
28. J. Boyd, A discourse on winning and losing [briefing slides], *Maxwell Air Force Base, AL: Air University Library. (Document No. MU 43947)* (1987)
29. B.S. Chandra, C.S. Sastry, S. Jana, Robust heartbeat detection from multimodal data via CNN-based generalizable information fusion. IEEE Trans. Biomed. Eng. **66**, 710–717 (2018)
30. S.P. Shashikumar, A.J. Shah, Q. Li, G.D. Clifford, S. Nemati, A deep learning approach to monitoring and detecting atrial fibrillation using wearable technology, in *2017 IEEE EMBS International Conference on Biomedical Health Informatics (BHI)* (2017), pp. 141–144
31. A.G. Bonomi, F. Schipper, L.M. Eerikäinen, J. Margarito, R.M. Aarts, S. Babaeizadeh, H.M. de Morree, L. Dekker, Atrial fibrillation detection using photo-plethysmography and acceleration data at the wrist, in *2016 Computing in Cardiology Conference (CinC)* (2016), pp. 277–280
32. T. Meng, X. Jing, Z. Yan, W. Pedrycz, A survey on machine learning for data fusion. Inf. Fusion **57**, 115–129 (2020)
33. W. Liu, O. Camps, M. Sznaier, Multi-camera multi-object tracking (2017). arXiv:1709.07065
34. I. Mcloughlin, Z. Xie, Y. Song, H. Phan, R. Palaniappan, Time–frequency feature fusion for noise robust audio event classification. Circuits Syst Signal Process. **39**, 1672–1687 (2020)
35. S.A. Rankawat, R. Dubey, Robust heart rate estimation from multimodal physiological signals using beat signal quality index based majority voting fusion method. Biomed. Signal Process. Control **33**, 201–212 (2017)
36. V. Nathan, I. Akkaya, R. Jafari, A particle filter framework for the estimation of heart rate from ECG signals corrupted by motion artifacts," in *2015 37th Annual International Conference of the IEEE Engineering in Medicine and Biology Society (EMBC)* (2015), pp. 6560–6565
37. T. Jeon, J. Yu, W. Pedrycz, Robust detection of heartbeats using association models from blood pressure and EEG signals. Biomed. Eng. Online **15**, 7 (2016)
38. S. Lapi, F. Lavorini, G. Borgioli, M. Calzolai, L. Masotti, M. Pistolesi, G.A. Fontana, Respiratory rate assessments using a dual-accelerometer device. Respir. Physiol. Neurobiol. **191**, 60–66 (2014)
39. D.A. Birrenkott, M.A. Pimentel, P.J. Watkinson, D.A. Clifton, A robust fusion model for estimating respiratory rate from photoplethysmography and electrocardiography. IEEE Trans. Biomed. Eng. **65**, 2033–2041 (2018)
40. E. Kenneth, A. Rajendra, N. Kannathal, C.M. Lim, Data fusion of multimodal cardiovascular signals," in *2005 IEEE Engineering in Medicine and Biology 27th Annual Conference* (2005), pp. 4689–4692

41. H. Huang, J. Liu, Q. Zhu, R. Wang, G. Hu, A new hierarchical method for inter-patient heartbeat classification using random projections and rr intervals. Biomed. Eng. Online **13**(06), 90 (2014)
42. T. Wartzek, C. BrÃser, M. Walter, S. Leonhardt, Robust sensor fusion of unobtrusively measured heart rate. IEEE J. Biomed. Health Inf. **18**(2), 654–660 (2014)
43. S. Jusoh, S. Almajali, A systematic review on fusion techniques and approaches used in applications. IEEE Access **8**, 14424–14439 (2020)
44. A. Rövid, V. Remeli, Towards raw sensor fusion in 3D object detection, in *2019 IEEE 17th World Symposium on Applied Machine Intelligence and Informatics (SAMI)* (2019), pp. 293–298
45. B. Shahian Jahromi, T. Tulabandhula, S. Cetin, Real-time hybrid multi-sensor fusion framework for perception in autonomous vehicles. Sensors **19** (2019)
46. J. Fayyad, M.A. Jaradat, D. Gruyer, H. Najjaran, Deep learning sensor fusion for autonomous vehicle perception and localization: a review. Sensors **20** (2020)
47. R.O. Chavez-Garcia, O. Aycard, Multiple sensor fusion and classification for moving object detection and tracking. IEEE Trans. Intell. Transp. Syst. **17**, 525–534 (2016)
48. V. John, S. Mita, RVNet: deep sensor fusion of monocular camera and radar for image-based obstacle detection in challenging environments, in *Image and Video Technology* (Springer International Publishing, Cham, 2019), pp. 351–364
49. R. Girshick, J. Donahue, T. Darrell, J. Malik, Rich feature hierarchies for accurate object detection and semantic segmentation (2014)
50. M.B. Alatise, G.P. Hancke, Pose estimation of a mobile robot based on fusion of IMU data and vision data using an extended Kalman filter. Sensors **17** (2017)
51. D. Luo, Y. Ma, X. Zhang, X. Wu, Multi-sensor fusion based robot self-activity recognition, in *2018 IEEE-RAS 18th International Conference on Humanoid Robots (Humanoids)* (2018), pp. 1–9
52. T. Nicosevici, R. Garcia, M. Carreras, M. Villanueva, A review of sensor fusion techniques for underwater vehicle navigation," in *Oceans'04 MTS/IEEE Techno-Ocean'04 (IEEE Cat. No.04CH37600)*, vol. 3 (2004), pp. 1600–1605
53. D. Novak, R. Riener, A survey of sensor fusion methods in wearable robotics. Robot. Auton. Syst. **73**, 155–170 (2015)
54. H.F. Nweke, Y.W. Teh, G. Mujtaba, M.A. Al-garadi, Data fusion and multiple classifier systems for human activity detection and health monitoring: review and open research directions. Inf. Fusion **46**, 147–170 (2019)
55. R. Gravina, P. Alinia, H. Ghasemzadeh, G. Fortino, Multi-sensor fusion in body sensor networks: state-of-the-art and research challenges. Inf. Fusion **35**, 68–80 (2017)
56. X. Xi, M. Tang, Z. Luo, Feature-level fusion of surface electromyography for activity monitoring. Sensors **18** (2018)
57. V. Bhateja, H. Patel, A. Krishn, A. Sahu, A. Lay-Ekuakille, Multimodal medical image sensor fusion framework using cascade of wavelet and contourlet transform domains. IEEE Sens. J. **15**, 6783–6790 (2015)
58. B. Yang, Z. Jing, H. Zhao, Review of pixel-level image fusion. J. Shanghai Jiaotong Univ. (Sci.) **15**, 6–12 (2010)
59. K.B. Ozanyan, Tomography defined as sensor fusion, in *2015 IEEE Sensors* (2015), pp. 1–4
60. C. Byington, M. Watson, H. Lee, M. Hollins, Sensor-level fusion to enhance health and usage monitoring systems, *Annual Forum Proceedings-AHS International*, vol. 2 (2008), pp. 1476–1485
61. J. Liu, Y. Hu, Y. Wang, B. Wu, J. Fan, Z. Hu, An integrated multi-sensor fusion-based deep feature learning approach for rotating machinery diagnosis. Meas. Sci. Technol. **29**, 055103 (2018)
62. M.H. Kamarudin, Z.H. Ismail, N.B. Saidi, Deep learning sensor fusion in plant water stress assessment: a comprehensive review. Appl. Sci. **11**(4) (2021)
63. P. Narkhede, R. Walambe, S. Mandaokar, P. Chandel, K. Kotecha, G. Ghinea, Gas detection and identification using multimodal artificial intelligence based sensor fusion. Appl. Syst. Innov. **4**(1)

64. G. Balakrishnan, F. Durand, J. Guttag, Detecting pulse from head motions in video, in *2013 IEEE Conference on Computer Vision and Pattern Recognition* (2013), pp. 3430–3437
65. H.B. Mitchell, *Data Fusion: Concepts and Ideas* (Springer Science & Business Media, 2012)
66. C. Orphanidou, *Signal Quality Assessment in Physiological Monitoring: State of the Art and Practical Considerations* (Springer International Publishing, 2018)
67. I. Silva, B. Moody, J. Behar, A. Johnson, J. Oster, G.D. Clifford, G.B. Moody, Robust detection of heart beats in multimodal data. Physiol. Meas. **36**(8), 549–552 (2015)
68. C. Diao, B. Wang, N. Cai, Data fusion of multivariate time series: application to noisy 12-lead ECG signals. Appl. Sci. **9**(1)
69. J. Pan, W.J. Tompkins, A real-time QRS detection algorithm. IEEE Trans. Biomed. Eng. **BME-32**, 230–236 (1985)
70. W.A.H. Engelse, C. Zeelenberg, A single scan algorithm for QRS-detection and feature extraction. Comput. Cardiol. **6**, 37–42 (1979)
71. W. Zong, G. Moody, D. Jiang, A robust open-source algorithm to detect onset and duration of QRS complexes. Comput. Cardiol. **2003**, 737–740 (2003)
72. W. Zhao, Y. Xu, C. Yan, J. Hu, D. Jia, X. Sun, T. You, A multilead fusion based QRS complex detection method on 12-lead electrocardiogram signals, in *2018 Computing in Cardiology Conference (CinC)*, vol. 45 (2018), pp. 1–4
73. L. Galeotti, C.G. Scully, J. Vicente, L. Johannesen, D.G. Strauss, Robust algorithm to locate heart beats from multiple physiological waveforms by individual signal detector voting. Physiol. Meas. **36**(8), 1705 (2015)
74. C.H. Antink, C. BrÃser, S. Leonhardt, Multimodal sensor fusion of cardiac signals via blind deconvolution: a source-filter approach, in *Computing in Cardiology 2014* (2014), pp. 805–808
75. A.E. Johnson, J. Behar, F. Andreotti, G.D. Clifford, J. Oster, R-peak estimation using multimodal lead switching. Comput. Cardiol. **2014**, 281–284 (2014)
76. M. Vollmer, Robust detection of heart beats using dynamic thresholds and moving windows. Comput. Cardiol. **2014**, 569–572 (2014)
77. M.J. Mollakazemi, F. Asadi, S. Ghiasi, S.H. Sadati, Applying quality index criterion for flexible multi-detection of heartbeat using features of multimodal data, in *2016 Computing in Cardiology Conference (CinC)* (2016), pp. 1065–1068
78. S.A. Rankawat, M. Rankawat, R. Dubey, Heart rate estimation from non-cardiovascular signals using slope sum function and teager energy," in *2015 International Conference on Industrial Instrumentation and Control (ICIC)* (2015), pp. 1534–1539
79. Q. Ding, Y. Bai, Y.B. Erol, R. Salas-Boni, X. Zhang, L. Li, X. Hu, Multimodal information fusion for robust heart beat detection. Comput. Cardiol. **2014**, 261–264 (2014)
80. C.H. Antink, E. Breuer, D.U. Uguz, S. Leonhardt, Signal-level fusion with convolutional neural networks for capacitively coupled ECG in the car, in *2018 Computing in Cardiology Conference (CinC)*, vol. 45 (2018), pp. 1–4
81. J.M. Warnecke, N. Boeker, N. Spicher, J. Wang, M. Flormann, T.M. Deserno, Sensor fusion for robust heartbeat detection during driving, in *2021 43rd Annual International Conference of the IEEE Engineering in Medicine Biology Society (EMBC)* (2021), pp. 447–450
82. B. Xie, H. Minn, Real-time sleep apnea detection by classifier combination. IEEE Trans. Inf Technol. Biomed. **16**(3), 469–477 (2012)
83. J. Di, J. Li, C. Liu, A Bayesian fusion model for heart rate annotations, in *2020 International Conference on Sensing, Measurement Data Analytics in the Era of Artificial Intelligence (ICSMD)* (2020), pp. 307–312
84. V. Nathan, R. Jafari, Particle filtering and sensor fusion for robust heart rate monitoring using wearable sensors. IEEE J. Biomed. Health Inform. **22**(6), 1834–1846 (2018)
85. G. De Morais Borges, V. Brusamarello, Bayesian fusion of multiple sensors for reliable heart rate detection, in *2014 IEEE International Instrumentation and Measurement Technology Conference (I2MTC) Proceedings* (2014), pp. 1310–1313
86. C. Brüser, J.M. Kortelainen, S. Winter, M. Tenhunen, J. Pärkkä, S. Leonhardt, Improvement of force-sensor-based heart rate estimation using multichannel data fusion. IEEE J. Biomed. Health Inf. **19**, 227–235 (2015)

87. A. Aygun, H. Ghasemzadeh, R. Jafari, Robust interbeat interval and heart rate variability estimation method from various morphological features using wearable sensors. IEEE J. Biomed. Health Inf. **24**, 2238–2250 (2020)
88. Q. Li, H. Alemzadeh, Z. Kalbarczyk, R.K. Iyer, A fault-tolerant hardware architecture for robust wearable heart rate monitoring, in *2015 9th International Conference on Pervasive Computing Technologies for Healthcare (PervasiveHealth)* (2015), pp. 185–192
89. Y. Huang, C. Sun, T. Jin, S. Yang, Z. Zhang, Unobtrusive inter-beat interval estimation from multichannel ballistocardiogram signal using Kalman filter, in *2020 42nd Annual International Conference of the IEEE Engineering in Medicine Biology Society (EMBC)* (2020), pp. 455–460
90. D. Garrido, J.J. Assioun, A. Keshishyan, M.A. Sanchez-Gonzalez, B. Goubran, Respiratory rate variability as a prognostic factor in hospitalized patients transferred to the intensive care unit. Cureus **10**(1), e2100 (2018)
91. T. Lauteslager, V. Dishakjian, L. Watson, J. Savarese, A.J. Williams, G.D. Leschziner, Detecting early signs of deterioration and preventing hospitalizations in skilled nursing facilities using remote respiratory monitoring. Respir. Med. Case. Rep. **50**, 102044 (2024)
92. A. Adami, R. Boostani, F. Marzbanrad, P.H. Charlton, A new framework to estimate breathing rate from electrocardiogram, photoplethysmogram, and blood pressure signals. IEEE Access **9**, 45832–45844 (2021)
93. X.-R. Ding, Y.-T. Zhang, H.K. Tsang, W. Karlen, A pulse transit time based fusion method for the noninvasive and continuous monitoring of respiratory rate, in *2016 38th Annual International Conference of the IEEE Engineering in Medicine and Biology Society (EMBC)* (2016), pp. 4240–4243
94. S. Nemati, A. Malhotra, G.D. Clifford, Data fusion for improved respiration rate estimation. EURASIP J. Adv. Signal Process. **2010**, 1–10 (2010)
95. P.H. Charlton, M. Villarroel, F. Salguiero, *Waveform Analysis to Estimate Respiratory Rate*, ch. 26 (Springer, Cham, 2016), pp. 377–390
96. S. Baker, W. Xiang, I. Atkinson, Determining respiratory rate from photoplethysmogram and electrocardiogram signals using respiratory quality indices and neural networks. PLoS One **16**, e0249843 (2021)
97. W. Karlen, S. Raman, J.M. Ansermino, G.A. Dumont, Multiparameter respiratory rate estimation from the photoplethysmogram. IEEE Trans. Biomed. Eng. **60**, 1946–1953 (2013)
98. Z. He, X. Chen, Z. Fang, T. Sheng, S. Xia, Fusion estimation of respiration rate from ECG and PPG signal based on android platform and wearable watch, in *2nd IET International Conference on Biomedical Image and Signal Processing (ICBISP 2017)* (2017), pp. 1–6
99. C. Orphanidou, T. Bonnici, P. Charlton, D. Clifton, D. Vallance, L. Tarassenko, Signal-quality indices for the electrocardiogram and photoplethysmogram: derivation and applications to wireless monitoring. IEEE J. Biomed. Health Inf. **19** (2015)
100. C. Orphanidou, Derivation of respiration rate from ambulatory ecg and ppg using ensemble empirical mode decomposition: comparison and fusion. Comput. Biol. Med. **81**, 45–54 (2017)
101. C. Ahlstrom, A. Johansson, T. Lanne, P. Ask, A respiration monitor based on electrocardiographic and photoplethysmographic sensor fusion, in *The 26th Annual International Conference of the IEEE Engineering in Medicine and Biology Society*, vol. 1 (Sept. 2004), pp. 2311–2314
102. O. Ronneberger, P. Fischer, T. Brox, U-net: convolutional networks for biomedical image segmentation, in *Medical Image Computing and Computer-Assisted Intervention–MICCAI 2015*, (Springer International Publishing, Cham, 2015), pp. 234–241
103. M. Chan, V. Ganti, O. Inan, Respiratory rate estimation using u-net-based cascaded framework from electrocardiogram and seismocardiogram signals. IEEE J. Biomed. Health Inf. **26**, 2481–2492 (2022)
104. P.E. Peppard, T. Young, J.H. Barnet, M. Palta, E.W. Hagen, K.M. Hla, Increased prevalence of sleep-disordered breathing in adults. Am. J. Epidemiol. **177**, 1006–1014 (2013)
105. B. Xie, H. Minn, Real-time sleep apnea detection by classifier combination. IEEE Trans. Inf Technol. Biomed. **16**, 469–477 (2012)

106. R. Atri, M. Mohebbi, Obstructive sleep apnea detection using spectrum and bispectrum analysis of single-lead ECG signal. Physiol. Meas. **36**(9), 1963–1980 (2015)
107. B. Sulistyo, N. Surantha, S.M. Isa, Sleep apnea identification using HRV features of ECG signals. Int. J. Electric. Comput. Eng. (IJECE) **8**, 3940–3948 (2018)
108. E. Urtnasan, Y.-J. Kim, J.-U. Park, E.-Y. Joo, K.-J. Lee, Deep learning approaches for automatic detection of sleep apnea events from an electrocardiogram. Comput. Methods Prog. Biomed. **180**, 105001 (2019)
109. L. Almazaydeh, K. Elleithy, M. Faezipour, Detection of obstructive sleep apnea through ECG signal features, in *2012 IEEE International Conference on Electro/Information Technology* (2012), pp. 1–6
110. C. Varon, A. Caicedo, D. Testelmans, B. Buyse, S. Van Huffel, A novel algorithm for the automatic detection of sleep apnea from single-lead ECG. IEEE Trans. Biomed. Eng. **62**(9), 2269–2278 (2015)
111. H.D. Nguyen, B.A. Wilkins, Q. Cheng, B.A. Benjamin, An online sleep apnea detection method based on recurrence quantification analysis. IEEE J. Biomed. Health Inf. **18**(4), 1285–1293 (2014)
112. X. Liang, X. Qiao, Y. Li, Obstructive sleep apnea detection using combination of CNN and LSTM techniques, in *2019 IEEE 8th Joint International Information Technology and Artificial Intelligence Conference (ITAIC)* (2019), pp. 1733–1736
113. A. Bernardini, A. Brunello, G.L. Gigli, A. Montanari, N. Saccomanno, AIOSA: an approach to the automatic identification of obstructive sleep apnea events based on deep learning. Artif. Intell. Med. **118**, 102133 (2021)
114. S.S. Mostafa, F. Mendonça, F. Morgado-Dias, A. Ravelo-Garcia, SpO2 based sleep apnea detection using deep learning, in *2017 IEEE 21st International Conference on Intelligent Engineering Systems (INES)* (2017), pp. 091–096
115. S.S. Mostafa, F. Mendonça, A.G. Ravelo-GarciÂa, F. Morgado-Dias, A systematic review of detecting sleep apnea using deep learning. Sensors **19**(22), 4934 (2019)
116. L. Cen, Z.L. Yu, T. Kluge, W. Ser, Automatic system for obstructive sleep apnea events detection using convolutional neural network, in *2018 40th Annual International Conference of the IEEE Engineering in Medicine and Biology Society (EMBC)* (2018), pp. 3975–3978
117. Y. Lin, H. Wu, C. Hsu, P. Huang, Y. Huang, Y. Lo, Sleep apnea detection based on thoracic and abdominal movement signals of wearable piezoelectric bands. IEEE J. Biomed. Health Inf. **21**(6), 1533–1545 (2017)
118. C. Avci, A. Akbaş, Sleep apnea classification based on respiration signals by using ensemble methods. IOS Press **26**, S1703–S1710 (2015)
119. H.-C. Chang, H.-T. Wu, P.-C. Huang, H.-P. Ma, Y.-L. Lo, Y.-H. Huang, Portable sleep apnea syndrome screening and event detection using long short-term memory recurrent neural network. Sensors **20**(21), 6067 (2020)
120. A. Otero, P. Felix, M.R. Alvarez, C. Zamarron, Fuzzy structural algorithms to identify and characterize apnea and hypopnea episodes, in *2008 30th Annual International Conference of the IEEE Engineering in Medicine and Biology Society* (2008), pp. 5242–5245
121. S. Gutta, Q. Cheng, H.D. Nguyen, B.A. Benjamin, Cardiorespiratory model-based data-driven approach for sleep apnea detection. IEEE J. Biomed. Health Inf. **22**(4), 1036–1045 (2018)
122. G. Memis, M. Sert, Multimodal classification of obstructive sleep apnea using feature level fusion, in *2017 IEEE 11th International Conference on Semantic Computing (ICSC)* (2017), pp. 85–88
123. R. Haidar, S. McCloskey, I. Koprinska, B. Jeffries, Convolutional neural networks on multiple respiratory channels to detect hypopnea and obstructive apnea events, in *2018 International Joint Conference on Neural Networks (IJCNN)* (2018), pp. 1–7
124. T. Van Steenkiste, D. Deschrijver, T. Dhaene, Sensor fusion using backward shortcut connections for sleep apnea detection in multi-modal data (2019)
125. X. Lv, J. Li, A multi-level features fusion network for detecting obstructive sleep apnea hypopnea syndrome, in *Algorithms and Architectures for Parallel Processing: 20th International Conference, ICA3PP 2020, New York City, NY, USA, Oct 2–4, 2020, Proceedings, Part III*, (Springer, Berlin, Heidelberg, 2020), pp. 509–519

126. Ö. Yildirim, P. Pławiak, R.-S. Tan, U.R. Acharya, Arrhythmia detection using deep convolutional neural network with long duration ECG signals. Comput. Biol. Med. **102**, 411–420 (2018)
127. G.D. Clifford, I. Silva, B. Moody, Q. Li, D. Kella, A. Shahin, T. Kooistra, D. Perry, R.G. Mark, The physionet/computing in cardiology challenge 2015: Reducing false arrhythmia alarms in the icu, in *2015 Computing in Cardiology Conference (CinC)* (2015), pp. 273–276
128. R. He, H. Zhang, K. Wang, Y. Yuan, Q. Li, J. Pan, Z. Sheng, N. Zhao, Reducing false arrhythmia alarms in the ICU using novel signal quality indices assessment method, in *2015 Computing in Cardiology Conference (CinC)* (2015), pp. 1189–1192
129. N. Sadr, J. Huvanandana, D. T. Nguyen, C. Kalra, A. McEwan, P. de Chazal, Reducing false arrhythmia alarms in the ICU by Hilbert QRS detection, in *2015 Computing in Cardiology Conference (CinC)* (2015), pp. 1173–1176
130. J. Su, S. Liu, Z. Sun, B. Sun, W. Ye, C. Rajagopalan, H. Xianliang, Real-time fusion of ECG and SpO2 signals to reduce false alarms, in *2018 Computing in Cardiology Conference (CinC)*, vol. 45 (2018), pp. 1–4
131. S. Fallet, S. Yazdani, J.-M. Vesin, A multimodal approach to reduce false arrhythmia alarms in the intensive care unit, in *2015 Computing in Cardiology Conference (CinC)* (2015), pp. 277–280
132. P. Couto, R. Ramalho, R. Rodrigues, Suppression of false arrhythmia alarms using ECG and pulsatile waveforms, in *2015 Computing in Cardiology Conference (CinC)* (2015), pp. 749–752
133. S.-K. Teo, J.C. Wong, B. Yang, F. Yang, L. Feng, T.W. Lim, Y. Su, Reducing false arrhythmia alarms in the ICU, in *2015 Computing in Cardiology Conference (CinC)* (2015), pp. 1177–1180
134. M. Liu, Z. Sun, W. Ye, X. He, H. Jiang, Y. Li, Y. Zhuang, Optimization strategies to reduce alarm fatigue in patient monitors, in *2020 Computing in Cardiology* (2020), pp. 1–4
135. C. Liu, L. Zhao, H. Tang, Reduction of false alarms in intensive care unit using multi-feature fusion method, in *2015 Computing in Cardiology Conference (CinC)* (2015), pp. 741–744
136. F. Plesinger, P. Klimes, J. Halamek, P. Jurak, False alarms in intensive care unit monitors: detection of life-threatening arrhythmias using elementary algebra, descriptive statistics and fuzzy logic, in *2015 Computing in Cardiology Conference (CinC)* (2015), pp. 281–284
137. W. Zong, Reduction of false critical ECG alarms using waveform features of arterial blood pressure and/or photoplethysmogram signals, in *2015 Computing in Cardiology Conference (CinC)* (2015), pp. 289–292
138. J.J. Gierałtowski, I. Grzegorczyk, K. Ciuchciński, K. Kośna, M. Soliski, P. Podziemski, Algorithm for life-threatening arrhythmias detection with reduced false alarms, in *2015 Computing in Cardiology Conference (CinC)* (2015), pp. 1201–1204
139. Q. Zhang, X. Chen, Z. Fang, S. Xia, False arrhythmia alarm reduction in the intensive care unit using data fusion and machine learning, in *2016 IEEE-EMBS International Conference on Biomedical and Health Informatics (BHI)* (2016), pp. 232–235
140. M. Caballero, G.M. Mirsky, Reduction of false cardiac arrhythmia alarms through the use of machine learning techniques, in *2015 Computing in Cardiology Conference (CinC)* (2015), pp. 1169–1172
141. P. de Chazal, M. O'Dwyer, R. Reilly, Automatic classification of heartbeats using ECG morphology and heartbeat interval features. IEEE Trans. Biomed. Eng. **51**(7), 1196–1206 (2004)
142. L.M. Eerikäinen, J. Vanschoren, M.J. Rooijakkers, R. Vullings, R.M. Aarts, Decreasing the false alarm rate of arrhythmias in intensive care using a machine learning approach, in *2015 Computing in Cardiology Conference (CinC)* (2015), pp. 293–296
143. C. Srivastava, S. Sharma, A. Jalali, A novel algorithm for reducing false arrhythmia alarms in intensive care units, in *2016 38th Annual International Conference of the IEEE Engineering in Medicine and Biology Society (EMBC)* (2016), pp. 2525–2528
144. C. Hoog Antink, S. Leonhardt, Reducing false arrhythmia alarms using robust interval estimation and machine learning, in *2015 Computing in Cardiology Conference (CinC)* (2015), pp. 285–288

145. V. Kalidas, L.S. Tamil, Enhancing accuracy of arrhythmia classification by combining logical and machine learning techniques, in *2015 Computing in Cardiology Conference (CinC)* (2015), pp. 733–736
146. M. Xu, J. Shen, H. Yu, Multimodal data classification using signal quality indices and empirical similarity-based reasoning, in *2015 Computing in Cardiology Conference (CinC)* (2015), pp. 1197–1200
147. T. Manna, A. Swetapadma, An improved method for detecting false alarm in arrhythmia ICU patients, in *2018 Fourth International Conference on Research in Computational Intelligence and Communication Networks (ICRCICN)* (2018), pp. 66–69
148. C. Daluwatte, L. Johannesen, J. Vicente, C.G. Scully, L. Galeotti, D.G. Strauss, Heartbeat fusion algorithm to reduce false alarms for arrhythmias, in *2015 Computing in Cardiology Conference (CinC)* (2015), pp. 745–748
149. F. Li, J. Wu, M. Jia, Z. Chen, Y. Pu, Automated heartbeat classification exploiting convolutional neural network with channel-wise attention. IEEE Access **7**, 122955–122963 (2019)
150. J. Zhang, D. Liang, A. Liu, M. Gao, X. Chen, X. Zhang, X. Chen, MLBF-Net: a multi-lead-branch fusion network for multi-class arrhythmia classification using 12-lead ECG. IEEE J. Transl. Eng. Health Med. **9**, 1–11 (2021)
151. R. Wang, Q. Yao, X. Fan, Y. Li, Multi-class arrhythmia detection based on neural network with multi-stage features fusion, in *2019 IEEE International Conference on Systems, Man and Cybernetics (SMC)* (2019), pp. 4082–4087
152. X. Ye, Q. Lu, Automatic classification of 12-lead ECG based on model fusion, in *2020 13th International Congress on Image and Signal Processing, BioMedical Engineering and Informatics (CISP-BMEI)* (2020), pp. 733–738
153. T. Hoang, N. Fahier, W.-C. Fang, Multi-leads ECG premature ventricular contraction detection using tensor decomposition and convolutional neural network, in *2019 IEEE Biomedical Circuits and Systems Conference (BioCAS)* (2019), pp. 1–4
154. J. Pellman, F. Sheikh, Atrial fibrillation: mechanisms, therapeutics, and future directions. Compr. Physiol. **5**, 649–665 (2015)
155. X. Fan, Q. Yao, Y. Cai, F. Miao, F. Sun, Y. Li, Multiscaled fusion of deep convolutional neural networks for screening atrial fibrillation from single lead short ECG recordings. IEEE J. Biomed. Health Inf. **22**(6), 1744–1753 (2018)
156. A.N. Hermans, M. Gawalko, L. Dohmen, R.M. van der Velden, K. Betz, D. Duncker, D.V. Verhaert, H. Heidbuchel, E. Svennberg, L. Neubeck, J. Eckstein, D.A. Lane, G.Y.H. Lip, H.J.G.M. Crijns, P. Sanders, J.M. Hendriks, N.A.H.A. Pluymaekers, D. Linz, Mobile health solutions for atrial fibrillation detection and management: a systematic review. Clin. Res. Cardiol. **111**, 479–491 (2021)
157. L. Zhu, V. Nathan, J. Kuang, J. Kim, R. Avram, J.E. Olgin, J. Gao, Atrial fibrillation detection and atrial fibrillation burden estimation via wearables. IEEE J. Biomed. Health Inf. 1–1 (2021)
158. L.M. Eerikäinen, A.G. Bonomi, F. Schipper, L.R.C. Dekker, H.M. de Morree, R. Vullings, R.M. Aarts, Detecting atrial fibrillation and atrial flutter in daily life using photoplethysmography data. IEEE J. Biomed. Health Inf. **24**(6), 1610–1618 (2020)

Chapter 3
Signal Quality Assessment Classifiers for Wearable IoT Sensors

Abstract This chapter presents a comparative study of artificial intelligence (AI)-based binary classifiers designed to assess the integrity of electrocardiogram (ECG) data, with a particular focus on implementation at the network edge. Ensuring data integrity at the edge allows for the early identification and exclusion of corrupted or low-quality signals, thereby reducing the volume of data that must be stored or transmitted. This, in turn, lowers the power and storage demands on Internet-of-Things (IoT) devices without compromising system performance. The chapter investigates various machine learning classifiers that evaluate ECG signal quality using low-complexity Signal Quality Indices (SQIs) derived from kurtosis and skewness measures. Experimental results reveal that a bagged ensemble of three neural networks offers the highest accuracy in detecting poor-quality ECG signals. Additionally, the chapter analyzes the computational complexity and energy consumption of individual classifier models as well as classifier fusion approaches, providing practical insights for resource-constrained environments. Together, these findings support the development of efficient, AI-driven edge solutions for real-time biomedical signal validation.

Keywords Signal quality assessment · Electrocardiogram · Skewness · Kurtosis · Feature extraction · Ensembles · Classifier fusion

3.1 Introduction

Continuous monitoring of patient health using data from Internet-of-Things (IoT) sensors, analyzed with machine learning techniques on cloud servers, enables early detection and allows for timely proactive interventions. For monitoring cardiovascular diseases (CVDs), a wearable cardiac sensor is typically used that can acquire a user's ECG data. Due to its ambulatory nature, the data acquired from wearable devices are often contaminated with various types of noises. Too high a noise level renders the data unusable for diagnostic applications and automated analysis when sent to a cloud server. The Physionet/CinC 2011 challenge attempted to address this issue by developing efficient real-time algorithms to run on a smartphone that can

A. John et al., *Deep Learning and Signal-Processing Methods for Multisensor Data Fusion*, https://doi.org/10.1007/978-3-031-96724-5_3

provide useful feedback to the user while acquiring an ECG signal [1]. However, there are still challenges such as the cost of transmitting massive amounts of data to the cloud or an intermediate node, latency incurred in communication and processing, and the complexity of processing a large quantity of data in the edge device. The performance of system-on-a-chip (SoC) technology used in IoT sensors has been steadily improving. Many SoCs now have sufficient computing power to perform complex computations locally, enabling processing at the edge of the network. Edge processing reduces data redundancies, prevents the transmission of corrupted or unusable data, and optimizes both storage and battery usage.

Various works on implementing power-saving architectures on wearable or implantable ECG processors have been discussed in the literature [2]. Techniques for efficient QRS peak detection in wearable ECG devices are discussed in [3]. However, these techniques assume that the data acquired from the IoT sensor is of acceptable quality, which is not always the case. In most of the existing biomedical SoCs, a simple lead-off detection mechanism is the only check to ensure the integrity of the acquired data. Therefore, in this chapter, this gap is addressed to extend the simple lead-off detection technique to a more comprehensive acceptability detection, using machine learning-based methods. This way, the sensor system can be alerted to avoid the acquisition and transmission of unusable data.

Methods for indicating signal quality as multiple classes or continuous values that specify the acceptability of ECG signals of diagnostic quality have been discussed in the literature for the multi-channel case [4, 5]. Clifford et al. [4] discussed the usage of six different SQIs as features for various machine learning algorithms and a couple of machine learning fusion methods for determining the signal quality of 12-channel ECG recordings into two classes: acceptable or unacceptable. The best performance was observed when a support vector machine (SVM) was used for classification, which exhibited a P Net score of 0.926 on the Physionet/ CinC 2011 challenge Test Set B. In [5], the authors discussed a hybridized rule-based engine that mimicked an SVM based on flat line detection, QRS peak detection, ECG signal standard deviation across all channels, etc. This work obtained a P Net Score of 0.92 on the Physionet/CinC 2011 challenge Test Set B.

In [6], the authors presented an extensive study on different signal quality indicators for the multi-class classification of ECG signals. They considered three different types of common noises—electrode motion, muscle artifacts, and baseline wander—at various SNR values and classified the signals into five different quality classes using SVMs. An accuracy of 80.26% and an overlapped accuracy of 98.60% were obtained on the test set by selecting 10 of the 14 SQIs for classification. Zhang et al. [7] discussed the deployment of Lempel-Ziv (LZ) complexity analysis in the context of ECG signal quality assessment. The data integrity of the signals was modified by adding noise at various SNRs and the LZ values were analyzed. It was observed that the LZ values increase in the presence of high-frequency noise and impulse noise at low SNRs. The method exhibited the highest sensitivity in identifying ECG signals corrupted by high-frequency noise with a sensitivity of 93.5%. Another work of significance in the single-lead ECG signal quality evaluation scenario is [8], where the authors proposed a fuzzy logic-based system to classify signals as acceptable or

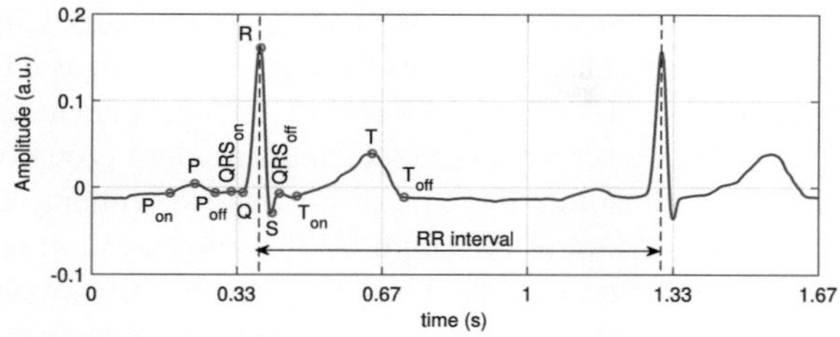

Fig. 3.1 ECG signal with fiducial characteristic points—the P, Q, R, S, and T waves

unacceptable based on a fuzzy evaluation post-fuzzy synthesis of the membership functions of six SQIs. This method exhibited a classification accuracy of 94.67% for the modified CinC 2017 challenge training dataset. In most of the works discussed in the literature, QRS detection-based SQIs or a large number of SQIs are used for signal quality assessment. However, to achieve accurate QRS detection or detect other fiducial points in ECG, i.e., the P, Q, R, S, or T waves (Fig. 3.1), the acquired signal has to be of acceptable quality [9]. Therefore, QRS detection-based SQIs or SQIs which require the detection of ECG fiducial points (F SQI) are not ideal to be used for determining the acceptability of the signal. Also, they are often computationally expensive due to the presence of digital filter blocks. This chapter proposes non-fiducial feature-based SQIs (NF SQIs) for signal acceptability classification of single-channel ECG data. The proposed approach doesn't require any signal pre-processing and therefore lowers the complexity for edge implementations.

The development of a binary classification technique for acceptability testing of the recorded ECG is developed in this chapter. The proposed technique uses only NF SQIs as input features to the classifier and is easier to implement compared to those which require QRS detection. The proposed technique extends the simple lead-off detection technique in wearable sensors using NF SQIs. Several existing machine learning classifiers were used to perform acceptability testing on a publicly available dataset—Physionet/ CinC 2017 challenge training set. The computational complexity and energy consumption of the classification task are estimated and reported. Three decision-level fusion-based classifiers that combine existing classifiers are discussed, and their energy consumption is estimated.

3.2 Method

3.2.1 Dataset

This work discusses how the clinical acceptability of a single-channel ECG lead can be carried out at the data acquisition end of an IoT device. The suitability and quality of the proposed methodology were tested on the Physionet/CinC 2017 challenge training dataset [10]. The dataset consists of single-lead ECG records collected using the AliveCor mobile measurement device, which was annotated by clinical experts as falling into four categories—normal rhythm, atrial fibrillation, other recordings, or too noisy to be classified. This dataset was chosen as it is a good representation of the data that would be obtained from wearable IoT devices. All the 284 noisy recordings in the training set were selected, and a random group of 284 normal recordings was chosen to constitute the dataset for this experiment. The recordings were sampled at 300 Hz and were of length 9000 samples (30 s) or 18000 samples (60 s). The signals were downsampled by four to reduce the number of computations but were not pre-processed before the feature generation stage to remove noises. However, the signal portions from 0 to 1 s were dropped for each record before feature generation to remove edge effects.

In order to assess the generalization performance of the classifier, a 10-fold cross-validation method is used. The dataset is first divided into ten equally sized mutually exclusive subsets. Nine subsets are used for training, and the last subset is used for evaluation. This process is repeated ten times, using a different subset for evaluation each time. The final test result is the average of the test results of all ten instances of testing.

3.2.2 *Method Outline*

The proposed methodology relies on the calculation and extraction of NF SQIs for each record. The extracted signal quality indicators are then used as features to train six different classifiers and ensembles of classifiers to determine whether a given record is of acceptable signal quality. The flow diagram is as shown in Fig. 3.2 with the details of the SQIs provided in Sects. 3.3.1 and 3.3.2.

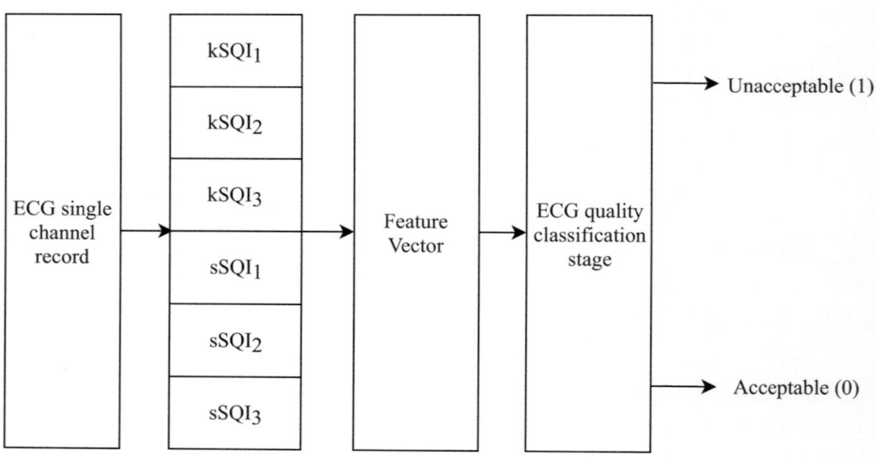

Fig. 3.2 Flow diagram of the proposed methodology with kSQI and sSQI indicating the signal quality indicators used in this study

3.3 Non-fiducial Signal Quality Indicators

SQIs play an important role in the assessment of the quality of data obtained from wearable devices that are prone to a wide variety of noises. For this experiment, NF SQIs that are discussed in the literature [9, 11] are used. The SQIs were empirically chosen for this application after multiple trials with the various SQIs discussed in the literature. The study found that the separability between the two classes was greater when a combination of these two SQIs was used.

3.3.1 Kurtosis SQI (kSQI)

Kurtosis is defined as the fourth standardized moment of a random variable. Kurtosis is a measure of how sharp the peak of the distribution is. A distribution with many outliers is expected to have a low value of kurtosis since its probability distribution will be flatter and will approach zero slower [11]. The commonly used equation for computing an estimate of the kurtosis, namely, kSQI is (as shown in Eq. (3.1)) [6]

$$kSQI = \frac{\frac{1}{N}\sum_i (s_i - \widehat{\mu})^4}{\widehat{\sigma}^4} \tag{3.1}$$

and $\widehat{\mu}$ and $\widehat{\sigma}$ are estimates of the mean and standard deviation of the set of samples s_i of size N. A kurtosis-based quality index has been used to assign the quality of signal segments in [12].

3.3.2 Skewness SQI (sSQI)

Skewness is defined as the third standardized moment of a random variable, an estimate of which is as per Eq. (3.2). Skewness measures the symmetry of the distribution. A distribution with no outliers will be more symmetric, and consequently have low values of skewness [11].

$$sSQI = \frac{\frac{1}{N}\sum_i \left(s_i - \hat{\mu}\right)^3}{\hat{\sigma}^3} \tag{3.2}$$

In [4], the sSQI along with other SQIs was used to train a classifier to detect signal quality.

In this approach, the nth record (of mainly 30 s or 60 s) is divided into three equal non-overlapping windows and the $kSQI^{(n)}$ and $sSQI^{(n)}$ for each window indexed by $i = 1, 2, 3$ is computed, resulting in a feature vector $x^{(n)} = [kSQI_1^{(n)}\ kSQI_2^{(n)}\ kSQI_3^{(n)}\ sSQI_1^{(n)}\ sSQI_2^{(n)}\ sSQI_3^{(n)}]$. The calculation of kSQI

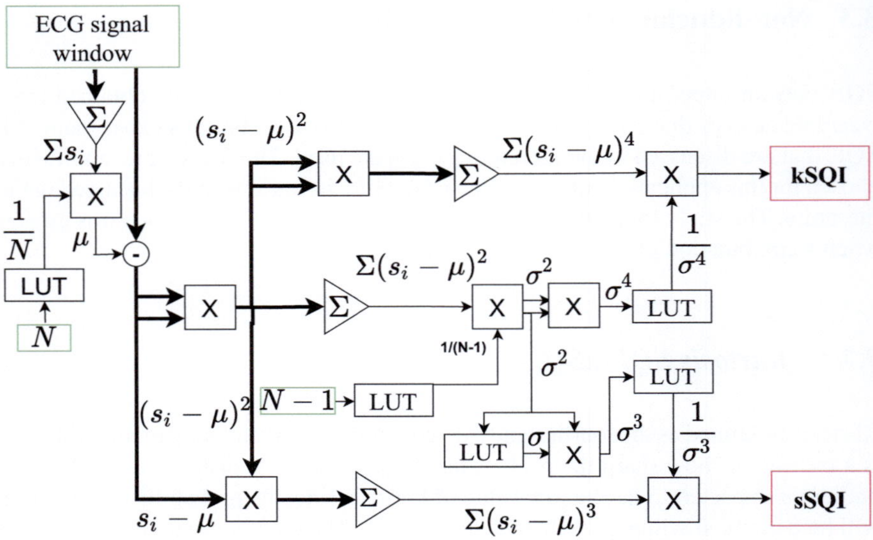

Fig. 3.3 Block diagram of the kSQI and sSQI calculation stage. The green boxes indicate inputs and the pink boxes indicate final outputs. LUTs stand for look-up tables. The bold connectors indicate vectors, and N indicates the number of samples per window. The details of kSQI and sSQI are discussed in detail in Sects. 3.3.1 and 3.3.2

and sSQI from an ECG signal window and the corresponding circuit diagram is as shown in Fig. 3.3. This method is used as a single SQI value will not be able to provide a good representation of the signal quality for the entire record. The number of windows was chosen based on the majority record length of 30 seconds, with the knowledge that at least a few ECG cycles are important to capture skewness and kurtosis in a record, for the task of ECG quality assessment. The window lengths should not be too large either, as too large a window would lead to smoothing. Smaller windows would lead to a larger number of features for the classification task, in addition to the issue of loss of relational information between samples. For comparison purposes, a record of acceptable quality and unacceptable quality and their corresponding feature vector is shown in Fig. 3.4. From Fig. 3.4, it can be observed how in a record of acceptable quality, all three values of kSQIs and sSQIs fall within a similar range, whereas in a record of unacceptable quality, the three values of kSQIs and sSQIs vary greatly. Once all the feature vectors in the training set were computed, the features were normalized and arranged into the matrix **X**:

$$X_k^{(n)} = \frac{x_k^{(n)} - \overline{x_k}}{\sigma_{x_k}} \tag{3.3}$$

where in Eq. (3.3), $\overline{x_k}$ and σ_{x_k} indicate the mean and standard deviation of each column indicated by k for the training set. This normalization ensures that no single

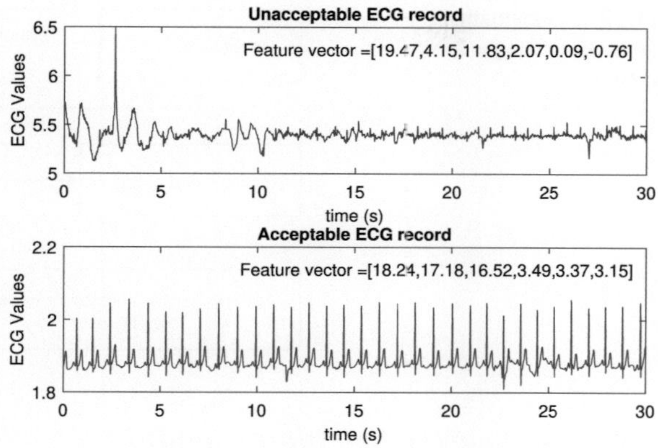

Fig. 3.4 A record of unacceptable signal quality and acceptable signal quality with the corresponding feature vectors

feature would become a major contributor to decision-making and to treat values that fall beyond the standard normal distribution as noise. This is referred to as standard normalization in this book.

3.4 Machine Learning (ML) for Signal Quality Assessment of ECG Signals

The SQIs obtained per record are used to train various ML algorithms to classify the data as being of acceptable (0) or unacceptable (1) signal quality. In this chapter, we focus on supervised learning, where class labels are provided during training. We compare the performance of six different classifiers in achieving this task:

- a *k*-Nearest Neighbor classifier (*k*-NN),
- a Decision Tree (DT),
- a Support Vector Machine (SVM),
- a Naïve Bayes (NB) classifier,
- a Linear Discriminant Analysis (LDA)-based classifier, and
- a Feedforward Neural Network (NN).

The parameters in each of the algorithms were fine-tuned by using a subset of the total dataset for training and validation. The best parameters were chosen for the 10-fold cross-validation experiment. The performances of low-cost ensembles are also evaluated for signal quality assessment.

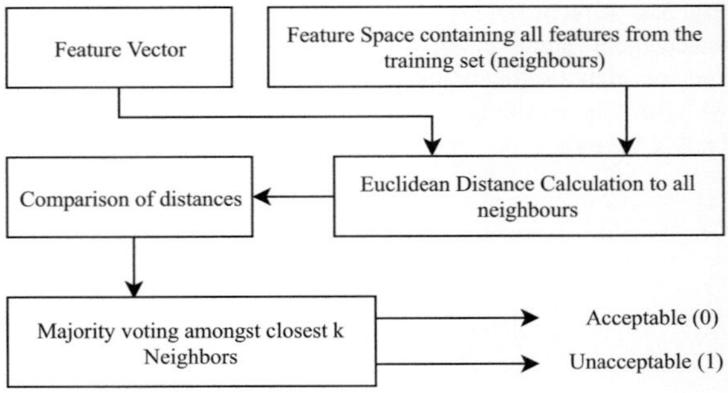

Fig. 3.5 Schematic of prediction using *k*-NN classifier

3.4.1 k-*Nearest Neighbor Classifier*

k-NN is a distance-based classifier wherein a query object is classified based on the classes of its *k* nearest neighbors, where *k* is a user-defined number and is, therefore, a non-parametric classifier. A majority voting approach is used to determine the class of the query object. The prediction stage using the *k*-NN model is as shown in Fig. 3.5. The *k*-NN classifier can be modified with new training data easily in an edge wearable device if required during update stages. In this experiment, the *k*-NN feature space is six-dimensional and populated with 512 objects (number of training examples) during the training phase. Euclidean distance was chosen as the distance measure. *k*-NN has been used for real-time ECG classification as discussed in [13]. In this experiment, the *k*-NN feature space is six-dimensional and populated with 512 objects (number of training examples) during the training phase.

3.4.2 Decision Trees

A decision tree (DT) is a learning method in which the important features that can be used to generally represent information are chosen, and decision nodes are created, with the leaf nodes yielding the classified results. The query object can be classified by analyzing a series of conditions and can be implemented easily in hardware. The node splitting criteria used was the Gini's diversity index. A figure indicative of the prediction process using a DT is as shown in Fig. 3.6, where $x1, x2, x3$.. indicate the features of the new query object, and the other constants are obtained during the training stage. DTs have been used for classifying ECG beat types in [14]. In this experiment, the DT model has 36 decision conditions in 18 decision nodes and 19 leaf nodes.

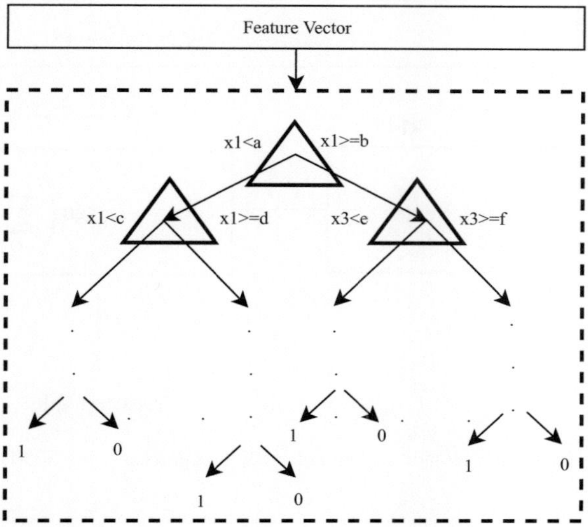

Fig. 3.6 Indicative schematic of prediction using DT

3.4.3 Support Vector Machines

SVMs are supervised learning models in which objects are classified by creating an optimal hyperplane that separates the objects during training. Oftentimes, the objects are transformed into a higher dimensional space using kernel functions to find the optimal hyperplane. The support vectors are the vectors that lie closest to the hyperplanes on either side of it. The SVM model calculates the offset b, coefficients c_i for each support vector and stores the corresponding class of each support vector as y_i. The prediction stage using the SVM model can be depicted as shown in Fig. 3.7. SVMs have been used to classify different ECG beat types in [15]. To classify acceptable ECG records from unacceptable, in this experiment, the SVM used a Gaussian radial basis kernel (Eq. (3.4)) with 112 support vectors. The Gaussian radial basis function has a transformation as shown in Eq. (3.4):

$$\mathbf{K}(\mathbf{x}, \mathbf{x}') = exp\left(-\frac{||\mathbf{x} - \mathbf{x}'||^2}{2\sigma^2}\right). \tag{3.4}$$

3.4.4 Naïve Bayes Classifier

NB classifier is a parametric probabilistic classifier that applies the Bayes theorem with the assumption that the features defining an object are independent. The probability of an object belonging to a particular class C_k given the feature vector of an object $\mathbf{x} = (x_1, \cdots, x_n)$ is written as $p(C_k|x_1, ..., x_n)$, which can be calculated using

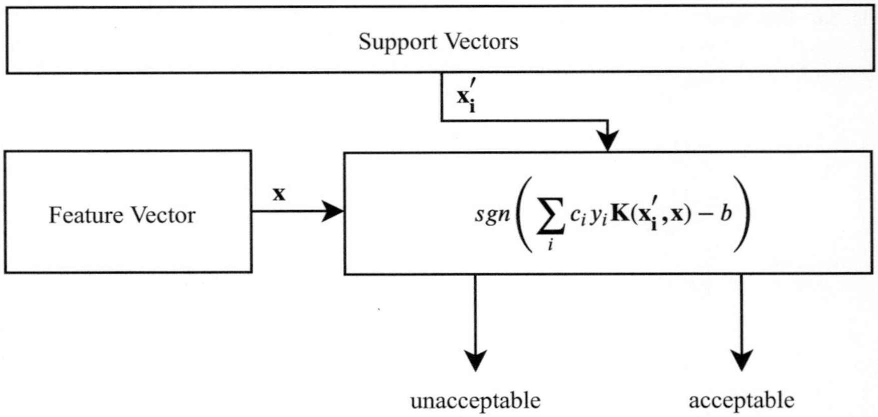

Fig. 3.7 Schematic diagram of prediction process using SVMs

Bayesian probability. For calculating Bayesian probabilities, prior probabilities need to be calculated, and the training stage is used for defining the prior probabilities and the probabilities of occurrence of a class $p(C_k)$. The Bayes classifier is the function that assigns a class label $\hat{y}$ as

$$\hat{y} = \underset{k\in\{1,\ldots,K\}}{\operatorname{argmax}} \; p(C_k) \prod_{i=1}^{n} p(x_i \mid C_k), \tag{3.5}$$

where k is the number of classes. Once the prior probabilities ($p(x_i|C_k$ ie., ($p(kSQI_1|C_1)$, $p(kSQI_2|C_1)$,$\cdots$, $p(sSQI_3|C_2)$, etc.) calculated during training are available, the class of a new record is evaluated as depicted in Fig. 3.8 for this experiment. Padmavathi et al. have used NB classifier for classifying ECG signals into signals having normal beats and abnormal beats [16].

3.4.5 Linear Discriminant Analysis

LDA finds a linear combination of features that characterizes or separates two or more classes of objects or events. The resulting combination can be used as a linear classifier for binary classification problems. Given a training set containing objects of a known class, LDA assumes that the conditional probability density functions $p(x|C_2)$ and $p(x|C_2)$ of feature vectors in training set $\mathbf{x}$ are normally distributed with mean vectors μ_0 and μ_1, respectively. The covariances Σ of both classes of features are considered the same in LDA. The prediction of the class of a new object is evaluated as depicted in Fig. 3.9 using LDA. In the two-class problem, the value of δ_k is used to predict the class Eq. (3.6).

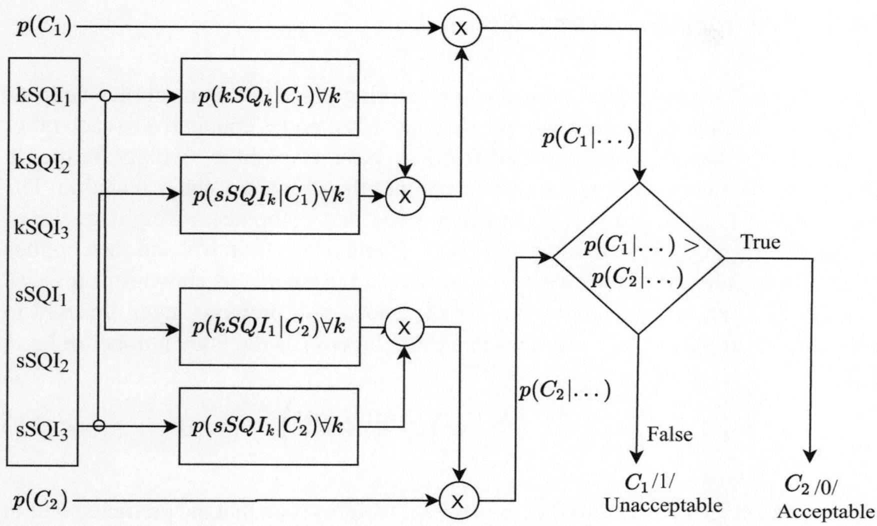

Fig. 3.8 Flow diagram of the prediction process using NB classifier

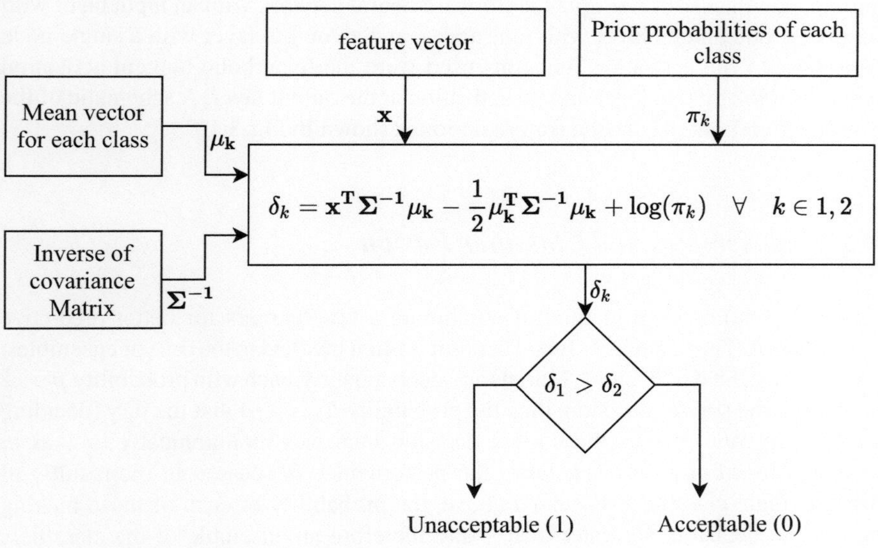

Fig. 3.9 Flow diagram of the prediction process using LDA

$$\delta_k = \mathbf{x}^{\mathbf{T}} \Sigma^{-1} \mu_{\mathbf{k}} - \frac{1}{2} \mu_{\mathbf{k}}^{\mathbf{T}} \Sigma^{-1} \mu_{\mathbf{k}} + log(\pi_k) \quad \forall \quad k \in 1, 2. \tag{3.6}$$

The object belongs to the class with the highest δ_k value. LDA has been previously used in literature for beat classification of ECG signals [17].

3.4.6 Feedforward Neural Network

A simple feedforward neural network is a learning algorithm that mimics neurons in the human brain. A neural network is composed of nodes connected to each other by links [18]. These links are directed and often have a weight w_i (weight for the ith input to a node) associated with it that determines the strength of the connection. The layer number is indicated by $[k]$, and each node first computes a weighted sum of its inputs $a^{[k-1]}$, i.e., the output of layer $[k-1]$ and a bias term $b^{[k]}$ and then applies an activation function g to this sum to derive the output $a^{[k]}$ as shown in Eq. (3.7). Equation (3.7) shows that the outputs of the nodes of a previous layer are used to compute the outputs at the nodes of the current layer (j is the node number in layer $[k]$).

$$a_j^{[k]} = g\left(\sum_{i=1}^{n} w_{i,j}^{[k]} a_i^{[k-1]} + b_j^{[k]}\right). \tag{3.7}$$

The training stage involves finding the optimal weights such that the predicted output matches closely with the desired outputs. Zhou *et al.* discussed the use of a convolutional neural network for signal quality assessment of ECG signals [19]. However, in this chapter, the use of a simple feedforward neural network with an input layer with six nodes, one hidden layer with four nodes, and an output layer with a single node is proposed. The activation functions used were the hyperbolic tangent activation function at the hidden layer and no activation at the output layer. A schematic of the neural network model chosen for prediction is shown in Fig. 3.10.

3.4.7 Ensembles and Classifier Fusion

Ensemble learning is a method of combining a few learners for better predictive performance. The Condorcet Jury Theorem is often invoked in the case of ensembles, which states that if $n = 2m + 1$ jurists act independently, each with probability $p > \frac{1}{2}$ of making the correct decision, then the probability $h_{2m+1}(p)$ that the jury (deciding by majority rule) makes the correct decision increases monotonically to 1 as m increases to infinity [20, 21]. From the performance (discussed in the results) of the individual classifiers discussed above, the probability of each of those making the correct decision is greater than $\frac{1}{2}$ and therefore an ensemble of the classifiers can be employed to improve the classification performance. Two common ensemble methods in ML, namely, bagging and boosting, are explored here. The base classifiers used in ensemble methods should be unstable classifiers such as DT and NN to introduce diversity in the ensemble [22, 23]. Stable classifiers (k-NN, SVM, NB and LDA here) that do not lead to major changes in the learned models with changes in training instances do not perform well when used in ensembles [22, 24].

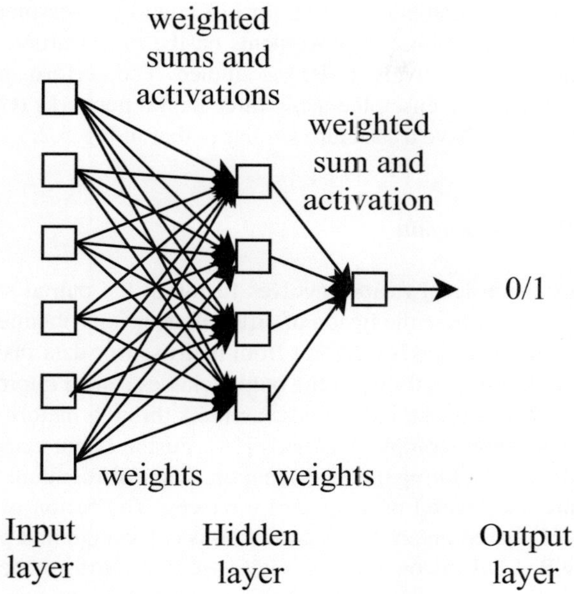

Fig. 3.10 Schematic of prediction using the NN

Bootstrap Aggregation (Bagging)

Bootstrap aggregation/bagging (B) is an ensemble method that introduces diversity in the various classifier models by random sampling of the training set with replacement [22]. The final prediction is based on a majority vote or a weighted majority vote. Mert et al. studied the effect of using bagged decision trees on ECG arrhythmia beat classification in [25]. In this chapter, the use of DTs and NNs in bagging for ECG quality assessment is studied. The work examines the results of ECG signal quality classification with a bagged ensemble of three DTs (B1) with 11, 11, and 9 decision nodes. The performance of an ensemble of three neural networks using bagging (B2) with a structure similar to that in Fig. 5.2 is also studied.

Gradient Boosting

Boosting is an ensemble technique in which a sequence of classifiers is trained, one after another, so that later classifiers are trained to better predict class labels for examples that earlier ones performed poorly on [26]. Boosting algorithms can be characterized as gradient descent on cost-functionals in an inner-product function space [27]. Gradient boosting (GB) is done by constructing additive models by sequentially fitting a simple base learner to current residuals by least-squares at each iteration [28]. A weighted eXtreme gradient boosting (XGBoost) method was used for ECG beat classification in [29]. In this chapter, the usage of DTs in non-weighted gradient boosting for ECG quality assessment with three decision trees in sequence making improvements on the base learner with 18, 15, and 12 decision nodes, respectively, (GB1) is studied. The performance of a non-weighted gradient boosting ensemble using three neural networks (GB2) is also studied. All three networks have a structure similar to that in Fig. 5.2.

Decision Fusion

Decision-level fusion involves fusion at the output stage as discussed in Chap. 2. However, here the fusion of different decisions obtained from a single sensor source is fused, which is different from multi-sensor data fusion discussed throughout this book. The reliability of the output classes can be improved through the fusion of the decisions of the individual classifiers through majority voting, weighted voting, or other more complex methods [30]. Fusion is not carried out during training and is done only during model evaluation. Three different methods of fusing the algorithms are used based on their performances. The fusion of the results of NB and LDA (F1) based on an if-else condition is carried out and performance is evaluated. The NB+LDA fusion was checked to see if the performance improves significantly even if the base classifiers that performed most poorly were used. The fusion of NB, DT, and LDA (F2) based on if-else-if conditions is discussed next. The NB+LDA+DT algorithm is checked as performance did not improve significantly with just NB+LDA and to check if the addition of a reasonably well-performing classifier might improve the performance. For this, DT was chosen due to the smaller number of computations required for classification. Finally, the fusion of all the learners (F3) based on majority voting is carried out to check for performance improvements. For breaking the tie, k-NN, SVM, and DT are used as they exhibited the best well-rounded performance (similar sensitivity and specificity) as a solo base classifier.

3.5 Results

3.5.1 Performance Analysis

The results for the various classification algorithms with dataset D using the three sets of kSQIs and sSQIs as features are shown in Table 3.1. Here the results are the micro-average of the results of the ten training and test sets from the 10-fold cross-validation method. The performance is indicated in terms of accuracy, sensitivity, specificity, precision/Positive Predictive Value (PPV), and F1-score which are calculated from the confusion matrix as shown in (3.8), (3.9), (3.10), (3.11), and (3.12), with a true positive (TP in the equations) indicating a correctly identified unacceptable record. The true negatives are indicated by TN, false positives as FP, and false negatives as FN in the equations:

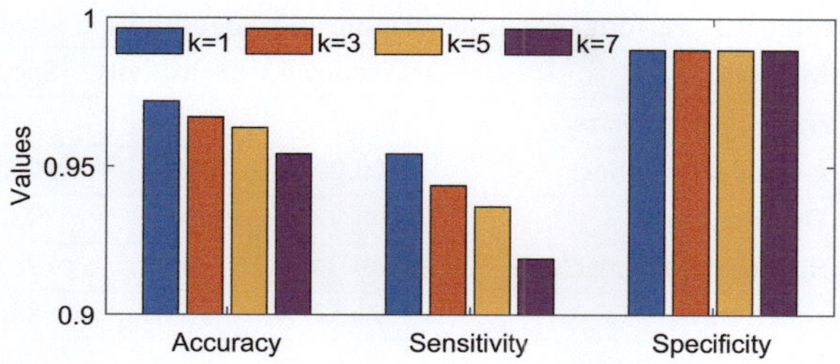

Fig. 3.11 Comparison bar plot of accuracy, sensitivity, and specificity of an unweighted k-NN classifier for $k = 1, 3, 5, 7$

$$\text{Accuracy (\%)} = \frac{\text{TP} + \text{TN}}{\text{TP} + \text{TN} + \text{FP} + \text{FN}}. \tag{3.8}$$

$$\text{Sensitivity (\%)} = \frac{\text{TP}}{\text{TP} + \text{FN}}. \tag{3.9}$$

$$\text{Specificity (\%)} = \frac{\text{TN}}{\text{TN} + \text{FP}}. \tag{3.10}$$

$$\text{PPV (\%)} = \frac{\text{TP}}{\text{TP} + \text{FP}}. \tag{3.11}$$

$$\text{F1- score (\%)} = \frac{2.\text{PPV}.\text{Se}}{\text{PPV} + \text{Se}} \times 100 = \frac{2 \times \text{TP}}{2 \times \text{TP} + \text{FP} + \text{FN}} \times 100. \tag{3.12}$$

From the table, it can be observed that during the training stage, k-NN exhibits 100% accuracy, which is expected. In the prediction stage, the k-NN classifier exhibited an accuracy of 97.36%. k-NN also exhibits the highest precision and specificity among the classifiers discussed but exhibits a lower sensitivity when compared to the other classifiers. The selection of the hyperparameter k for this experiment was chosen based on the performance of various k values for voting as shown in Figs. 3.11 and 3.12, respectively. For the weighted voting-based k-NN, the weights were taken as the inverse of the distance of the neighbors from the query object. The performances of the different methods were analyzed in terms of accuracy (Acc), sensitivity (Se), and specificity (Sp), which are calculated from the confusion matrix. From the figures, it is observed that the classifier has the highest accuracy and sensitivity when $k = 1$ (therefore no weighting is required), and therefore the k value is set to 1 for this experiment, and an exhaustive search method was used to find the nearest neighbor.

As discussed in Sect. 3.4.1, k-NN is slow during testing and needs to store all the training objects in memory. Therefore, unless the k-NN algorithms yield very high accuracy, it wouldn't make sense to employ them for signal classification on an edge SoC. The performance of k-NN can be improved by using a larger and more diverse training set, but, at the same time, the number of objects to be stored in the feature space for testing also increases. Moreover, k-NN can be made faster by performing a k-dimensional tree search instead of an exhaustive search for the nearest neighbors.

Table 3.1 Performance of the classification algorithms in identifying unacceptable signal records

Method	Accuracy	Sensitivity	Specificity	Precision	F1-score
Training performance					
k-nearest neighbor	**100.00**	**100.00**	**100.00**	**100.00**	**100.00**
Decision tree	98.92	99.3	98.55	98.56	98.93
Support vector machine	99.04	99.14	98.94	98.95	99.04
Naïve Bayes	91.57	96.83	86.31	87.61	91.99
Linear discriminant	87.97	89.40	86.54	86.92	88.14
Neural network	96.83	96.95	96.71	96.72	96.83
Bagging (3 DTs)	98.67	98.87	98.47	98.48	98.67
Gradient boosting (3 DTs)	99.65	99.88	99.41	99.42	99.65
Bagging (3 NNs)	**99.75**	99.49	**100.00**	**100.00**	**99.75**
Gradient boosting (3 NNs)	99.49	98.98	**100.00**	**100.00**	99.49
Prediction performance					
k-nearest neighbor	97.36	95.77	**98.94**	**98.91**	97.32
Decision tree	95.25	95.07	95.42	95.41	95.24
Support vector machine	**98.24**	**98.24**	98.24	98.24	**98.24**
Naïve Bayes	91.55	96.83	86.27	87.58	91.97
Linear discriminant	87.50	88.38	86.62	86.85	87.61
Neural network	95.42	94.72	96.13	96.07	95.39
Bagging (3 DTs)	97.36	97.18	97.54	97.53	97.35
Gradient boosting (3 DTs)	95.77	95.42	96.13	96.10	95.76
Bagging (3 NNs)	**99.47**	**98.94**	**100.00**	**100.00**	**99.47**
Gradient boosting (3 NNs)	98.94	97.89	**100.00**	**100.00**	98.93
ML fusion (NB+LDA)	95.60	92.25	**98.94**	98.87	95.45
ML fusion (DT+NB+LDA)	95.95	96.48	95.42	95.47	95.97
ML fusion (majority vote)	97.54	96.83	98.24	98.21	97.52

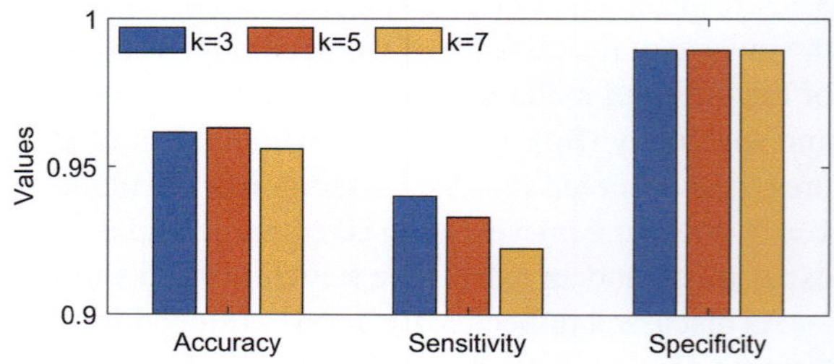

Fig. 3.12 Comparison bar plot of accuracy, sensitivity, and specificity of a weighted k-NN classifier for $k = 3, 5, 7$

The performance of DT shows good accuracies such as 98.92% and 95.25% in both the training and prediction stages, respectively. The DT model in this experiment has 36 decision conditions, 19 leaf nodes with the height of the lowest leaf node being 7. The advantage of a DT is that classification is fast with small memory overhead, and the current DT model is suitable for deployment on an SoC if power constraints can be met, which will be discussed in Sect. 3.5.2.

SVMs exhibit good accuracy levels in the training stage of 99.04% as well as an accuracy of 98.24% in the prediction stage. It exhibits the highest sensitivity, accuracy, and F1-scores among the classifiers discussed. The advantage of SVMs in this scenario is that sensitivities and specificities are on par with each other, reducing alarm fatigue or recording of corrupted data by mistake.

The NB classifier and LDA demonstrate limited accuracy in both the training and evaluation phases. This can be attributed to the selection of suboptimal parameters based on the input features during the training phase. These low accuracies stem from the large number of false positives that the models classify, i.e., clean records are often classified as noisy by these algorithms, with systems ending up discarding signals of good quality. Regardless, the NB classifier exhibits a sensitivity that is better than the sensitivities of the DT and *k*-NN classifier, implying that artifact-ridden data can be identified easily and discarded and a new recording can be collected.

The neural network exhibits an accuracy of 96.83% in the training stage and an accuracy of 95.25% in the test stage. This indicates that although neural networks tend to overfit, the neural network model obtained here can generalize. The neural network is more specific than sensitive and has better overall accuracy than the DT model.

Bagging and gradient boosting with DTs exhibit good performance in the testing stage. Bagging exhibits better performance than boosting in the training stage in terms of all parameters considered. The gradient boosting method exhibits a very high accuracy and F1-score in the training stage, but these values drop considerably during testing, and therefore the possibility of the model overfitting to the training data is high. The tendency of gradient boosting with DTs to overfit can be reduced by pruning the decision trees. The results shown are after repeatedly testing the performance of various tree heights on the training set. Bagging of NNs is found to exhibit good performance during both training and testing, and similar performances in both stages indicate that the model is not overfitting to the training data. B2 exhibits the best performance during testing for all the parameters considered with 99.47% accuracy. The gradient boosting method of ensembling NNs is found to be highly specific and precise, with a slight drop in sensitivity compared to the training stage. However, the GB2 method is only second to the B2 method with an accuracy of 98.94%.

The fusion of NB and LDA models shows a better performance than when NB and LDA are considered separately. The specificity of this method is on par with the *k*-NN method which exhibits the highest specificity but has low sensitivity. The fusion of DT, NB, and LDA exhibits better performance than all the learners considered separately in the case of all parameters considered, except in the case of specificity, where the NB learner outperforms this method. The fusion of all the learners shows

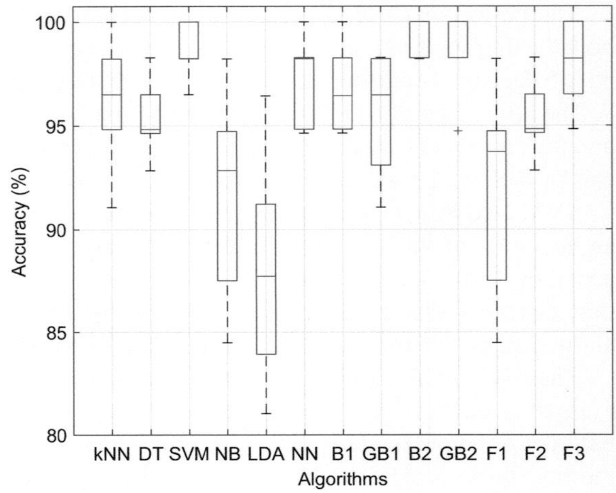

Fig. 3.13 Box plot of accuracy values during testing for all learners over 10-fold cross-validation

good performance but does not outperform the best-performing solo base classifiers, i.e., SVM and *k*-NN. Fusion methods are usually found to perform better, as when one source of information/decision fails, the other sources of information or decision might be able to correct for this error. F1 and F2 were found to improve the overall accuracy, specificity, precision, and F1-score than the individual classifiers used, although there has been a drastic drop in sensitivity in the case of F1, and a slight drop in sensitivity in the case of F2 when compared against the NB classifier. SVM and *k*-NN were still found to outperform F3 in all measures of performance. Since none of the classifiers complemented each other in this scenario, i.e., correcting for each other for all instances, a good rule for complementary fusion or cooperative fusion could not be established, and the lack of performance improvement could be attributed to that.

The box plot of accuracy values over 10-fold cross-validation of the various classifiers during testing is as shown in Fig. 3.13. From the figure, it can be seen that some of the learners are consistent with the results, while a few others show greater variances.

The performance comparison in terms of accuracy with different training set sizes is as shown in Fig. 3.14. It can be seen that for all the learners, except NB and F1, performance improves gradually with an increase in number of training samples. The performance of the NB classifier drops with the size of training samples as outliers to the mean features of each class have a greater probability of getting sampled into the training set and thereby shifting the parameters of the normal distribution learned by the NB classifier. Thereby, NB classifier performs poorly for the average features which are large in number compared to the outliers. The performance of the F1 method drops with the increasing number of training samples as well due to its dependence on the NB classifier, while F2 and F3 do not follow this trend as dependence on NB is much lower in F2 and F3 due to the presence of DT in F2 and all the other classifiers in F3.

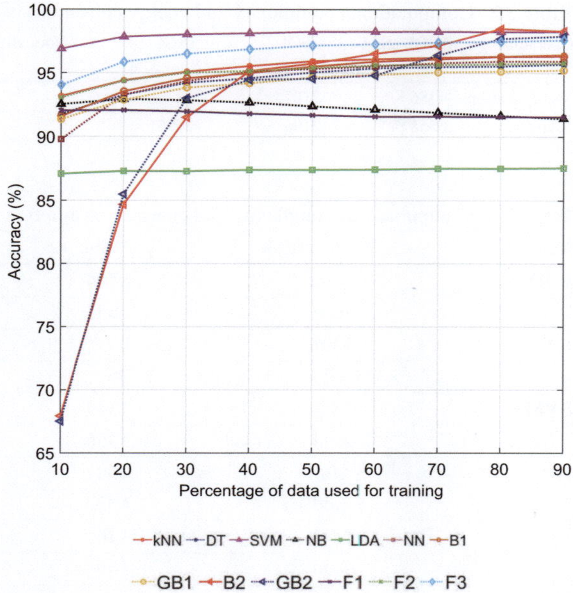

Fig. 3.14 Accuracy versus training data size

3.5.2 Complexity Calculation

To estimate the energy requirement during prediction, the computational complexity of the proposed method for prediction is calculated in terms of the number of multiplications and additions required for each prediction, as shown in Tables 3.2 and 3.3. The computational complexity of the SQI feature extraction and normalization has been evaluated to indicate the approximate total energy requirement and is as shown in Table 3.2. Table 3.2 shows the number of multiplications and additions required for records of length of 30 s. Computations of kSQI and sSQI can share a few multipliers and adders, as shown in Fig. 3.2. The shared number of adders was 6558, and the shared number of multipliers was 4386. Computation of kSQI alone using the shared values required 2187 multiplications and 2196 additions. To carry out sSQI specific calculations, 12 multipliers and 2196 adders are required. The normalizing stage requires six multiplication and six addition operations assuming that the inverse of the standard deviation is calculated prior to prediction. For all the calculations, all Taylor series approximations have been carried out to the second order ($\sqrt{x}$, e^x, etc.). Common division operations are assumed to be carried out using multiplication of the stored inverses ($\frac{1}{\sqrt{2\pi}}$, $\frac{1}{2}$, etc.) or through Taylor series expansions. Table 3.3 shows the computational complexity during the prediction stage of each of the classifiers discussed. Since the proposed algorithms are for implementation on IoT edge devices, the implementation using fixed-point representation and corresponding performance needs to be considered. Various ML implementations using fixed-point

Table 3.2 Computational complexity and energy consumption of the SQIs and normalization

Operation	kSQI, sSQI	kSQI	sSQI	Normalization	Total	Energy total (nJ)
Multiplications	4386	2187	12	6	6591	4.273
Additions	6558	2196	2196	6	10956	

Table 3.3 Computational complexity and energy consumption of the classifiers during prediction

Method	Operation	Value	Energy total (nJ)
k-NN	Mult	8704	3.394
	Add	6655	
DT	Mult	0	0.014
	Add	36	
SVM	Mult	1344	0.568
	Add	1456	
NB	Mult	168	0.065
	Add	133	
LDA	Mult	170	0.066
	Add	146	
NN	Mult	33	0.013
	Add	29	
B1	Mult	0	0.024
	Add	62	
GB1	Mult	0	0.018
	Add	46	
B2	Mult	99	0.039
	Add	90	
GB2	Mult	99	0.039
	Add	91	
F1	Mult	338	0.131
	Add	281	
F2	Mult	338	0.131
	Add	316	
F3	Mult	10419	4.120
	Add	8458	

representation were found to reduce hardware complexity without a significant drop in accuracy (less than 1%) [31, 32].

For estimating the computational complexity of the *k*-NN classifier, the Euclidean distance calculation to each object in the training set is carried out. 512 feature vectors were used for training, and the estimation was carried out by counting the operations involved in Euclidean distance calculation between two six-dimensional vectors as well as comparisons. k-NN requires approximately 8704 multiplications and 6655 additions for prediction. Since k-NNs are distance-based metrics that merely use the calculated distances for comparisons within the training set, accurate distance computations are not important. Therefore, for this classifier, the computational cost can be further reduced by using approximate multipliers for computations as discussed in Appendix A.1.

In the DT, each decision node involves two comparisons. Each comparison is assumed to be equivalent to an addition operation, and a total of 36 additions are carried out for the classification of each record. This method is well suited for implementation on an SoC with regard to energy efficiency.

The SVM classifier had identified 112 support vectors and a Gaussian kernel. So the computational complexity involved transformation into the kernel space and corresponding calculations and comparisons. The SVM required approximately 1344 multiplications and 1456 additions for the prediction of the quality of a record.

The total number of operations for prediction in the NB model constitutes the probability calculation from the probability distribution functions and comparisons. In this classifier model, 168 multiplications and 133 additions are required for the classification of a single record.

170 multiplications and 146 additions are required for the LDA classifier model for the signal quality prediction of a record. The values are calculated by assuming that the class covariance inverse (Σ^{-1}) is stored in the model.

The neural network model with one hidden layer required 33 multiplication and 29 additions for the classification of a single record. This involves the multiplication of inputs with the weights and additions as per Eq. (3.7) and the cost of activations.

The computational complexity of bagging and gradient boosting with decision trees is the cost of prediction of all the decision trees used in each of the ensemble models. The bagging model chosen required 62 additions for the decisions, while the boosting model with each subsequent tree being smaller than the previous tree required 46 additions for the decisions and aggregation. The B2 and GB2 methods use three neural networks of similar structures each, and both methods require 99 multiplications. The bagging of neural networks requires 90 additions, and gradient boosting requires 91 additions. The total energy requirements of both algorithms are nearly the same. The computational complexity of ensemble methods is lower than that of k-NN, SVM, NB, and LDA, thereby showing that a single best-performing model need not be the best solution for a task and ensembles of various weak learners can be used for ECG quality assessment tasks. Therefore, ensembles can be considered for very low-cost implementations of ECG quality assessment.

The cost of the decision fusion methods considered is the sum of the costs of its component learners and the cost of the aggregation method and is as shown in Table 3.3. The fusion of NB and LDA as well as DT, NB, and LDA is less expensive than that of k-NN and SVM classifiers individually and can be considered for low-cost implementations.

In Tables 3.2 and 3.3, the total energy requirement in terms of nJ is indicated. This is done by assuming that the energy required is 0.39 pJ for a 16-bit by 16-bit multiplier accumulator (MAC) in 28 nm fully depleted silicon on insulator (FD-

Table 3.4 Comparison of the best-performing classifier with state-of-the-art ECG record quality assessment algorithms

Article	Clifford et al. [4]	Tat et al. [5]	Li et al. [6]	Zhang et al. [7]	Zhao et al. [8]	This work	
Dataset	CinC 2011 challenge: Set B	CinC 2011 challenge: Set B	MITDB (simulated noise)	MITDB (simulated noise)	Subset of CinC'17 challenge: training	Subset of CinC '17 challenge: training	Subset of CinC '17 challenge: training
Number of records	500	500	48×2^a	48×2^a	300[a]	512[b]	
Number of channels	12	12	1	1	1	1	
Number of classes	2	2	5	2	2	2	
Method	F SQIs and NF SQIs; ML and fusion	Flat-line and QRS complex based conditions	F SQIs and NF SQIs, SVM	LZ values; threshold-ing	F SQIs and NF SQIs; Fuzzy logic	NF SQIs; ML	NF SQIs; ML fusion
Accuracy	92.6%	92.6%	80.26%	–	94.67%	99.47%	95.95%
Sensitivity	–	–	–	93.5% (high-frequency noise)	90.33%	98.24%	96.48%

[a]Two-channel records used as single-channel records
[b]10-fold cross-validation method test results

SOI) technology [33, 34]. It is observed that the decision tree required the least energy for prediction in Table 3.3. However, as the prediction accuracy of DT is not sufficiently high, the use of DT on the wearable IoT device is not preferable. The parametric models such as NB and LDA fail to obtain good accuracies even though the computational complexity involved in prediction is low. From the values in Table 3.1, it can be observed that the neural network ensemble using bagging exhibits the best performance in terms of accuracy, sensitivity, specificity, precision, and F1-score. The computational requirements of the bagging method are low and can be employed for online record quality assessment on wearable IoT devices (Table 3.3).

The performance of the discussed NF SQI-based NN bagging classifier is compared against the performance of state-of-the-art ECG record quality assessment algorithms in Table 3.4, and it can be observed that the proposed algorithm exhibits the best performance. However, direct comparison is not advisable due to the different datasets used for analysis.

3.6 Discussions

The findings of this chapter provide valuable insights into the performance of various binary classifiers for data integrity detection in wearable IoT edge devices. The results suggest that ensemble methods, particularly bagged neural networks, outperform traditional classifiers in terms of accuracy and robustness. This aligns with existing research emphasizing the advantages of combining multiple models (through multi-model fusion) to improve generalization and mitigate the effects of noise in physiological data. A key strength of the proposed approach is its reliance on non-fiducial signal quality indices (NF SQIs), eliminating the need for complex feature extraction techniques like QRS detection. This significantly enhances computational efficiency, making it well suited for resource-constrained IoT edge devices. Additionally, the study demonstrates that decision-level fusion methods improve classification performance, suggesting that leveraging multiple classifiers in combination can be an effective strategy for signal quality assessment in noisy environments. However, a notable limitation of the binary classification approach is that it inherently leads to discarding data deemed as low quality. Collecting physiological data, especially from wearable devices, requires significant effort, time, and resources, and rejecting data due to classification errors or binary classification thresholds could result in the loss of potentially useful information. This limitation can be overcome through soft classification approaches or confidence-based weighting methods to allow for data correction, reconstruction, or weighted fusion (which will be discussed in subsequent chapters) rather than outright rejection.

Beyond this, several practical challenges remain prior to real-world implementation. While the proposed models perform well under controlled conditions, their robustness to real-world variability—such as motion artifacts, sensor placement differences, and diverse patient populations—requires further investigation. Future research should consider adaptive learning techniques, allowing models to refine their predictions over time based on individual user characteristics. Additionally, testing these models on diverse datasets and real-time implementations will be crucial for validating their effectiveness beyond simulated conditions. By addressing these factors, machine learning-based data integrity detection can significantly enhance the reliability of wearable IoT-based healthcare monitoring while minimizing unnecessary data loss.

3.7 Conclusions

The chapter explored various machine learning algorithms that can be used in single-channel ECG quality detection where signal segments are marked as acceptable/unacceptable. The machine learning classification-based method discussed is found to be suitable for estimating the quality of ECG signals obtained from wearable IoT devices based on performance accuracy as well as computational complexity. To improve the reliability of various inference tasks, records of poor signal quality need to be discarded and re-recorded, calling for algorithms that are highly sensitive to poor signal quality. In this chapter, two signal quality indicators, namely, kSQI and sSQI—-with each of them calculated for three non-overlapping windows of a signal record—were used as features for classification, and an ensemble of neural networks was found to perform best in terms of all performance parameters considered. Two of the fusion methods (NB+LDA and NB+LDA+DT) studied were found to outperform their component base classifiers in terms of accuracy, specificity, precision, and F1-score, and were not as computationally expensive as the individual k-NN or SVM classifiers. The ensemble methods using decision trees and neural networks were shown to exhibit better performance than individual learners and were found to be computationally less expensive than k-NN, SVM, NB, and LDA. The ensemble of neural networks that exhibited the best performance during testing is of low computational complexity and is suited for deployment on an energy-efficient IoT device. The results indicate that reliable accuracy levels in signal integrity assessment can be achieved through the use of SQIs as features in conjunction with simple machine learning algorithms. Energy-efficient deep neural networks may be deployed on IoT devices for more reliable signal quality assessment at a later stage when sufficient training data is available.

References

1. I. Silva, G.B. Moody, L. Celi, Improving the quality of ECGs collected using mobile phones: the PhysioNet/computing in cardiology challenge 2011, in *2011 Computing in Cardiology* (Sept. 2011), pp. 273–276
2. X. Liu, J. Zhou, Y. Yang, B. Wang, J. Lan, C. Wang, J. Luo, W.L. Goh, T.T.-H. Kim, M. Je, A 457 nW near-threshold cognitive multi-functional ECG processor for long-term cardiac monitoring. IEEE J. Solid-State Circuits **49**(11), 2422–2434 (2014)
3. F. Zhang, Y. Wei, Y. Lian, Efficient QRS detection in wearable ECG devices for body sensor network, in *2008 5th International Summer School and Symposium on Medical Devices and Biosensors* (June 2008), pp. 289–292
4. G.D. Clifford, D. Lopez, Q. Li, I. Rezek, Signal quality indices and data fusion for determining acceptability of electrocardiograms collected in noisy ambulatory environments, in *2011 Computing in Cardiology* (Sept. 2011), pp. 285–288
5. T.H.C. Tat, C. Xiang, LE. Thiam, Physionet challenge 2011: improving the quality of electrocardiography data collected using real time QRS-complex and T-wave detection, in *2011 Computing in Cardiology* (Sept. 2011), pp. 441–444
6. Q. Li, C. Rajagopalan, G.D. Clifford, A machine learning approach to multi-level ECG signal quality classification. Comput. Methods Programs Biomed. **117**(3), 435–447 (2014)
7. Y. Zhang, S. Wei, C.D. Maria, C. Liu, Using Lempel-Ziv complexity to assess ECG signal quality. J. Med. Biol. Engi. **36**(5), 625–634 (2016)
8. Z. Zhao, Y. Zhang, SQI quality evaluation mechanism of single-lead ECG signal based on simple heuristic fusion and fuzzy comprehensive evaluation. Front. Physiol. **9**, 727 (2018)
9. U. Satija, B. Ramkumar, M.S. Manikandan, A review of signal processing techniques for electrocardiogram signal quality assessment. IEEE Rev. Biomed. Eng. **11**, 36–52 (2018)
10. G.D. Clifford, C. Liu, B. Moody, L.-W.H. Lehman, I. Silva, Q. Li, A.E. Johnson, R.G. Mark, AF classification from a short single lead ECG recording: the physionet/computing in cardiology challenge 2017, in *2017 Computing in Cardiology (CinC)* (2017), pp. 1–4

11. C. Orphanidou, *Signal Quality Assessment in Physiological Monitoring: State of the Art and Practical Considerations* (Springer International Publishing, 2018)
12. Q. Li, R.G Mark, G.D Clifford, Robust heart rate estimation from multiple asynchronous noisy sources using signal quality indices and a Kalman Filter. Physiol. Meas. **29**, 15–32 (2008)
13. A. Walinjkar, J. Woods, Personalized wearable systems for real-time ECG classification and healthcare interoperability: real-time ECG classification and FHIR interoperability, in *2017 Internet Technologies and Applications (ITA)* (Sept. 2017), pp. 9–14
14. L. Zhang, H. Peng, C. Yu, An approach for ECG classification based on wavelet feature extraction and decision tree, in *2010 International Conference on Wireless Communications Signal Processing (WCSP)* (2010), pp. 1–4
15. S. Shahbudin, S.N. Shamsudin, H. Mohamad, Discriminating ECG signals using Support Vector Machines, in *2015 IEEE Symposium on Computer Applications Industrial Electronics (ISCAIE)* (2015), pp. 175–180
16. S. Padmavathi, E. Ramanujam, Naive Bayes classifier for ECG abnormalities using multivariate maximal time series motif. Proc. Comput. Sci. **47**, 222–228 (2015). (Graph Algorithms, High Performance Implementations and Its Applications, ICGHIA 2014)
17. M.H. Song, J. Lee, H.D. Park, K.J. Lee, Classification of heartbeats based on linear discriminant analysis and artificial neural network, in *2005 IEEE Engineering in Medicine and Biology 27th Annual Conference* (Jan 2005), pp. 1151–1153
18. S. Russell, P. Norvig, *Artificial Intelligence: A Modern Approach*, 3rd ed. (Prentice Hall Press, USA, 2009)
19. X. Zhou, X. Zhu, K. Nakamura, N. Mahito, ECG quality assessment using 1D-convolutional neural network," in *2018 14th IEEE International Conference on Signal Processing (ICSP)* (Aug 2018), pp. 780–784
20. P.J. Boland, Majority systems and the Condorcet Jury theorem. J. R. Stat. Soc. Ser. D (The Stat.) **38**(3), 181–189 (1989)
21. T.G. Dietterich, Ensemble methods in machine learning, in *Multiple Classifier Systems. MCS 2000. Lecture Notes in Computer Science* (Springer, Berlin, Heidelberg, 2000), pp. 1–15
22. L. Breiman, Bagging predictors. Mach. Learn. **24**, 123–140 (1996)
23. L. Breiman, Heuristics of instability and stabilization in model selection. Ann. Stat. **24**(6), 2350–2383 (1996)
24. L. Breiman, Bias, variance and arcing classifiers, *Technical report 460, Statistics department, University of California*, vol. 11 (1996)
25. A. Mert, N. Kılıç, A. Akan, Evaluation of bagging ensemble method with time-domain feature extraction for diagnosing of arrhythmia beats. Neural Comput. Appl. **24**, 317–326 (2014)
26. Y. Freund, Boosting a weak learning algorithm by majority. Inf. Comput. **121**(2), 256–285 (1995)
27. L. Mason, J. Baxter, P. Bartlett, M. Frean, Boosting algorithms as gradient descent, in *Advances in Neural Information Processing Systems*, vol. 12 (MIT Press, 1999)
28. J.H. Friedman, Stochastic gradient boosting. Comput. Stati. Data Anal. **38**(4), 367–378 (2002). (Nonlinear Methods and Data Mining)
29. H. Shi, H. Wang, Y. Huang, L. Zhao, C. Qin, C. Liu, A hierarchical method based on weighted extreme gradient boosting in ECG heartbeat classification. Comput. Methods Programs Biomed. **171**, 1–10 (2019)
30. B.V. Dasarathy, Sensor fusion potential exploitation-innovative architectures and illustrative applications. Proc. IEEE **85**, 24–38 (1997). (Jan.)
31. P. Peng, Y. Mingyu, X. Weisheng, Running 8-bit dynamic fixed-point convolutional neural network on low-cost ARM platforms, in *2017 Chinese Automation Congress (CAC)* (2017), pp. 4564–4568
32. D. Anguita, A. Ghio, L. Oneto, X. Parra, J. Reyes-Ortiz, Human activity recognition on smartphones using a multiclass hardware-friendly support vector machine, in *International Workshop on Ambient Assisted Living* (Dec. 2012), pp. 216–223
33. H. Reyserhove, N. Reynders, W. Dehaene, Ultra-low voltage datapath blocks in 28nm UTBB FD-SOI, in *2014 IEEE Asian Solid-State Circuits Conference (A-SSCC)* (Nov. 2014), pp. 49–52

34. R. Taco, I. Levi, M. Lanuzza, A. Fish, An 88-fJ/40-MHz [0.4 V]–0.61-pJ/1-GHz [0.9 V] dual-mode logic 8 × 8 bit multiplier accumulator with a self-adjustment mechanism in 28-nm FD-SOI. IEEE J. Solid-State Circuits **54**, 560–568 (2019)

Chapter 4
A Multi-sensor Fusion Technique for Heart Rate Estimation

Abstract This chapter investigates the role of multi-sensor data fusion in enhancing the accuracy and robustness of heart rate estimation, particularly under conditions involving significant noise interference on the signal. By combining signals from multiple sensor channels, data fusion techniques can mitigate the impact of noise and improve the reliability of the estimated heart rate. The chapter introduces and compares a range of state-of-the-art and novel fusion strategies, with a focus on their performance using the MIT-BIH Arrhythmia Database supplemented by noise from the MIT Noise Stress Test Database. Special attention is given to scenarios with low signal-to-noise ratios (SNR), where two approaches—Kalman fusion and α-trimmed mean filtering—demonstrate superior performance. Kalman fusion excels when both input channels are noisy, whereas α-trimmed mean filtering is more effective when at least one clean channel is available. To leverage the strengths of both methods, a hybrid algorithm is presented that dynamically switches between them based on a signal quality indicator (SQI) serving as a proxy for SNR. This adaptive fusion strategy achieves improved heart rate estimation accuracy and outperforms several leading techniques, offering valuable insights for designing resilient, real-time health monitoring systems.

Keywords Data fusion · Electrocardiogram · Heart rate estimation · Signal quality indicators · Bayesian filtering · Kalman fusion · Algorithm switching

4.1 Introduction

The analysis of the ECG signal is widely applied in the diagnosis of various heart disorders [1]. Heart rate is one of the vital signs indicating disturbances to the cardiovascular system [2], and a wearable cardiac sensor that can acquire a user's ECG data is typically used for continuous physiological monitoring. A clean ECG contains a regular cycle of the P wave corresponding to the depolarization of the auricle, the QRS complex indicating the depolarization of the ventricles, and the T wave corresponding to the repolarization of the ventricles. Extracting any one of the repetitive complexes and waveforms can be used to calculate the heart rate. With develop-

A. John et al., *Deep Learning and Signal-Processing Methods for Multisensor Data Fusion*, https://doi.org/10.1007/978-3-031-96724-5_4

ments in IoT architectures enabling the deployment of wearable health monitoring devices, research in health monitoring like cardiovascular health and human activity recognition is fast developing [3, 4].

ECG signals can be easily corrupted by noise such as baseline wander, muscle artifacts, and electrode motion [5, 6]. The effects of these noises corrupting the ECG signals can lead to drastic variations in the heart rate calculated compared to the actual heart rate, leading to false alarms in the intensive care units as well as in wearable heart rate monitoring devices, which would further lead to alarm fatigue [7]. In other cases, a single lead can fail, leading to a loss of information. In this chapter, the possibility of fusing information from multiple ECG leads is considered because as long as the noise is not correlated between sensors and has a zero mean and constant variance, fusing similar sensors like this reduces the variance of the combined noise by a factor of the number of sensors [8]. Therefore, data fusion can be used to improve the performance of the heart rate estimation algorithms. Although data fusion is a very commonly used solution for heart rate estimation, no conclusive methods of usage or methodology have been established for data fusion of ECG leads for heart rate calculation, as seen in Chap. 2. This chapter explores several decision-level fusion methodologies that utilize RR intervals derived from QRS detection algorithms. Their performance is compared against single-lead ECG heart rate calculation algorithms to assess their effectiveness.

4.1.1 Related Works

In literature, as seen in Chap. 2, most methods on heart rate estimation focus on the detection of the QRS peak in electrocardiogram signals for calculating heart rate, as the QRS peak is usually easy to detect due to it contributing to 90% of the signal energy in a beat cycle. The Physionet/CinC 2014 challenge aimed to improve heartbeat detection from multimodal data consisting of ECG and other non-cardiovascular signals. All major results from the challenge have been discussed in [9], and the top score (as of May 2015) on the hidden test data set was 93.64%. A novel beat SQI-based majority voting fusion algorithm for robust heart rate (HR) estimation from cardiovascular and non-cardiovascular signals was proposed in [10]. A probabilistic method to estimate the R-peak locations of an ECG signal using a particle filter was discussed in [11], which exhibited a mean absolute heart rate estimation error of 5.044 beats-per-minute (bpm) at –6 dB SNR on the MIT-BIH Arrhythmia database. A probabilistic model to synthesize the heart rate annotations from multiple annotators (QRS detection algorithms) from ECG signals was proposed in [12]. An unsupervised Bayesian framework to synthesize the heart rate through fusion from multiple annotators was proposed, which exhibited a mean squared error of 14.37 bpm on the Physionet/CinC 2014 challenge 2014 dataset [13]. For beat detection from ECG and other non-cardiovascular signals, a modified slope sum function and the Teager-Kaiser energy operator method were used. The performance of the majority voting fusion method was evaluated on the Physionet/CinC 2014 challenge hidden test set and was found to achieve a score of 90.89% [14]. In [15], a novel fusion algorithm of 12-lead ECG signals is described based on the idea of a local weighted linear prediction algorithm for obtaining a better estimate of the ECG signal from 12 noisy channels taken from the Physionet/ CinC 2011 challenge database, which can then be used for heart rate detection.

As discussed in previous chapters, SQIs are an important consideration in data fusion algorithms as only those signal leads that are reliable should be used for fusion. SQI-based fusion of physiological signals for heartbeat detection has been discussed in [16, 17]. Methods for indicating signal quality as multiple classes or continuous values that specify the acceptability of ECG signals of diagnostic quality have been discussed previously in the literature [5, 18, 19]. In the previous chapter, signal quality assessment for a record of 30/60 s was discussed. However, for temporal inferences like heart rate estimation, where events of interest occur once every second or less, methods for instantaneous signal quality assessment become necessary.

In this chapter, various data fusion techniques to fuse R peak to R peak (RR) intervals obtained from 2 lead ECG signals are proposed, explored, and compared against each other. Some of these fusion algorithms make use of the instantaneous signal quality indices of the signals for fusion or decision-making, which are developed in this chapter. Although multiple channels may intuitively seem superior to a single channel, the real challenge is determining effective methods to harness this potential. This chapter focuses on identifying the optimal algorithms for accurately calculating RR intervals using state-of-the-art R-peak detection techniques. The flow diagram of the proposed methods and study is shown in Fig. 4.1. An SQI-based fusion algorithm selection model is proposed in this chapter, which uses a signal quality indicator as a surrogate for signal SNR to selectively switch between the various fusion algorithms. The extraction of RR intervals is discussed in Sect. 4.2. Signal quality indicators for determining the signal quality of the ECG signals are discussed in Sect. 4.3, followed by the discussion on Bayesian elimination of improbable RR intervals in Sect. 4.4. The discussion on the various fusion algorithms explored in this work is discussed in Sect. 4.5, the performance comparison is carried out in Sect. 4.6, and the final results are covered in Sect. 4.7.

4.1.2 Math Notation

- $x(t)$ is a continuous signal.
  $x[n]$ is a discrete-time sequence such that $x[n] = x(nT_s)$, where T_s is the sample interval and $T_s = \frac{1}{F_s}$, with F_s indicating the sampling frequency in Hz.
- $\widehat{\mu}$ is used to represent an estimate of μ and $\widetilde{\mu}$ is used to represent a moving estimate of μ.
- $x[n-]$ is used to represent the set of past samples of the discrete time sequence x at $x[n]$.
- $\lfloor w \rfloor$ indicates the floor function applied to a real number w.

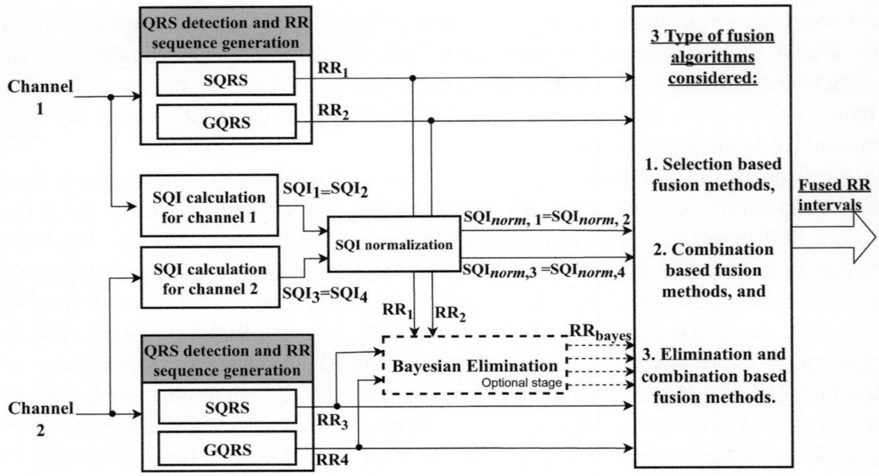

Fig. 4.1 Flow diagram of the proposed fusion methodology for heart rate estimation

- The methodology involved uses a low pass filter at various stages, and here this low pass filter is defined as, LPF_α.

$$y = LPF_\alpha(x);$$

$$y[n] = \alpha y[n-1] + (1-\alpha)x[n] \tag{4.1}$$

Typically $\alpha = 0.99$ and from the transfer function, the 3 dB cutoff frequency is calculated as $0.0016 \times F_s$, where F_s is the sampling frequency. For a sampling frequency of 360 Hz, the 3 dB cutoff frequency $F_{3dB} \approx 0.6$ Hz.

4.1.3 Dataset

The study uses 2-channel ECG recordings for algorithm evaluation. The recordings are taken from the MIT-BIH arrhythmia database [20, 21]. The MIT-BIH arrhythmia database contains 48 half-hour excerpts of two-channel ambulatory ECG recordings, obtained from 47 subjects studied at Boston's Beth Israel Hospital Arrhythmia Laboratory. Pre-recorded noise (baseline wander, muscle artifacts, and electrode motion noise) available from the MIT-BIH Noise Stress Test Database (NSTDB) [20–22] is used for algorithm evaluation. The noise recordings were obtained from physically active volunteers and standard ECG recorders and electrodes. The electrodes were placed on the limbs in positions in which the subjects' ECGs were not visible.

Therefore, it should be noted that the noise recordings used were not simulated, but rather taken from recordings obtained from physically active volunteers to simulate ambulatory noise that may appear in ambulatory ECG recordings.

4.2 Single-Channel QRS Complex Detection and RR Sequence Calculation

The QRS complex locations were extracted using 2 widely used QRS detection algorithms made available in the WFDB toolbox [20, 23]: GQRS and SQRS [24]; from which 2 streams of RR intervals per ECG signal channel are calculated: $RR_1, RR_2...RR_4$. The RR sequence is generated by taking the interval between two consecutive beat annotations. The RR interval at a given time sample is derived from the beat annotation prior to that time sample and the beat annotation that is followed immediately. The details can be observed in Fig. 4.2. The performance of the fusion algorithms discussed in this chapter are compared against the performance of the single-channel RR intervals calculated from these standard QRS detection algorithms. The QRS algorithms may not exhibit the same performance over all the noise conditions and may not be linearly dependent on the ECG signal's quality, which may lead to difficulties in analyzing the performance of the fusion algorithms. Since this chapter focuses on reviewing and analyzing various fusion techniques for unimodal data across different noise levels, it does not discuss or evaluate the performance of QRS detection algorithms.

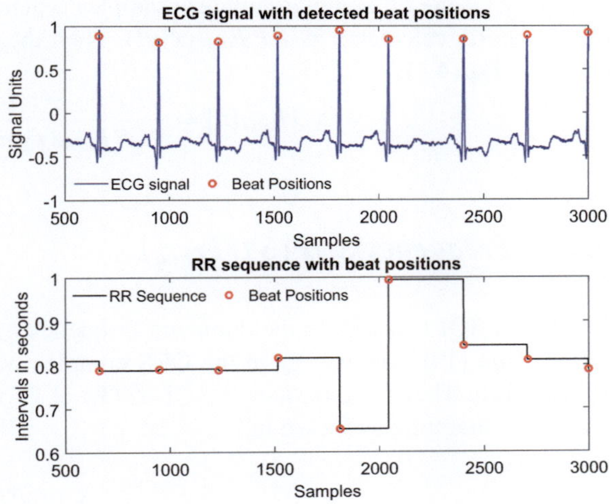

Fig. 4.2 Figure detailing how the RR sequence is extracted from the beat annotations

4.3 Signal Quality Indicators (SQIs)

With the emergence of telemedicine and continuous monitoring of vital signs inside and outside clinical settings, the data acquired from wearable devices are often contaminated with various types of noises such as electrode motion noise, muscle artifacts, and motion artifacts. At times, the acquired data may lack sufficient diagnostic quality, making it unsuitable for further processing. This is where SQIs play a crucial role. Over the years, their integration has proven essential for the successful deployment of monitoring devices.

4.3.1 Kurtosis SQI (SQI_{kurt})

As discussed in Chap. 3, Kurtosis is defined as the fourth standardized moment of a probability distribution [18]. The sample kurtosis for N samples of X denoted by x_i is calculated as $kSQI \approx \frac{\frac{1}{N}\sum(x_i-\widehat{\mu})^4}{\widehat{\sigma}^4}$, where $\widehat{\mu}$ and $\widehat{\sigma}$ are the empirical estimates of the mean and standard deviation of the distribution (the ECG signal analysis window), which provides a single kurtosis measure for the analysis window. Therefore, to provide an instantaneous estimate of the kurtosis SQI, the following estimation method is proposed:

$$x_1[n] = (x[n] - \widetilde{\mu[n]})^4 \tag{4.2}$$

$$x_2[n] = (x[n] - \widetilde{\mu[n]})^2, \tag{4.3}$$

where $x[n]$ is the ECG signal sample at sample point n and $\widetilde{\mu[n]}$ is the moving mean of the record, calculated as $LPF_{0.99}(x[n])$. Then the kurtosis SQI is calculated as shown in Eq. (4.4).

$$SQI_{kurt}[n] = \frac{LPF_{0.99}(x_1[n])}{(LPF_{0.99}(x_2[n]))^2} \tag{4.4}$$

4.3.2 Frequency SQI (SQI_{freq})

Frequency SQI is based on the dominant frequency band that contains ECG signal information [19]. The energy in the QRS complex is centered at 10 Hz and is 10 Hz in width. Therefore, frequency SQI (fSQI) of ECG signal x is mathematically defined in literature as shown in Eq. (4.5):

$$fSQI = \frac{\int_{f=5Hz}^{f=15Hz} S_x(f)df}{\int_{f=5Hz}^{f=50Hz} S_x(f)df}, \tag{4.5}$$

where S_x is the power spectral density of x. In the equation for fSQI, the numerator represents the energy of the QRS wave, and the denominator represents the overall energy of the ECG signal. To calculate an fSQI at each signal sample point per record, the method proposed is as shown in Eq. (4.6):

$$SQI_{freq}[n] = \frac{LPF_{0.99}(|(h_1 * x)(n)|^2)}{LPF_{0.99}(|(h_2 * x)(n)|^2)}, \tag{4.6}$$

where $h_1[n]$ and $h_2[n]$ are the impulse responses of bandpass filters with cutoff frequencies 5 Hz and 15 Hz corresponding to the QRS peak frequency range, and 5 Hz and 50 Hz corresponding to the ECG signal frequency range respectively. The outputs of these two filters are squared to calculate the energy in the desired frequency ranges and are then subsequently filtered using the low pass filter $LPF_{0.99}$.

4.3.3 Total SQI

The SQI_{kurt} and SQI_{freq} are combined to form the total SQI per channel as

$$SQI[n] = SQI_{kurt}[n].SQI_{freq}[n] \tag{4.7}$$

Figure 4.3a–c shows the output total SQI at varying SNRs for when the ECG signal is corrupted by (a) electrode motion noise, (b) muscle artifacts, and (c) baseline wander noise respectively. Here, the total SQI is averaged over a signal record of 20 seconds in length per sample per SNR. Ideally, the SQI values must vary linearly with SNR, but since this is not possible as signal quality can vary from 0 to ∞, a monotonically increasing function is a sufficiently good estimate of the signal quality. From Fig. 4.3, it can be concluded that the total SQI, which is a product-based combination of kurtosis SQI and frequency SQI is a sufficiently good indicator of signal quality as the function is monotonically increasing against the SNR values. The total SQI for an ECG signal window can be seen in Fig. 4.4c and d. QRS detection-based SQIs or SQIs which require the detection of ECG fiducial points (F SQI) are not ideal for use in determining the acceptability of the signal as discussed in Chap. 3. Both kurtosis and frequency SQIs do not require the estimation of the fiducial points (ie., the QRS peak locations).

4.3.4 The SQI per RR Interval Stream

For each channel, the signal quality indicator is calculated as shown in Eq. (4.7). Each RR interval stream obtained per channel (1 or more QRS detectors can act on a single channel of ECG signal to obtain one or more RR interval streams) is assigned the signal quality indicator for that channel, which then forms $SQI_k[n]$ where k

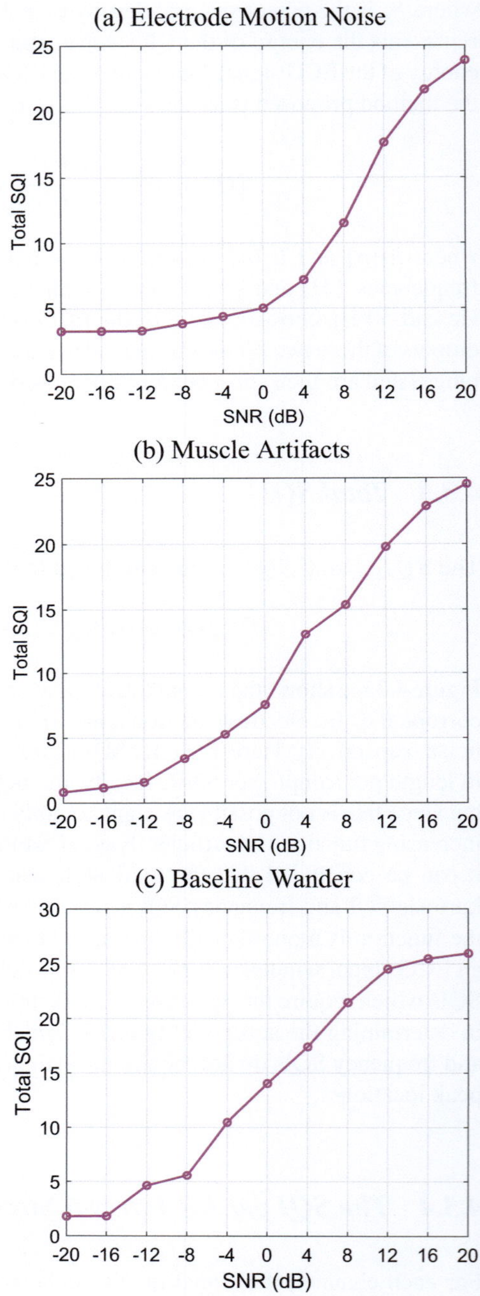

Fig. 4.3 The total SQI versus SNR plot for when the ECG signal is corrupted by **a** electrode motion noise, **b** muscle artifacts, and **c** baseline wander noise

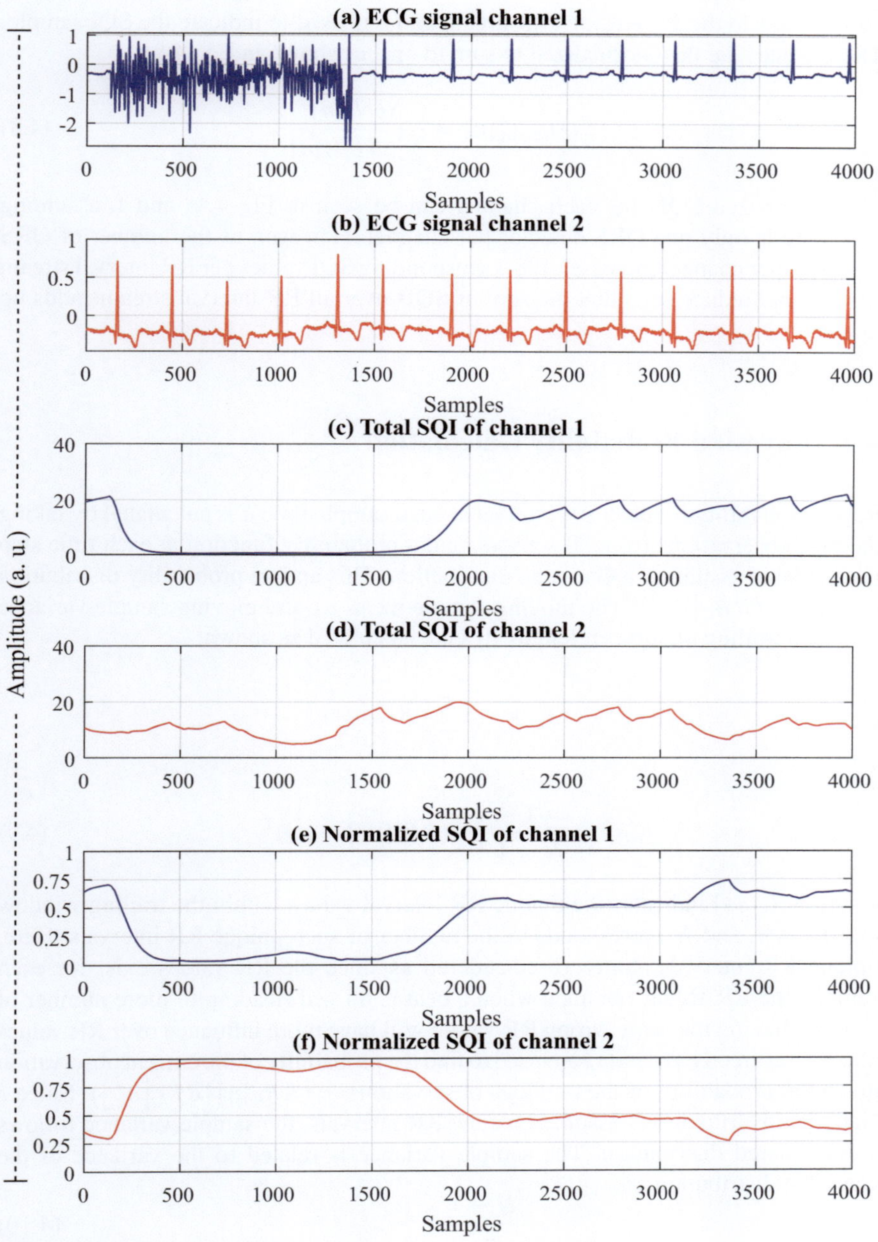

Fig. 4.4 SQI Variations with noise in channel 1

corresponds to the kth RR interval stream and n is used to indicate the SQI sample. These values are then normalized to sum to one, as shown in Eq. (4.8):

$$SQI_{norm_k}[n] = \frac{SQI_k[n]}{\sum_k SQI_k[n]} \tag{4.8}$$

The normalized SQIs for each channel can be seen in Fig. 4.4e and f, assuming that there is only one QRS detector per channel. However, as the number of QRS detectors per channel increases, the corresponding SQI values per RR interval stream per channel reduce such that the sum of SQIs over all RR interval streams adds up to unity.

4.4 Bayesian Probability Calculation

Bayes probability for each RR interval at each sample time n is calculated by taking the past observations from all sensors, and a probability function at each time step is calculated assuming a Gaussian distribution. This apriori probability distribution is called $p\left(RR_k\left[n^-\right]\right)$. The moving sample mean μ_k and moving sample variance s_k^2 over a trailing window of length 10 s are calculated as shown:

$$\mu_k = \frac{1}{N}\sum RR_k[n*]$$

$$s_k^2 = \frac{1}{N-1}\sum (RR_k[n*] - \mu_k)^2, \tag{4.9}$$

where $RR_k[n*]$ indicates the unique RR interval values within the trailing window of 10 s at n, and N corresponds to the number of such unique RR intervals. Here, unique RR intervals alone are considered as since the RR values exist for each sample, the RR values obtained when a beat is missed (leading to more number of samples having the same wrong RR value) will have more influence over RR values obtained correctly. μ_k and s_k^2 are used to find the probability of the current observation given the probability of the previous observations, $p\left(RR_k\left[n\right] | RR_k\left[n^-\right]\right)$. Since a Gaussian distribution is assumed for the RR intervals, the sample variance follows a chi-squared distribution. The sample variance is related to the variance of the Gaussian distribution as

$$\frac{N-1}{\sigma_k^2} s_k^2 \sim \chi_N^2 \tag{4.10}$$

Therefore, the variance of the distribution can be estimated as

$$s_k^2 \frac{N-1}{\chi^2_{N-1,0.005}} < \sigma_k^2 < s_k^2 \frac{N-1}{\chi^2_{N-1,0.995}}, \tag{4.11}$$

with a confidence of 99%, and the lower limit is taken as the estimate of variance. Therefore, the standard deviation estimate is calculated as shown in Eq. (4.12):

$$s_l^2 \triangleq s_k^2 . \frac{N-1}{\chi^2_{N-1,0.005}}, \tag{4.12}$$

where N is the total number of unique RR intervals in the window of length 10 s and $\chi^2_{N,0.005}$ is the value of the inverse complementary χ^2 cumulative distribution function with N observations and a confidence interval of 0.005. From this, $p\left(RR_k\,[n]\,|RR_k\left[n^-\right]\right)$ is calculated as shown in Eq. (4.13):

$$p(RR_k[n]|RR_k[n^-]) = \frac{1}{\sqrt{2\pi s_l^2}}.e^{\frac{-1}{2}\frac{(\mu_k - RR_k[n])^2}{s_l^2}} \tag{4.13}$$

With that, a grouping of probabilities per RR interval is obtained. The probabilities at each sample/RR interval index are normalized such that the probabilities add up to 1.

4.5 Fusion Algorithms

The RR intervals calculated per detector per channel are fused to estimate the correct RR interval using some common fusion algorithms discussed in this section. Three broad categories of fusion methods are considered:

1. selection of a single most probable RR value,
2. calculation of most probable value by combination, and
3. calculation of the most probable value through elimination and combination.

These fusion methods are detailed in Sects. 4.5.1, 4.5.2, and 4.5.3.

4.5.1 Selection of a Single Most Probable Value

Median

The median method extracts the median RR interval at each sample point.

$$RR_{median}\,[n] = Median_k\,(RR_k\,[n]) \tag{4.14}$$

RR Best SQI

In the Best SQI-based RR calculation, the RR interval obtained from either channel is chosen depending on which channel has the Best SQI. From the QRS algorithms, the SQRS algorithms were found to exhibit the best average performance for this particular data set when the signals are noisy, and therefore, the RR interval obtained from the SQRS algorithm is chosen from the channel (all RR intervals from a single channel have the same SQI) with the highest SQI.

$$\begin{aligned} RR_{bestSQI}[n] &= RR_{k*} \\ where \quad k* &= argmax_k(SQI_{norm_k}[n]) \end{aligned} \tag{4.15}$$

Best Bayes

The Bayesian probabilities obtained post-Bayesian filtering can be used to select the most probable RR interval as

$$\begin{aligned} RR_{BestBayes}[n] &= RR_{k*}[n] \\ where \quad k* &= argmax_k\left(p(RR_k[n]|RR_k[n^-])\right). \end{aligned} \tag{4.16}$$

In the situation where multiple RR channels have equal probabilities, the preference order is RR interval from SQRS, followed by RR interval from GQRS.

4.5.2 Calculation of Most Probable Value by Combination

Simple Weighted Average

In the simple weighted averaging (SWA) method of fusion, the RR intervals obtained from the multiple detectors and channels are fused by adding the RR intervals weighted by the normalized SQIs (as per Eq. (4.17)).

$$RR_{SWA}[n] = \sum_k RR_k[n].SQI_{norm_k}[n] \tag{4.17}$$

SWA with Bayesian Probabilities

Although the Bayesian elimination stage is optional, the feasibility of substituting the SQIs with the Bayesian probabilities obtained in the Bayesian elimination is analyzed. The Bayes probabilities can be used to indicate the suitability of the RR interval streams obtained per detector per channel for fusion. The probabilities can be normalized such that they add up to 1:

$$SQI_{Bayes_k}[n] = \frac{p(RR_k[n] | RR_k[n^-])}{\sum_k p(RR_k[n] | RR_k[n^-])} \tag{4.18}$$

From the Bayesian probabilities based SQI obtained in (Eq. (4.18)), the SWA using SQI_{Bayes_k} can be calculated as shown in (Eq. (4.19)):

$$RR_{SWA_{Bayes}}[n] = \sum_k RR_k[n].SQI_{Bayes_k}[n] \tag{4.19}$$

Kalman Fusion

Kalman fusion is a well-known technique in data fusion based on Kalman filtering [25], as seen in Chap. 2. Kalman filter creates an estimate of the current state of a system and then updates the state based on observations at the current time step. The estimate of the current state is then used to update the initial prediction for the state at the next time step. Prior to the Kalman fusion stage, the algorithm first checks for possibly erroneous RR intervals from the input RR intervals RR_k by checking if the RR interval exceeds 1.7 s or falls below 0.3 s. If such values are observed, they are replaced by the closest upper or lower limit values to obtain the pre-processed RR intervals $RR_{pre,k}$. These corrected interval sequences are used to find the RR interval sequence estimate $RR_{kal}[n]$, or the state of interest, which is a scalar at time step n (in general Kalman filters have a state vector which is what the filter learns to estimate and which is continuously updated, and in this case, a scalar state variable $\triangleq RR_{kal}[n]$) is employed. The observation vector will be the RR intervals calculated from the QRS detection algorithms $RR_{pre,k}[n]$ at time step n, which is the input to the Kalman filter. For performing Kalman filtering, some important parameters are

- $\sigma_k^2[n]$, the apparent noise variance of each individual RR interval detector per channel, which is represented as a row vector of length k, and
- $\sigma_{RR_{kal}}^2[n]$, the apparent noise variance of the previous estimation (at time step $n-1$), which is a scalar.

The noise $R[n]$ is defined as the difference between the sensor values and the apparent actual values. These values can be initialized to be any value and is defined as $R[n] = diag([\sigma_k^2[n]])\forall k$; corresponding to each set of sensor observations $RR_{pre,k}[n]$ at time step n [25]. A weighting system A is created such that

$$A[n] = \sigma_{RR_{kal}}^2[n].\left[\sigma_{RR_{kal}}^2[n]J + R[n]\right]^{-1} V, \tag{4.20}$$

where V is a column vector of ones, and J is a matrix of ones of dimension $k \times k$. Once all the requisite parameters at n are available, the RR interval estimate $RR_{kal}[n+1]$ can be estimated through the Kalman filtering process as [25]:

$$RR_{kal}\,[n+1] = RR_{kal}\,[n] \,.\, \left(1 - A[n]^T V\right) + A[n]^T RR_{pre,k}[n], \tag{4.21}$$

where $RR_{kal}[1] = 0$. Next, the new noise variances $\sigma_k^2[n+1]$ and $\sigma_{RR_{kal}}^2[n+1]$ based on their errors in comparison to the actual value is calculated. In real-life applications, no actual value of ground truth is available to use as a control. So in its stead the RR interval sequence obtained from the Simple Weighted Averaging ($RR_{SWA}[n]$) is used:

$$\sigma_k^2[n+1] = \sigma_k^2[n] + \frac{(RR_{pre,k}[n] - RR_{SWA}[n])^2 - \sigma_k^2[n]}{j+1}, \tag{4.22}$$

where j is the number of updates that have already occurred (here $j = n$, which corresponds to the time step). Noise variance from the previous estimate, $\sigma_{RR_{kal}}^2$ is updated as

$$\sigma_{RR_{kal}}^2[n+1] = \sigma_{RR_{kal}}^2[n] + \frac{(RR_{SWA}[n] - RR_{SWA}[n-1])^2 - \sigma_{RR_{kal}}^2[n]}{M}, \tag{4.23}$$

where the parameter M controls how quickly the values of the output variance should change, and was set as 5 empirically.

4.5.3 Calculation of Most Probable Value Through Elimination and Combination

σ-Trimmed Mean Filter

This filtering method makes use of the $RR_{SWA}\,[n]$ computed in Eq. (4.17) to filter out RR interval streams that deviate by more than half of the standard deviation of all RR interval streams from the $RR_{SWA}\,[n]$. The RR intervals that satisfy the following inequality Eq. (4.24):

$$|RR_k[n] - RR_{SWA}[n]| > \sigma[n]/2, \tag{4.24}$$

are discarded to yield a new set of RR intervals having indices k' at n, and a new average of the remaining RR intervals is calculated, where $\sigma\,[n]$ is the standard deviation of RR intervals from all k detectors at sample n. This filtering stage would remove highly unlikely RR intervals. In this context, out of the 4 RR interval channels, 1 or 2 RR intervals would be removed at each time step n. The RR interval calculated after the σ–trimmed mean filtering stage is calculated as shown in Eq. (4.25):

$$RR^{(\sigma)}[n] = \frac{\sum_{k \in k'} RR_k[n]}{|k'|}, \tag{4.25}$$

where $|k'|$ is the cardinality of the set k' at n.

α-Trimmed Mean Filter

α-trimmed mean filter is a method of filtering out $\lfloor \alpha K \rfloor$ channels or streams that are furthest below and above the median value of all K RR interval streams $RR_k\,[n]$ where $k = 1, 2, ..K$. The values that are furthest below and above the median are discarded, resulting in a new set of streams having indices k'. For, e.g., if $K = 4$, and stream index 2 is removed then $\{k'\} = \{1, 3, 4\}$. The SQIs are then renormalized using this subset of streams to obtain $SQI^{(\alpha)}_{norm_k}$ as shown in Eq. (4.26):

$$SQI^{(\alpha)}_{norm_k}[n] = \frac{SQI_{norm_k}[n]}{\sum_{u \in k'} SQI_u[n]} \quad \forall \;\; k \in k'. \tag{4.26}$$

The modified simple weighted average is then calculated as shown in Eq. (4.27) to obtain the RR interval estimate from the α-trim mean filtering algorithm.

$$RR^{(\alpha)}_{SWA}[n] = \sum_{k \in k'} RR_k[n].SQI^{(\alpha)}_{norm_k}[n] \tag{4.27}$$

4.5.4 Bayes-Eliminated Fusion Algorithms

Bayesian elimination is an optional stage used in eliminating any improbable RR intervals, obtained per channel per QRS detector algorithm, based on the probability distribution generated from the previous RR intervals as discussed in Sect. 4.4. The probabilities obtained per RR interval are normalized at each sample/ RR interval index such that the probabilities add up to 1. Now, the intervals with relevant probabilities below 0.05 are discarded. With the remaining intervals, the fused results from the methods discussed in Sects. 4.5.1, 4.5.2, and 4.5.3 are recalculated with the potential benefit of removing the influence of erroneous RR intervals. The fusion methods discussed in Sects. 4.5.1, 4.5.2, and 4.5.3 after the Bayesian elimination stage is called Bayes-eliminated fusion algorithms in subsequent portions of this chapter.

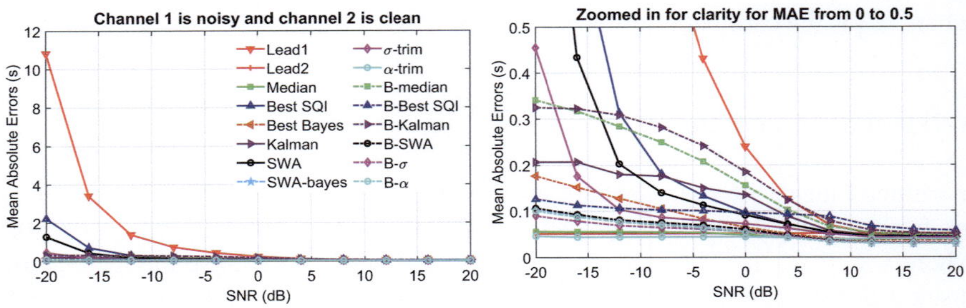

Fig. 4.5 The plot of MAE versus SNR when channel 1 is corrupted by muscle artifacts and channel 2 is clean

4.6 Performance Comparison

4.6.1 Mean Absolute Error

For the performance evaluation of the various fusion methods discussed in Sect. 4.5, the mean of the absolute errors (E_{abs}) ie., the MAE between the estimated RR intervals ($\widehat{RR[n]}$) and the ground truth RR interval ($RR[n]$) is used as a metric. E_{abs} is calculated per record as shown in Eq. (4.28), where N is the length of the record. The RR intervals have a unit of seconds and, therefore, E_{abs} has a unit of seconds. To aid in the comparison process, a table of steps used for each of the fusion algorithms is provided (Table 4.1).

$$E_{abs} = \frac{1}{N} \sum |\widehat{RR[n]} - RR[n]| \tag{4.28}$$

Three different scenarios are simulated with three different types of noise added to channel 1 alone, channel 2 alone, and both channels such that the signal-to-noise ratio is varied from –20 to 20 dB, and the performance of the fusion algorithms in terms of E_{abs} is as shown in Tables 4.2, 4.3, and 4.4 respectively. Here, the E_{abs} values indicate the mean of the E_{abs} of all records in the MIT-BIH Arrhythmia database (except records 101, 117, and 200 that were dropped due to the poor performance of SQRS and GQRS algorithms when the signals were clean on channel 2). The algorithms exhibiting the best performance at each SNR are highlighted in bold and underlined in the Tables. The algorithms that exhibit better performance than the clean individual lead are highlighted in bold. The performance of the algorithms in terms of E_{abs} when muscle artifacts corrupt channel 1, channel 2 and both channels from from –20 to 20 dB SNR is shown Figs. 4.5, 4.6, and 4.7 respectively.

Among the selection algorithms, the median method discussed in section "Median" exhibits the best performance, which is close to the performance of the non-corrupted lead in the scenario where only one lead is corrupted by noise in the high SNR regions. In the scenario where both leads are corrupted by noise, performance in the low SNR

Table 4.1 Steps involved for RR interval estimation for each algorithm

Steps	Algorithm						
	Median	Best SQI	Best Bayes	Kalman[a]	SWA	SWA-bayes	σ-trim
SQI calculation		✓			✓		
SQI normalization					✓		
Removal of RR intervals $\notin [0.3, 1.7]s$				✓			
Bayes probability calculation			✓			✓	
Bayes probability normalization						✓	
Bayes probability elimination							
Steps	Algorithm						
	α-trim	Bayes median	Bayes best SQI	Bayes Kalman[a]	Bayes SWA	Bayes σ-trim	Bayes α-trim
SQI calculation	✓		✓		✓		✓
SQI normalization	✓		✓		✓		✓
Removal of RR intervals $\notin [0.3, 1.7]s$				✓			
Bayes probability calculation		✓	✓	✓	✓	✓	✓
Bayes probability normalization		✓	✓	✓	✓	✓	✓
Bayes probability elimination		✓	✓	✓	✓	✓	✓

[a]Kalman fusion algorithm and the Bayes Kalman algorithm used in this work uses the RR interval estimate from the SWA algorithm as the control and therefore would require the SQI calculation and SQI normalization steps. However, to carry out Kalman fusion, any other RR interval estimate can be used as control, and therefore the SQI calculation and SQI normalization steps are not checked in this table

Table 4.2 Mean absolute errors (MAEs) in seconds indicating the performance of fusion algorithms when lead 1 is corrupted by noise and lead 2 is clean

SNR (dB)	Pre Fusion[a]		No Bayesian elimination								With Bayesian Elimination					
	Lead 1	Lead 2	Median	Best SQI	Best Bayes	Kalman	SWA	SWA-bayes	σ-trim	α-trim	Median	Best SQI	Kalman	SWA	σ-trim	α-trim
Electrode motion																
–20	0.694	0.051	0.119	0.285	0.195	0.271	0.184	0.145	0.125	**0.091**	0.287	0.123	0.330	0.138	0.117	0.128
–16	0.495	0.051	0.114	0.222	0.180	0.261	0.144	0.137	0.111	**0.089**	0.274	0.121	0.328	0.134	0.116	0.124
–12	0.391	0.051	0.108	0.220	0.160	0.252	0.130	0.126	0.110	**0.083**	0.263	0.116	0.318	0.129	0.114	0.119
–8	0.325	0.051	0.092	0.213	0.137	0.218	0.121	0.113	0.105	**0.077**	0.246	0.112	0.295	0.119	0.107	0.110
–4	0.254	0.051	0.088	0.177	0.112	0.174	0.108	0.096	0.095	**0.069**	0.208	0.124	0.251	0.102	0.093	0.093
0	0.175	0.051	0.065	0.127	0.085	0.123	0.089	0.074	0.080	**0.060**	0.150	0.115	0.188	0.078	0.072	0.069
4	0.107	0.051	**0.045**	0.083	0.060	0.079	0.068	0.052	0.062	**0.045**	0.095	0.125	0.122	0.055	0.052	**0.047**
8	0.064	0.051	**0.039**	0.056	**0.043**	0.053	0.052	**0.036**	**0.048**	**0.040**	0.063	0.085	0.074	**0.038**	**0.037**	**0.035**
12	0.049	0.051	**0.039**	**0.047**	**0.038**	**0.045**	**0.046**	**0.031**	**0.044**	**0.040**	0.051	0.065	0.056	**0.032**	**0.032**	**0.031**
16	0.046	0.051	**0.039**	**0.046**	**0.037**	**0.044**	**0.045**	**0.030**	**0.042**	**0.039**	0.049	0.060	0.052	**0.031**	**0.031**	**0.030**
20	0.044	0.051	**0.039**	0.045	**0.036**	**0.043**	**0.044**	**0.029**	**0.042**	**0.039**	0.048	0.058	0.050	**0.031**	**0.030**	**0.030**
Muscle artifact																
–20	10.819	0.051	0.056	2.220	0.177	0.207	1.273	0.102	0.455	**0.046**	0.341	0.127	0.325	0.107	0.090	0.101
–16	3.408	0.051	0.055	0.715	0.152	0.207	0.434	0.091	0.176	**0.044**	0.317	0.113	0.322	0.093	0.078	0.088
–12	1.377	0.051	0.055	0.313	0.127	0.180	0.203	0.079	0.103	**0.044**	0.284	0.106	0.308	0.081	0.069	0.077
–8	0.733	0.051	0.054	0.181	0.105	0.176	0.140	0.071	0.086	**0.044**	0.250	0.102	0.282	0.075	**0.064**	**0.070**
–4	0.431	0.051	0.053	0.133	0.083	0.150	0.113	0.064	0.079	**0.044**	0.208	0.101	0.242	0.069	**0.061**	**0.063**
0	0.240	0.051	**0.048**	0.097	0.064	0.135	0.091	0.055	0.072	**0.044**	0.156	0.096	0.185	0.060	0.055	0.053
4	0.126	0.051	**0.043**	0.072	**0.050**	0.088	0.071	**0.044**	0.062	**0.042**	0.102	0.093	0.124	**0.048**	**0.045**	**0.042**
8	0.069	0.051	**0.039**	0.052	**0.041**	0.057	0.054	**0.034**	**0.050**	**0.039**	0.066	0.088	0.078	**0.036**	**0.035**	**0.033**
12	0.050	0.051	**0.039**	**0.046**	**0.037**	**0.047**	**0.046**	**0.030**	**0.043**	**0.039**	0.052	0.067	0.058	**0.032**	**0.031**	**0.031**
16	0.046	0.051	**0.039**	**0.046**	**0.036**	**0.044**	**0.045**	**0.029**	**0.042**	**0.039**	0.049	0.060	0.052	**0.031**	**0.030**	**0.030**
20	0.044	0.051	**0.039**	0.046	**0.036**	**0.043**	**0.044**	**0.029**	**0.042**	**0.039**	0.048	0.056	0.049	**0.030**	**0.030**	**0.029**

(continued)

Table 4.2 (continued)

SNR (dB)	Pre Fusion[a]		No Bayesian elimination								With Bayesian Elimination					
	Lead 1	Lead 2	Median	Best SQI	Best Bayes	Kalman	SWA	SWA-bayes	σ-trim	α-trim	Median	Best SQI	Kalman	SWA	σ-trim	α-trim
Baseline wander																
−20	0.320	0.051	**0.050**	0.132	0.077	0.117	0.085	0.055	0.061	**0.041**	0.138	0.091	0.162	0.056	**0.050**	**0.051**
−16	0.143	0.051	**0.044**	0.077	0.055	0.082	0.061	**0.042**	0.051	**0.041**	0.095	0.081	0.109	**0.044**	**0.040**	**0.040**
−12	0.075	0.051	**0.041**	0.057	**0.044**	0.059	0.052	**0.035**	**0.047**	**0.041**	0.069	0.079	0.077	**0.037**	**0.035**	**0.034**
−8	0.053	0.051	**0.040**	**0.048**	**0.038**	**0.047**	**0.048**	**0.031**	**0.045**	**0.041**	0.055	0.070	0.060	**0.033**	**0.032**	**0.031**
−4	0.047	0.051	**0.040**	**0.047**	**0.037**	**0.043**	**0.046**	**0.030**	**0.044**	**0.040**	0.050	0.062	0.053	**0.031**	**0.031**	**0.030**
0	0.045	0.051	**0.040**	0.046	**0.036**	**0.043**	**0.045**	**0.029**	**0.043**	**0.040**	0.048	0.059	0.051	**0.030**	**0.030**	**0.029**
4	0.044	0.051	**0.040**	0.046	**0.035**	**0.043**	**0.044**	**0.029**	**0.042**	**0.039**	0.048	0.057	0.049	**0.030**	**0.030**	**0029**
8	0.044	0.051	**0.040**	0.046	**0.035**	**0.043**	**0.044**	**0.029**	**0.042**	**0.039**	0.048	0.056	0.049	**0.030**	**0.030**	**0.029**
12	0.044	0.051	**0.039**	0.046	**0.035**	**0.043**	**0.044**	**0.029**	**0.042**	**0.039**	0.048	0.056	0.049	**0.030**	**0.030**	**0.029**
16	0.044	0.051	**0.039**	0.046	**0.035**	**0.042**	**0.044**	**0.029**	**0.042**	**0.039**	0.048	0.056	0.049	**0.030**	**0.030**	**0.029**
20	0.044	0.051	**0.038**	0.046	**0.035**	**0.042**	**0.044**	**0.029**	**0.042**	**0.038**	0.048	0.056	0.049	**0.030**	**0.030**	**0.029**

[a]Pre-Fusion lead 1 and lead 2 indicate the average of the MAEs of the RR intervals obtained from the 2 QRS algorithms per channel

Table 4.3 Mean absolute errors in seconds indicating the performance of fusion algorithms when lead 2 is corrupted by noise and lead 1 is clean

SNR (dB)	Pre fusion[a]		No Bayesian elimination								With Bayesian elimination					
	Lead 1	Lead 2	Median	Best SQI	Best Bayes	Kalman	SWA	SWA-bayes	σ-trim	α-trim	Median	Best SQI	Kalman	SWA	σ-trim	α-trim
Electrode motion																
–20	0.044	0.687	0.112	0.203	0.185	0.284	0.159	0.135	0.098	**0.076**	0.281	0.124	0.332	0.128	0.107	0.118
–16	0.044	0.484	0.104	0.155	0.168	0.280	0.124	0.126	0.088	**0.072**	0.268	0.120	0.329	0.123	0.105	0.114
–12	0.044	0.377	0.103	0.141	0.150	0.266	0.108	0.116	0.084	**0.071**	0.256	0.117	0.318	0.117	0.102	0.109
–8	0.044	0.310	0.083	0.118	0.130	0.236	0.098	0.103	0.078	**0.061**	0.235	0.112	0.294	0.107	0.094	0.099
–4	0.044	0.242	0.082	0.099	0.105	0.193	0.087	0.087	0.070	**0.061**	0.192	0.104	0.250	0.091	0.080	0.082
0	0.044	0.169	0.058	0.075	0.081	0.141	0.073	0.068	0.061	**0.049**	0.136	0.097	0.190	0.070	0.063	0.061
4	0.044	0.106	**0.042**	0.060	0.058	0.090	0.059	0.049	0.052	**0.041**	0.086	0.085	0.125	0.051	0.047	**0.043**
8	0.044	0.072	**0.040**	0.052	0.045	0.062	0.050	**0.038**	0.045	**0.040**	0.062	0.077	0.083	**0.038**	**0.036**	**0.035**
12	0.044	0.057	**0.040**	0.048	**0.039**	0.049	0.045	**0.031**	**0.042**	**0.040**	0.051	0.062	0.062	**0.032**	**0.031**	**0.030**
16	0.044	0.052	**0.039**	0.046	**0.036**	**0.044**	**0.044**	**0.029**	**0.041**	**0.040**	0.049	0.058	0.054	**0.031**	**0.030**	**0.030**
20	0.044	0.051	**0.038**	0.046	**0.036**	**0.043**	**0.044**	**0.029**	**0.041**	**0.038**	0.048	0.057	0.051	**0.031**	**0.030**	**0.030**
Muscle artifact																
–20	0.044	10.578	0.049	1.643	0.165	0.257	0.932	0.096	0.287	**0.040**	0.330	0.103	0.318	0.101	0.086	0.095
–16	0.044	3.359	0.049	0.304	0.142	0.250	0.333	0.084	0.132	**0.040**	0.305	0.097	0.316	0.086	0.073	0.081
–12	0.044	1.383	0.049	0.176	0.118	0.239	0.156	0.072	0.081	**0.040**	0.272	0.096	0.302	0.073	0.063	0.069
–8	0.044	0.735	0.049	0.147	0.096	0.214	0.107	0.064	0.066	**0.039**	0.237	0.093	0.279	0.066	0.057	0.061
–4	0.044	0.430	0.049	0.104	0.077	0.190	0.088	0.058	0.060	**0.039**	0.192	0.092	0.239	0.062	0.054	0.056
0	0.044	0.240	0.045	0.079	0.061	0.159	0.074	0.051	0.056	**0.038**	0.138	0.089	0.191	0.054	0.049	0.048
4	0.044	0.129	**0.040**	0.063	0.050	0.106	0.061	**0.042**	0.051	**0.038**	0.090	0.086	0.138	0.045	**0.042**	**0.040**
8	0.044	0.076	**0.040**	0.050	**0.042**	0.068	0.050	**0.035**	0.045	**0.038**	0.062	0.081	0.091	**0.036**	**0.034**	**0.033**
12	0.044	0.058	**0.039**	0.047	**0.037**	0.051	0.046	**0.031**	**0.043**	**0.038**	0.052	0.067	0.065	**0.032**	**0.031**	**0.030**
16	0.044	0.053	**0.038**	0.046	**0.036**	0.045	**0.044**	**0.029**	**0.042**	**0.038**	0.049	0.059	0.055	**0.031**	**0.030**	**0.030**
20	0.044	0.051	**0.038**	0.046	**0.036**	**0.043**	**0.044**	**0.029**	**0.041**	**0.037**	0.048	0.058	0.051	**0.030**	**0.030**	**0.030**

(continued)

Table 4.3 (continued)

SNR (dB)	Pre fusion[a]		No Bayesian elimination								With Bayesian elimination					
	Lead 1	Lead 2	Median	Best SQI	Best Bayes	Kalman	SWA	SWA-bayes	σ-trim	α-trim	Median	Best SQI	Kalman	SWA	σ-trim	α-trim
Baseline wander																
−20	0.044	0.309	0.044	0.097	0.072	0.132	0.069	0.051	0.049	**0.041**	0.125	0.084	0.166	0.051	0.045	0.046
−16	0.044	0.143	**0.041**	0.073	0.053	0.095	0.053	**0.040**	0.044	**0.041**	0.086	0.073	0.118	**0.041**	**0.037**	**0.037**
−12	0.044	0.082	**0.041**	0.056	0.044	0.070	0.047	**0.034**	**0.042**	**0.041**	0.064	0.064	0.085	**0.035**	**0.033**	**0.032**
−8	0.044	0.060	**0.041**	0.047	**0.038**	0.053	0.044	**0.031**	**0.042**	**0.041**	0.054	0.059	0.065	**0.032**	**0.031**	**0.030**
−4	0.044	0.053	**0.040**	0.046	**0.036**	0.046	0.044	**0.029**	**0.042**	**0.040**	0.050	0.058	0.055	**0.031**	**0.030**	**0.030**
0	0.044	0.052	**0.040**	0.046	**0.036**	0.044	**0.044**	**0.029**	**0.041**	**0.040**	0.048	0.056	0.052	**0.030**	**0.030**	**0.030**
4	0.044	0.051	**0.040**	0.046	**0.036**	**0.043**	**0.044**	**0.029**	**0.041**	**0.038**	0.048	0.056	0.050	**0.030**	**0.030**	**0.030**
8	0.044	0.051	**0.040**	0.046	**0.036**	**0.043**	**0.044**	**0.029**	**0.041**	**0.038**	0.048	0.056	0.050	**0.030**	**0.030**	**0.030**
12	0.044	0.051	**0.039**	0.046	**0.036**	**0.043**	**0.043**	**0.029**	**0.041**	**0.036**	0.048	0.056	0.050	**0.030**	**0.030**	**0.030**
16	0.044	0.051	**0.039**	0.046	**0.036**	**0.043**	**0.043**	**0.029**	**0.041**	**0.036**	0.049	0.056	0.050	**0.030**	**0.030**	**0.030**
20	0.044	0.051	**0.038**	**0.044**	**0.036**	**0.043**	**0.043**	**0.029**	**0.040**	**0.036**	0.049	0.055	0.050	**0.030**	**0.030**	**0.030**

[a]Pre-Fusion lead 1 and lead 2 indicate the average of the MAEs of the RR intervals obtained from the 2 QRS algorithms per channel

Table 4.4 Mean absolute errors in seconds indicating the performance of fusion algorithms when both lead 1 and lead 2 are corrupted by noise

SNR (dB)	Pre fusion[a]		No Bayesian elimination								With Bayesian Elimination					
	Lead 1	Lead 2	Median	Best SQI	Best Bayes	Kalman	SWA	SWA-bayes	σ-trim	α-trim	Median	Best SQI	Kalman	SWA	σ-trim	α-trim
Electrode motion																
–20	0.694	0.687	0.462	0.482	0.425	0.281	0.476	0.422	0.480	0.462	0.478	0.477	0.339	0.386	0.387	0.385
–16	0.495	0.484	0.359	0.369	0.405	0.281	0.363	0.397	0.365	0.359	0.434	0.437	0.338	0.362	0.363	0.361
–12	0.391	0.377	0.313	0.320	0.375	0.277	0.315	0.365	0.316	0.313	0.396	0.406	0.329	0.340	0.340	0.337
–8	0.325	0.310	0.274	0.279	0.329	0.261	0.277	0.318	0.277	0.274	0.351	0.363	0.302	0.301	0.301	0.298
–4	0.254	0.242	0.216	0.227	0.255	0.217	0.222	0.246	0.221	0.216	0.285	0.302	0.257	0.236	0.236	0.232
0	0.175	0.169	0.144	0.158	0.168	0.158	0.154	0.158	0.152	0.145	0.203	0.229	0.199	0.154	0.153	0.149
4	0.107	0.106	0.083	0.098	0.090	0.102	0.095	0.084	0.093	0.083	0.127	0.158	0.138	0.086	0.085	0.080
8	0.064	0.072	0.052	0.061	0.054	0.066	0.061	0.047	0.058	0.053	0.079	0.100	0.091	0.049	0.047	0.045
12	0.049	0.057	0.043	0.049	0.041	0.050	0.048	0.034	0.046	0.044	0.057	0.068	0.064	0.035	0.034	0.034
16	0.046	0.052	0.041	0.046	0.037	0.045	0.045	0.031	0.043	0.042	0.052	0.062	0.055	0.032	0.031	0.031
20	0.044	0.051	0.040	0.045	0.037	0.043	0.044	0.030	0.041	0.041	0.050	0.059	0.052	0.031	0.031	0.030
Muscle artifact																
–20	10.819	10.578	9.052	10.235	0.413	0.221	10.286	0.595	10.302	9.140	0.727	0.776	0.385	0.693	0.693	0.690
–16	3.408	3.359	2.737	3.037	0.380	0.203	3.047	0.483	3.055	2.743	0.599	0.624	0.339	0.527	0.527	0.526
–12	1.377	1.383	1.068	1.118	0.336	0.185	1.115	0.369	1.118	1.068	0.471	0.480	0.294	0.379	0.380	0.380
–8	0.733	0.735	0.533	0.546	0.280	0.167	0.543	0.278	0.543	0.533	0.369	0.374	0.252	0.280	0.280	0.282
–4	0.431	0.430	0.310	0.313	0.208	0.148	0.311	0.196	0.309	0.313	0.285	0.290	0.210	0.199	0.199	0.201
0	0.240	0.240	0.180	0.184	0.129	0.122	0.182	0.124	0.180	0.182	0.209	0.218	0.168	0.131	0.130	0.131
4	0.126	0.129	0.101	0.109	0.073	0.092	0.106	0.072	0.103	0.102	0.144	0.156	0.127	0.079	0.077	0.077
8	0.069	0.076	0.059	0.066	0.047	0.064	0.064	0.043	0.061	0.059	0.088	0.102	0.089	0.046	0.045	0.045
12	0.050	0.058	0.045	0.050	0.039	0.050	0.049	0.033	0.046	0.045	0.062	0.072	0.065	0.035	0.035	0.034
16	0.046	0.053	0.042	0.046	0.037	0.045	0.046	0.030	0.043	0.042	0.053	0.062	0.055	0.032	0.031	0.031
20	0.044	0.051	0.040	0.044	0.036	0.043	0.044	0.029	0.041	0.041	0.050	0.059	0.051	0.031	0.030	0.030

(continued)

Table 4.4 (continued)

SNR (dB)	Pre fusion[a]		No Bayesian elimination								With Bayesian Elimination					
	Lead 1	Lead 2	Median	Best SQI	Best Bayes	Kalman	SWA	SWA-bayes	σ-trim	α-trim	Median	Best SQI	Kalman	SWA	σ-trim	α-trim
Baseline wander																
–20	0.320	0.309	**0.232**	**0.264**	**0.147**	**0.123**	**0.258**	**0.143**	**0.256**	**0.233**	**0.204**	**0.224**	**0.167**	**0.142**	**0.141**	**0.138**
–16	0.143	0.143	**0.107**	**0.125**	**0.088**	**0.087**	**0.119**	**0.082**	**0.116**	**0.107**	**0.128**	**0.141**	**0.116**	**0.082**	**0.081**	**0.079**
–12	0.075	0.082	**0.062**	**0.070**	**0.057**	**0.064**	**0.068**	**0.050**	**0.066**	**0.062**	0.086	0.098	0.085	**0.052**	**0.051**	**0.049**
–8	0.053	0.060	**0.047**	**0.052**	**0.042**	**0.051**	**0.051**	**0.036**	**0.049**	**0.047**	0.064	0.073	0.065	**0.037**	**0.037**	**0.036**
–4	0.047	0.053	**0.042**	**0.046**	**0.038**	**0.045**	**0.046**	**0.031**	**0.044**	**0.043**	0.055	0.064	0.056	**0.033**	**0.032**	**0.032**
0	0.045	0.052	**0.041**	0.046	**0.036**	**0.043**	**0.044**	**0.029**	**0.042**	**0.042**	0.051	0.059	0.053	**0.031**	**0.030**	**0.030**
4	0.044	0.051	**0.041**	0.046	**0.036**	**0.043**	**0.044**	**0.029**	**0.042**	**0.041**	0.049	0.057	0.051	**0.030**	**0.030**	**0.030**
8	0.044	0.051	**0.041**	0.046	**0.036**	**0.043**	**0.044**	**0.029**	**0.042**	**0.042**	0.049	0.057	0.050	**0.030**	**0.030**	**0.030**
12	0.044	0.051	**0.041**	0.046	**0.036**	**0.043**	**0.044**	**0.029**	**0.042**	**0.041**	0.049	0.056	0.050	**0.030**	**0.030**	**0.029**
16	0.044	0.051	**0.041**	0.046	**0.036**	**0.043**	**0.044**	**0.029**	**0.042**	**0.041**	0.049	0.057	0.050	**0.030**	**0.030**	**0.029**
20	0.044	0.051	**0.041**	**0.044**	**0.036**	**0.043**	**0.044**	**0.029**	**0.042**	**0.041**	0.049	0.056	0.050	**0.030**	**0.030**	**0.030**

[a] Pre-Fusion lead 1 and lead 2 indicate the average of the MAEs of the RR intervals obtained from the 2 QRS algorithms per channel

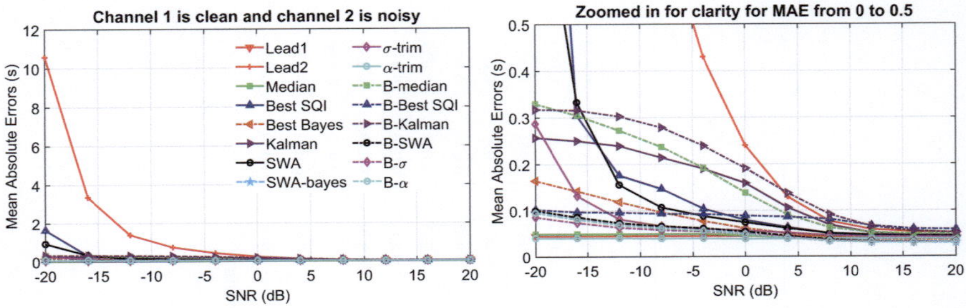

Fig. 4.6 The plot of MAE versus SNR when channel 2 is corrupted by muscle artifacts and channel 1 is clean

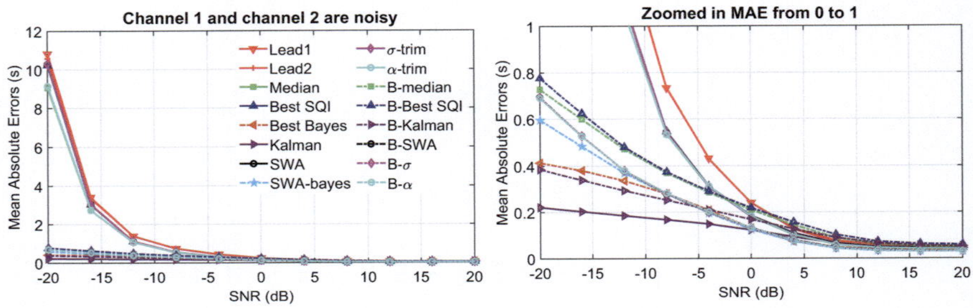

Fig. 4.7 The plot of MAE versus SNR when both channels 1 and 2 are corrupted by muscle artifacts

regions (below 8 dB) is close to the best-performing selection algorithm, and is better than the performance of the individual leads without fusion. However, in the case of the Bayes-eliminated median selection, when only one channel is corrupted by noise, the Bayes-eliminated median method does not exhibit a performance better than that of the individual leads prior to fusion. The Bayes-eliminated median method could be performing poorly when Bayesian elimination is carried out as due to the Bayesian elimination stage, the median interval which is selected in the simple median algorithm (which is from the reliable channel) is eliminated occasionally. This is also an indicator of the unreliability of the QRS detection algorithms.

The best SQI method discussed in section "RR Best SQI" exhibits good performance in the low SNR regions, but is outperformed by the median method in all scenarios. In the high SNR regions, its performance is comparable to the cleanest single channel. Bayes-eliminated best SQI method exhibits better performance compared to the best SQI method. This could be attributed to the Bayesian elimination stage successfully eliminating RR intervals and the renormalized SQIs ensuring that the RR interval from the non-noisy channel is selected, unlike in the Bayes-eliminated median selection where the median value is directly selected.

The best Bayes algorithm discussed in section "Best Bayes" exhibits a performance similar to the median method in the low SNR regions. The errors in performance could be simply attributed to high Bayes probabilities being assigned to beat values detected from a channel when the channel is noisy, because a few wrong RR intervals in the initial stages of Bayesian elimination could lead to subsequent RR intervals further away from the ground truth being assigned higher probabilities. This could also be due to the algorithm choosing just 1 RR interval based on the Bayesian probabilities without fusing all the RR intervals by averaging or weighted averaging. In the high SNR regions, it exhibits a mean absolute error comparable to or even less than the cleanest individual lead. In the scenario where both leads are corrupted by muscle artifacts, the best Bayes method exhibits a performance that is significantly better than the individual leads, indicating that the Bayes probabilities are most reliable when both signals are corrupted by noise.

The Simple Weighted Average (SWA) method discussed in section "Simple Weighted Average" exhibits good performance when only one channel is corrupted by noise, as this method takes into account the signal quality. In the case of the Bayes-eliminated SWA method, the performance further improves. This is as expected, as discussed in the case of the best SQI method. In the scenario where both channels are corrupted by noise, both SWA and Bayes-eliminated SWA provide a better estimate of the RR interval sequence compared to the individual leads.

The simple weighted average with Bayes probabilities discussed in section "SWA with Bayesian Probabilities" outperforms the SQI-based SWA in the low SNR regions when only one channel is corrupted by noise. This is expected, as the Bayes probabilities would be high for RR intervals from clean channels. This is consistent with what is observed of the best SQI method. The best SQI method, however, considers only one channel of information which may not be the most reliable, although the Bayes probability indicates it is. This issue is solved in the case of SWA with Bayes probabilities, and it is observed that this method outperforms the best SQI method when only one channel is corrupted by noise. In the high SNR regions, this method exhibits the best performance among all the methods considered and outperforms the performance of the average single-channel method.

Among the fusion methods that use combination-based fusion, Kalman fusion discussed in section "Kalman Fusion" does not perform very well in the low SNR regions when only one channel is corrupted by noise. This is because the algorithm treats all RR intervals equally and as if calculated from 2 signals of the same quality, even though RR_{SWA} is taken as the control sequence. In the scenario where both channels are corrupted by noise, Kalman fusion exhibits the best performance in the low SNR regions as both channels are now of similar reliability. The Bayes-eliminated Kalman fusion shows a further drop in performance compared to Kalman fusion, indicating that sometimes the Bayesian elimination stage eliminates RR segments that are reliable, in favor of unreliable RR segments as discussed in the case of median filtering. The performance of the Kalman fusion algorithm could be improved upon by using improved estimates of initial conditions and the update parameters like M in Eq. (4.23). However, since the Kalman fusion method is the best-performing method in the low SNR regions when both signals are corrupted by noise and the MAE is significantly lower compared to the second-best performing algorithm, Kalman fusion is a clear winner for implementation in noisy scenarios when both channels are corrupted by noise.

With regard to the two elimination and combination algorithms, the σ-trim mean filtering method discussed in section "σ-Trimmed Mean Filter" can be thought of as an elimination algorithm followed by a simple averaging. Therefore, its performance is very close to the performance of the SQI-based simple weighted average method in the low SNR regions when only one channel is corrupted by noise (we assume that the eliminated RR sequences are from the noisy channels). However, when a channel is corrupted by muscle artifacts, leading to very large variations in input RR intervals (as indicated by the MAE of lead 1 in Table 4.2 and MAE of lead 2 in Table 4.3 for muscle artifacts at –20 dB), the performance is much better compared to the SWA method. This could be due to significantly erroneous RR intervals being eliminated, compared to SQI-based averaging, where even the channel with very poor SQI does contribute a little to the weighted average. The Bayes-eliminated σ-trim method closely follows the performance of the σ-trim method. When a single channel is corrupted by noise, both the σ-trim mean filtering method and the Bayes-eliminated σ-trim mean filtering method perform better than the non-corrupted lead in the high SNR regions. When both channels are corrupted by noise, both the σ-trim mean filtering method and the Bayes-eliminated σ-trim mean filtering method perform better than the individual leads in estimating the RR sequence at all SNRs.

The α-trim mean filtering method discussed in section "α-Trimmed Mean Filter" exhibits the best performance, when only one channel is corrupted by noise, in the low SNR regions among all the algorithms considered. This indicates that the median-based elimination and SQI-based combination work well in the low SNR regions as unreliable RR intervals are eliminated, and the remnant RR intervals are fused on the basis of the renormalized SQIs. The Bayes α-trim method performs poorly compared to the simple α-trim method, as good RR intervals are probably eliminated in the low SNR regions. However, in the high SNR region, Bayes-eliminated α-trim mean filtering exhibits the best performance along with SWA using Bayes probabilities, indicating that Bayes probability-based methods are highly reliable in the high SNR regions. The α-trim mean filtering method is consistently found to be either the best-performing algorithm or exhibits a better performance than the single non-corrupted lead in all scenarios and is, therefore, a suitable fusion algorithm when only one channel is corrupted by noise.

4.6.2 Bland Altman Plots

Figures 4.8, 4.9, and 4.10 show the Bland Altman plots for when the ECG signal channel 1 is corrupted by electrode motion noise and channel 2 is clean, channel 2 is corrupted by baseline wander and channel 1 is clean, and both channels are corrupted by muscle artifacts at –20 and 20 dB respectively. The Bland Altman plot has the reference RR on the x-axis instead of the mean of the reference RR and calculated RR from the algorithm for ease of comparison [26]. The y-axis shows Δ, which is calculated as

$$\Delta = \widehat{RR[n]} - RR[n], \tag{4.29}$$

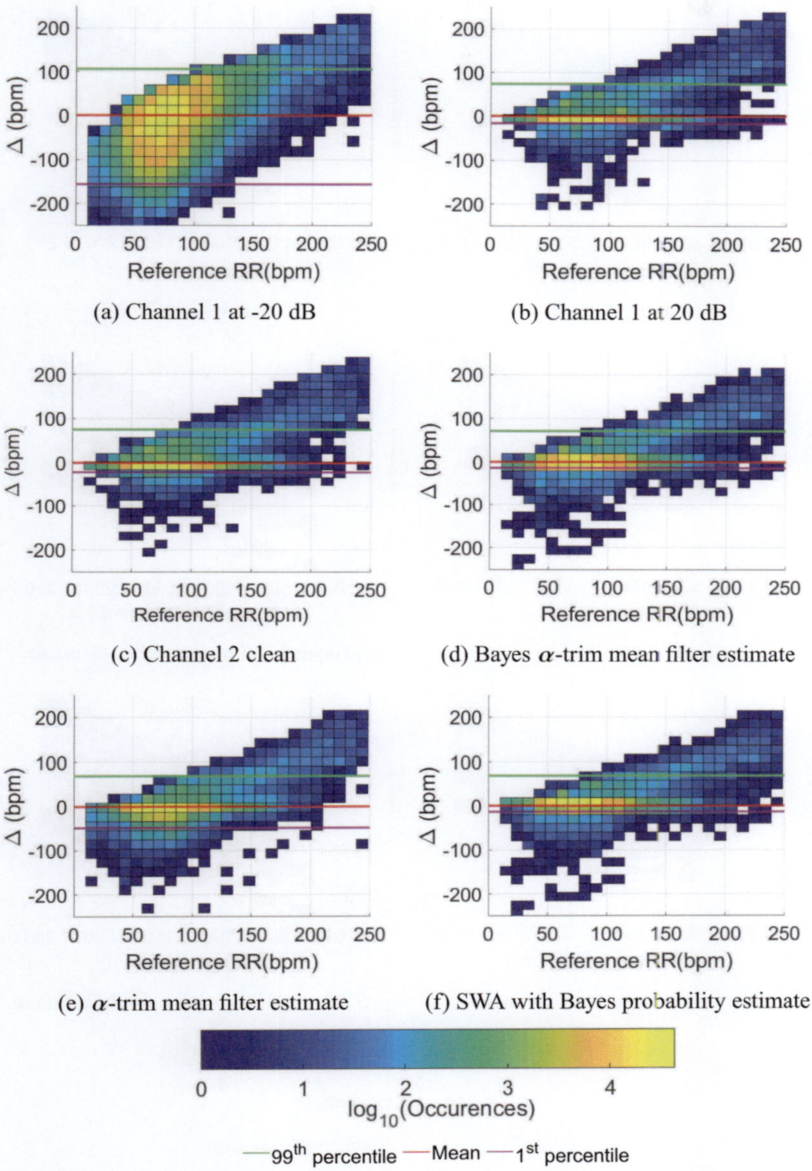

(a) Channel 1 at -20 dB
(b) Channel 1 at 20 dB
(c) Channel 2 clean
(d) Bayes α-trim mean filter estimate
(e) α-trim mean filter estimate
(f) SWA with Bayes probability estimate

Fig. 4.8 Bland Altman plot showing the agreement between the RR interval information obtained from **a** Channel 1, **c** Channel 2, and **e** α-trim mean filtering when channel 1 is corrupted by electrode motion noise at –20 dB and channel 2 is clean, and **b** Channel 1, **d** Bayes α-trim mean filtering, and **f** SWA with Bayes probabilities when Channel 1 is corrupted by electrode motion noise at 20 dB and when channel 2 is clean, with the reference RR interval for all records. The limits of agreements of the differences are shown through P_{99} and P_1 in the figure to indicate where 98% of the estimates lie. The values recorded are in beats per minute (The y-axis limits of the plots have been adjusted for ease of comparison and, therefore, some outliers are not visible in the figures)

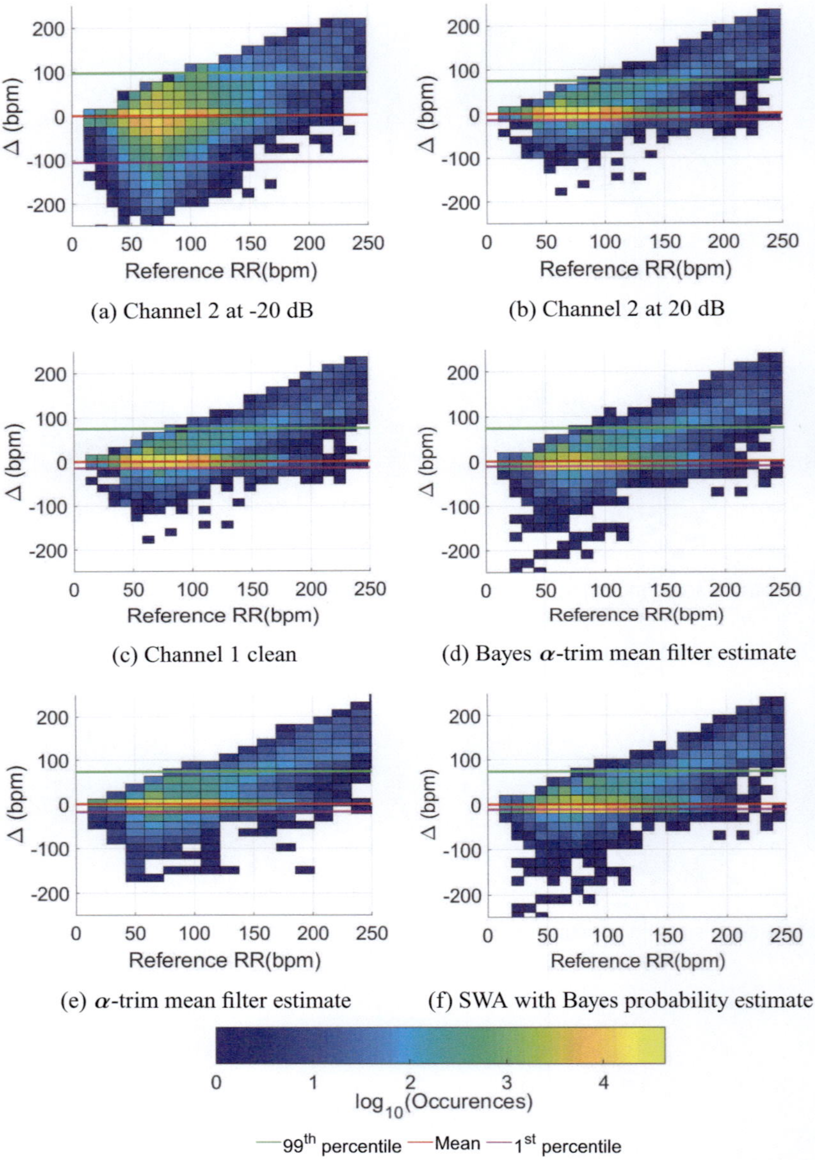

Fig. 4.9 Bland Altman plot showing the agreement between the RR interval information obtained from **a** Channel 2, **c** Channel 1, and **e** α-trim mean filtering when channel 2 is corrupted by baseline wander noise at –20 dB and channel 1 is clean, and **b** Channel 2, **d** Bayes α-trim mean filtering, and **f** SWA with Bayes probabilities when Channel 2 is corrupted by baseline wander noise at 20 dB and when channel 2 is clean, with the reference RR interval for all records. The limits of agreements of the differences are shown through P_{99} and P_1 in the figure to indicate where 98% of the estimates lie. The values recorded are in beats per minute (The y-axis limits of the plots have been adjusted for ease of comparison and, therefore, some outliers are not visible in the figures)

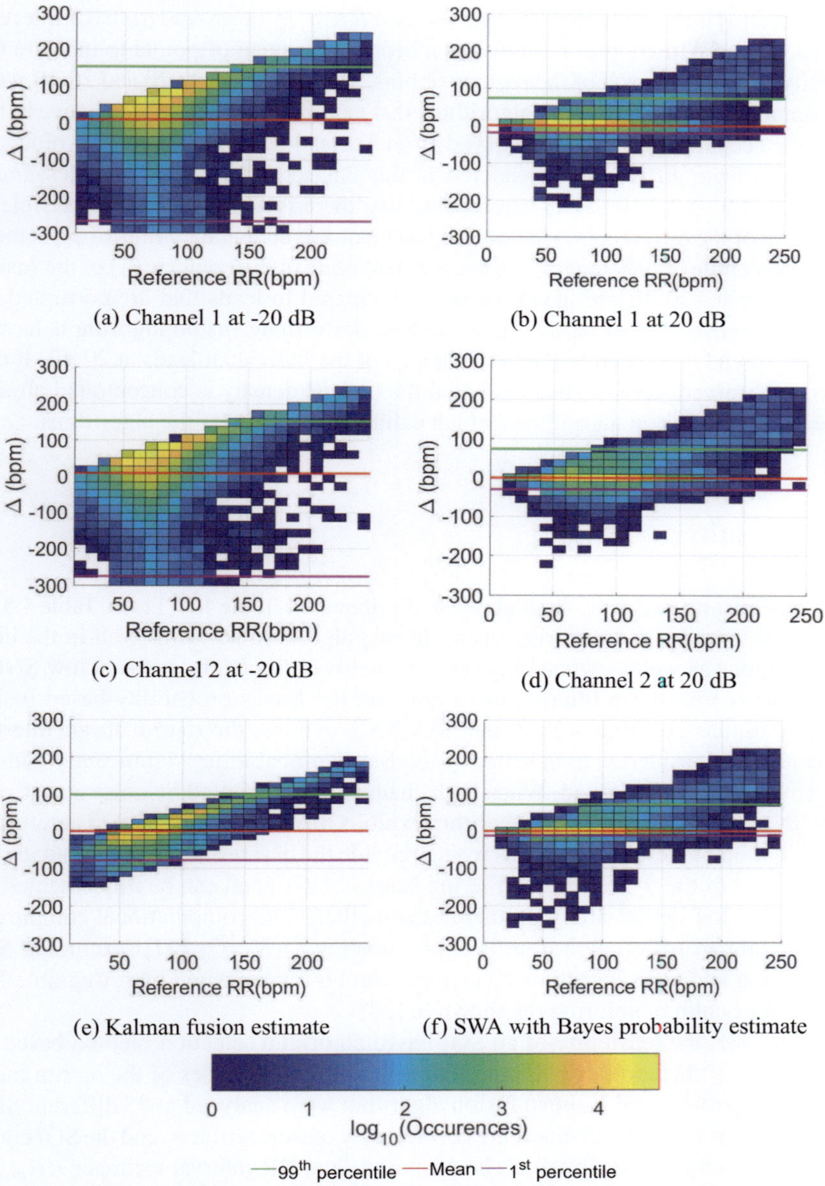

Fig. 4.10 Bland Altman plot showing the agreement between the RR interval information obtained from **a** Channel 1, **c** Channel 2, and **e** Kalman fusion when both channel 1 and channel 2 are corrupted by muscle artifacts at –20 dB, and **b** Channel 1, **d** Channel 2, and **f** SWA with Bayes probabilities when both channel 1 and channel 2 is corrupted by muscle artifacts at 20 dB, with the reference RR interval for all records. The limits of agreements of the differences are shown through P_{99} and P_1 in the figure to indicate where 98% of the estimates lie. The values recorded are in beats per minute (The y-axis limits of the plots have been adjusted for ease of comparison and, therefore, some outliers are not visible in the figures)

where $\widehat{RR[n]}$ is the estimated RR intervals and $RR[n]$ is the ground truth RR interval.

The Bland Altman plot is shown as a heat map instead of points to indicate the density of points in each of the regions or bins. The scenarios at –20 and 20 dB were chosen to showcase the fusion algorithms that exhibited the best performance in the extreme scenarios, as can be observed from Tables 4.2, 4.3, and 4.4. The plots are generated from all records considered in this dataset for every 10 samples. From Figs. 4.8, 4.9, and 4.10, it can be observed that there is no noticeable bias visible in the mean of the differences. However, it can be observed that the limits of agreement (99th percentile of differences, P_{99} and 1st percentile of differences, P_1) of the fusion algorithms at –20 dB are much narrower compared to leads that are corrupted by noise. Moreover, it can be seen that the best-performing fusion algorithms have a narrow spread compared to the performance of the individual leads at 20 dB. It can also be observed from the heat map that the highest density is concentrated around $\Delta = 0$, for the fusion algorithms, which exhibits the merits of the algorithms.

4.7 Results

The key results and best strategies are discussed in Table 4.5. From Table 4.5, it can be observed that the α-trim mean filtering algorithm performs well in the high SNR regions as well as when only one channel is corrupted by noise at low SNRs. Since the α-trim mean filtering algorithm and the Bayes probability-based fusion exhibit similar performances in the high SNR regions, the α-trim mean filtering algorithm is considered, as calculating the Bayes probabilities significantly adds to the computational overhead. When both channels are corrupted by noise and at very low SNRs, the Kalman fusion algorithm exhibits the best performance. Therefore, a combination of these two methods would provide the best results. The computational complexity of the α- trim mean filtering stage is $\mathcal{O}(N)$ and can be implemented on a wearable IoT device after optimizing the method. The computational complexity of the Kalman fusion algorithm is approximately $\mathcal{O}(N^{2.376})$ [27] (excluding SQI calculation and QRS detection stages) and can be implemented on a wearable IoT device for health monitoring (as shown in [28]).

Therefore, the feasibility of an SQI-based algorithm selection method based on an SQI cutoff threshold is analyzed. For this, the performances of the α-trim mean filtering algorithm and Kalman fusion algorithm were analyzed at 25 different SNR combinations when the channels are corrupted by muscle artifacts, and the SQI cutoff levels were chosen empirically. The proposed final RR interval estimate RR_{pro} at each sample is selected based on the SQIs of both leads at each timestamp as shown in Eq. (4.30).

$$RR_{pro}[n] = \begin{cases} RR_{kal}[n] & \text{if } SQI_1[n] < 5.2 \ \& \\ & SQI_2[n] < 5.2 \\ RR_{\alpha}[n] & \end{cases} \tag{4.30}$$

Table 4.5 Strategies for fusion methods for different scenarios for robust heart rate estimation

Scenario		Strategy/method
Low SNR regions	Single channel corrupted by noise	Elimination and combination with SQIs- α-trim mean filtering
	Both channels corrupted by noise	Combinations with equal weights for both channels (no SQI)-Kalman Fusion
High SNR regions		Elimination and combination with SQI -α-trim mean filtering or Bayes probability based fusion SWA-Bayes/ Bayes α-trim mean filtering

We further analyze the efficacy of this method by corrupting the signal leads at 25 different SNR combinations per sample per signal in the dataset with electrode motion noise. The performance in terms of mean absolute error and the root mean square error (RMSE) are as shown in Table 4.6 (RMSE is used at this stage to provide comparisons with other works easily). The RMSE is calculated as

$$E_{rms} = \sqrt{\frac{1}{N}\sum(\widehat{RR[n]} - RR[n])^2} \tag{4.31}$$

From Table 4.6, it can be observed that in the high SNR regions, the algorithm selection method provides an estimate which is close to that obtained from the α-trim mean filtering algorithm, while in the low SNR regions, the final estimate is influenced by the Kalman fusion algorithm to provide a better estimate than the α-trim mean filtering algorithm. It is important to note that while symmetric scenarios are expected to yield similar performance, variations in the performance of QRS detectors on lead 1 and lead 2—particularly in the clean signal scenario (as previously discussed)—result in slight differences. When both the signal channels are corrupted by noise such that the signal-to-noise ratio is 20 dB, the proposed fusion algorithm exhibits a performance improvement of 53.91% compared to the RR interval estimate obtained by simple averaging of the RR intervals from both leads. When both the signal channels are corrupted by noise such that the signal-to-noise ratio is –20 dB (low SNR), the proposed fusion algorithm exhibits a performance improvement of 20.83% compared to the mean RR interval estimate from both channels. Since, in the low SNR regions, the algorithm performance provides a much better estimate than that obtained from the individual channels, this method of algorithm selection can be considered to be quite effective. Moreover, in a very low power setting, the Kalman fusion algorithm can be eliminated to bring down the computational complexity as the Kalman fusion algorithm is desirable only in the low SNR regions, and the α-

Table 4.6 SQI-based algorithm selection to obtain reliable RR interval estimate

SNR (dB)		Algorithm performance in MAE (seconds)				Proposed algorithm performance (RMSE) (s)
Lead 1	Lead 2	Average of leads 1 and 2	α-trim	Kalman	Proposed algorithm	
–20	–20	0.690	0.462	0.277	0.318	0.378
–20	–8	0.502	0.293	0.258	0.253	0.348
–20	0	0.432	0.186	0.256	0.203	0.311
–20	8	0.383	0.092	0.255	0.130	0.263
–20	20	0.373	0.068	0.254	0.103	0.226
–8	–20	0.506	0.302	0.240	0.258	0.353
–8	–8	0.317	0.274	0.231	0.231	0.338
–8	0	0.247	0.204	0.219	0.201	0.296
–8	8	0.198	0.117	0.216	0.134	0.249
–8	20	0.188	0.088	0.214	0.109	0.215
0	–20	0.431	0.191	0.238	0.201	0.319
0	–8	0.242	0.201	0.218	0.203	0.300
0	0	0.172	0.145	0.159	0.150	0.269
0	8	0.124	0.081	0.127	0.091	0.226
0	20	0.113	0.060	0.126	0.071	0.199
8	–20	0.375	0.085	0.262	0.112	0.266
8	–8	0.187	0.100	0.228	0.121	0.250
8	0	0.116	0.071	0.137	0.081	0.226
8	8	0.068	0.052	0.067	0.053	0.202
8	20	0.057	0.041	0.053	0.042	0.182
20	–20	0.365	0.062	0.269	0.087	0.212
20	–8	0.177	0.076	0.234	0.095	0.200
20	0	0.106	0.051	0.140	0.060	0.187
20	8	0.058	0.039	0.062	0.040	0.179
20	20	0.048	0.041	0.043	0.038	0.183

trim mean filtering can be chosen as the fusion algorithm for all scenarios with good confidence.

The performance[1] of the SQI-based fusion algorithm selection method is compared against other state-of-the-art RR interval estimation algorithms (note that there

[1] For the purpose of comparison, each RR-interval estimate error is extrapolated to a beats-per-minute (bpm) error in Table 4.7. However, it should be noted that this is somewhat misleading as this bpm error would only occur if the RR-interval error in bpm persisted for an entire minute which

Table 4.7 Comparison with state-of-the-art RR interval estimation methods

Article	Nathan et al. [11]	This work	Xie et al. [12]	This work	Di et al. [13]	This work
Dataset	MIT-BIH Arrhythmia database	MIT-BIH Arrhythmia database	2014 Physionet/ CinC challenge database	MIT-BIH Arrhythmia database	2014 Physionet/ CinC challenge database	MIT-BIH Arrhythmia database
Signals	ECG	ECG	ECG	ECG	ECG	ECG
Channels	1-channel	2-channels	1-channel	2-channels	1-channel	2-channels
Fusion Method	–	RR interval fusion	RR interval fusion	RR interval fusion	RR interval fusion	RR interval fusion
SNR	–6 dB	–6 dB	Clean	Clean	Clean	Clean
MAE (bpm)	5.044	2.745	–	1.387	–	1.387
RMSE (bpm)	–	13.357	13.96	12.11	14.37	12.11

are various other heartbeat detection algorithms in the literature, but here in this chapter, the focus is on RR interval estimation) and is as shown in Table 4.7. The comparison with Nathan et al. [11] is carried out when the signals are at -6 dB SNR, the same subset of the MIT-BIH Arrhythmia database as used in [11] is used, and the proposed fusion method exhibits an MAE of 2.745 bpm. The comparison with Xie et al. [12] and Di et al. [13] is carried out when the signals are clean. It can be observed that in all scenarios the proposed method outperforms the state-of-the-art. However, direct comparison with Xie et al. [12] and Di et al. [13] is not advisable due to the different datasets used for analysis.

4.8 Discussions

This chapter demonstrates the effectiveness of ECG data fusion algorithms in improving heart rate estimation for wearable IoT sensors. Selection-based methods, such as the median approach, showed consistent performance in noisy scenarios when at least one lead remained clean, while Bayesian elimination further refined accuracy by eliminating unreliable RR intervals. The Kalman fusion method, utilizing past observations, exhibited superior performance in scenarios where both leads were corrupted, making it a strong candidate for real-time physiological monitoring. Combination-based approaches, particularly simple weighted averaging with SQI normalization, dynamically adapted to noise conditions, outperforming single-lead estimation. The integration of the Bayesian probabilities into SWA further enhanced

is a practical impossibility as the estimates are unbiased; hence the use of the MAE and RMSE metrics in seconds elsewhere in this work.

performance, especially in low SNR conditions. Elimination-based methods, such as the α-trimmed mean filtering, effectively eliminated outlier RR intervals and improved estimation accuracy. Overall, the results highlight the importance of selecting appropriate fusion techniques based on signal quality and noise conditions. While selection-based methods are useful when one lead is relatively clean, combination and elimination approaches offer greater estimation quality in noisy environments.

4.9 Conclusions

The chapter explores various fusion algorithms to accurately estimate the RR interval by using the beats detected from two simultaneously recorded ECG channels obtained from two state-of-the-art QRS detection algorithms. The performance of the fusion algorithms on signals corrupted by muscle artifacts, electrode motion, and baseline wander indicate the suitability of the fusion algorithm for deployment on a wearable IoT device. It was observed that the Kalman fusion algorithm exhibits the best performance when both channels have signals with low SNRs, and the α-trim mean filtering algorithm is suitable for all the other noise scenarios. Consequently, the feasibility of using SQIs as a surrogate for signal SNR for algorithm selection is proposed, and it was observed that the strategy of using Kalman fusion when SQIs of both channels is lower than a threshold and α-trim mean filtering elsewhere in the input signal space is effective in obtaining the best estimate of the RR interval. When both the signal channels are corrupted by noise such that the signal-to-noise ratio is 20 dB, the proposed fusion algorithm exhibits a performance improvement of 53.91% compared to the RR interval estimate obtained by simple averaging of the RR intervals obtained from the individual leads. When both the signal channels are corrupted by noise such that the signal-to-noise ratio is –20 dB (low SNR), the proposed fusion algorithm exhibits a performance improvement of 20.83% compared to the mean RR interval estimate from both channels. Moreover, since the Kalman fusion algorithm is desirable only in the low SNR regions, the computational complexity can be further reduced by employing just the α-trim mean filtering for all scenarios with good confidence. The proposed fusion algorithm is an example of redundant, decision-level, and DEI-DEO fusion. The proposed fusion strategies are suitable for the fusion of unimodal signals while the fusion of multimodal signals requires an intermediate transformation into a common representational format. One such method is proposed in the next chapter.

References

1. W.-L. Chin, C.-C. Chang, C.-L. Tseng, Y.-Z. Huang, T. Jiang, Bayesian real-time QRS complex detector for healthcare system. IEEE Internet Things J. **6**(3), 5540–5549 (2019)

2. C. Liu, X. Zhang, L. Zhao, F. Liu, X. Chen, Y. Yao, J. Li, Signal quality assessment and lightweight QRS detection for wearable ECG smartvest system. IEEE Internet Things J. **6**(2), 1363–1374 (2019)
3. W. Lu, F. Fan, J. Chu, P. Jing, S. Yuting, Wearable computing for internet of things: a discriminant approach for human activity recognition. IEEE Internet Things J. **6**(2), 2749–2759 (2019)
4. T. Wu, F. Wu, C. Qiu, J.-M. Redouté, M.R. Yuce, A rigid-flex wearable health monitoring sensor patch for IoT-connected healthcare applications. IEEE Internet Things J. **7**(8), 6932–6945 (2020)
5. U. Satija, B. Ramkumar, M. Sabarimalai Manikandan, Real-time signal quality-aware ECG telemetry system for IoT-based health care monitoring. IEEE Internet of Things J. **4**(3), 815–823 (2017)
6. T. Wartzek, C. Brüser, M. Walter, S. Leonhardt, Robust sensor fusion of unobtrusively measured heart rate. IEEE J. Biomed. Health Inform. **18**(2), 654–660 (2014)
7. M.-C. Chambrin, Alarms in the intensive care unit: how can the number of false alarms be reduced? Crit. Care **5**(4), 184–188 (2001)
8. Matlab, Understanding sensor fusion and tracking, part 1: What is sensor fusion? (2019)
9. I. Silva, B. Moody, J. Behar, A. Johnson, J. Oster, G.D. Clifford, G.B. Moody, Robust detection of heart beats in multimodal data. Physiol. Meas. **36**(8), 549–552 (2015)
10. S.A. Rankawat, R. Dubey, Robust heart rate estimation from multimodal physiological signals using beat signal quality index based majority voting fusion method. Biomed. Signal Process. Control **33**, 201–212 (2017)
11. V. Nathan, I. Akkaya, R. Jafari, A particle filter framework for the estimation of heart rate from ECG signals corrupted by motion artifacts, in *2015 37th Annual International Conference of the IEEE Engineering in Medicine and Biology Society (EMBC)* (2015), pp. 6560–6565
12. Y. Xie, J. Li, T. Zhu, C. Liu, Continuous-valued annotations aggregation for heart rate detection. IEEE Access **7**, 37664–37671 (2019)
13. J. Di, J. Li, C. Liu, A Bayesian fusion model for heart rate annotations, in *2020 International Conference on Sensing, Measurement Data Analytics in the era of Artificial Intelligence (ICSMD)* (2020), pp. 307–312
14. S.A. Rankawat, M. Rankawat, R. Dubey, Heart rate estimation from non-cardiovascular signals using slope sum function and teager energy, in *2015 International Conference on Industrial Instrumentation and Control (ICIC)* (2015), pp. 1534–1539
15. C. Diao, B. Wang, N. Cai, Data fusion of multivariate time series: application to noisy 12-lead ECG signals. Appl. Sci. **9**(1) (2019)
16. M.A. Pimentel, M.D. Santos, D.B. Springer, G.D. Clifford, Hidden semi-Markov model-based heartbeat detection using multimodal data and signal quality indices. Comput. Cardiol. **2014**, 553–556 (2014)
17. A.E. Johnson, J. Behar, F. Andreotti, G.D. Clifford, J. Oster, R-peak estimation using multimodal lead switching. Comput. Cardiol. **2014**, 281–284 (2014)
18. C. Orphanidou, *Signal Quality Assessment in Physiological Monitoring: State of the Art and Practical Considerations* (Springer International Publishing, 2018)
19. Q. Li, C. Rajagopalan, G.D. Clifford, A machine learning approach to multi-level ECG signal quality classification. Comput. Methods Programs Biomed. **117**(3), 435–447 (2014)
20. A.L. Goldberger, L.A. Amaral, L. Glass, J.M. Hausdorff, P.C. Ivanov, R.G. Mark, J.E. Mietus, G.B. Moody, C.K. Peng, H.E. Stanley, Physiobank, physiotoolkit and physionet: components of a new research resource for complex physiologic signals. Circulation **101**, e215–e220 (2000). (Jun.)
21. G.B. Moody, R.G. Mark, The impact of the MIT-BIH arrhythmia database. IEEE Eng. Med. Biol. Mag. **20**, 45–50 (2001). (May)
22. G. Moody, W. Muldrow, R. Mark, A noise stress test for arrhythmia detectors. Comput. Cardiol. **11**, 381–384 (1984)
23. I. Silva, G. Moody, An open-source toolbox for analysing and processing physionet databases in MATLAB and Octave. J. Open Res. Softw. **2**, e27 (2014). (Sep.)

24. W.A.H. Engelse, C. Zeelenberg, A single scan algorithm for QRS-detection and feature extraction. Comput. Cardiol. **6**, 37–42 (1979). (Nov.)
25. J.M. Feldman, M.H. Ebrahim, I. Bar-Kana, Robust sensor fusion improves heart rate estimation: clinical evaluation. J. Clin. Monit. **13**, 379–384 (1997). (Nov)
26. M. Cecconi, A. Rhodes, J. Poloniecki, G.D. Rocca, R.M. Grounds, Bench-to-bedside review: the importance of the precision of the reference technique in method comparison studies-with specific reference to the measurement of cardiac output. Crit. Care **13**, 201 (2009)
27. C. Montella, The Kalman filter and related algorithms: a literature review (05 2011)
28. J. He, M. Zhou, X. Wang, Y. Han, Application of Kalman filter and k-NN classifier in wearable fall detection device, in *2017 IEEE SmartWorld, Ubiquitous Intelligence Computing, Advanced Trusted Computed, Scalable Computing Communications, Cloud Big Data Computing, Internet of People and Smart City Innovation (SmartWorld/SCALCOM/UIC/ATC/CBDCom/IOP/SCI)* (2017), pp. 1–7

Chapter 5
A Multi-sensor and Multimodal Data Fusion Technique for Heartbeat Detection

Abstract Accurate heartbeat detection is an essential requirement in modern healthcare monitoring, offering critical insights into cardiac health and cardiorespiratory fitness. This chapter introduces a novel multimodal data fusion technique that leverages discrete wavelet transform (DWT) to enhance beat detection accuracy by integrating electrocardiogram (ECG) and photoplethysmogram (PPG) signals—especially in ambulatory and noise-prone environments. Key signal characteristics are extracted in the wavelet domain and merged into a unified feature signal through a weighted averaging process, where the weights are determined using a signal quality index tailored for periodic or quasi-periodic signals with varying morphologies. The final fused signal is then used for heartbeat peak detection. The performance of the algorithm is rigorously evaluated under diverse noise conditions and signal-to-noise ratios (SNRs), ranging from –30 to 50 dB. The method consistently achieves high sensitivity (99.69%), positive predictive value (99.64%), and low relative error in beat-to-beat interval detection, significantly outperforming state-of-the-art single-channel approaches. This chapter highlights the algorithm's potential for real-time, robust heart rate monitoring in wearable and IoT-based healthcare systems, where conventional methods may fail due to noisy or incomplete data.

Keywords Multimodal data fusion · Data fusion · Electrocardiogram · Photoplethysmogram · Discrete stationary wavelet transform · Common representation format · Continuous signal quality indicators · Signal-quality aware fusion

5.1 Introduction

One of the fundamental tasks associated with wearable physiological health monitoring is to accurately detect the subject's heartbeat. Severe motion artifacts, electrode motion noise, and electromyographic noise can affect ambulatory data obtained from electrodes in biosensors, which is a huge challenge in carrying out accurate heartbeat detection in wearables. An excessive false alarm rate from incorrect feature extraction may lead to alarm fatigue and subsequently slow down response times to

A. John et al., *Deep Learning and Signal-Processing Methods for Multisensor Data Fusion*, https://doi.org/10.1007/978-3-031-96724-5_5

adverse events [1]. Data fusion methods are found to achieve performance improvements without significant variations in existing data acquisition hardware and with minimal additional computational and power consumption costs.

The electrocardiogram (ECG) is a recording of the electrical activity of the heart, and the QRS complex in the ECG signal is used as the signal feature for most heartbeat detectors. Several algorithms have been developed to detect the QRS complex location and identify patients' conditions, cardiac activities, arrhythmia, and heart rate variability (HRV) [2–5], as discussed in Chap. 4. Robust and low-power heartbeat detection algorithms suitable for wearable devices and telehealth monitoring have been discussed in the literature [6, 7]. Most straightforward methods include the detection of QRS complexes in ECG signals with time-domain-based thresholding methods. However, the time-domain-based methods may miss the occasional low amplitude QRS peaks. On the other hand, frequency-domain analyses assess a variety of frequency components of an ECG signal containing noise and may give inaccurate results concerning heart rate [8]. In recent years, smartwatches have started employing photoplethysmogram (PPG) sensors to detect a user's heart rate [9]. A PPG signal is an optically obtained blood volume change measurement associated with each cardiac cycle. It is comparatively inexpensive and more convenient for heart rate detection than ECG recordings. Several algorithms to calculate heart rate from PPG signals have been developed [10–12]. Traditional algorithms that exclusively use a single sensor data source, i.e., ECG or PPG, for detecting heartbeats are highly susceptible to motion artifacts, baseline wander, and power-line interference and produce inferior accuracy in ambulatory conditions, and this chapter aims to address this issue through data fusion [13]. In this chapter, ECG and PPG signals are fused for heartbeat detection. The multimodal nature of these signals requires that the signals be transformed into a common representational format prior to fusion for accurate inferences. Therefore, in this chapter, a novel multimodal technique of fusing ECG and PPG signals using the discrete stationary wavelet transform (DSWT) for ambulatory heartbeat detection is presented.

5.1.1 Related Works

The fusion of ECG and PPG signals and other signal sources in the context of heart rate estimation is well explored in the literature [14–16], as discussed in Chap. 4. The fusion of ECG and PPG signals in the context of respiration rate estimation is also discussed in the literature [17]. However, this is the first attempt at fusing ECG and PPG signals alone to improve heartbeat detection accuracy. Various algorithms aimed at improving heartbeat detection from discontinuous data in bedside monitors using data fusion were developed as part of the Physionet/CinC 2014 challenge. The Physionet/CinC 2014 challenge aimed to improve heartbeat detection in multimodal data. The challenge dataset included signals like ECG, blood pressure (BP), stroke volume (SV), electrooculogram (EOG), and others. All the major results from the challenge are discussed in [18]. The algorithm that obtained the top score in the challenge with a gross sensitivity of 95.64% and positive predictive value (PPV) of 93.48% used a combination of ECG signals and other pulsatile signals for peak identification with slope and peak-sensitive band-pass digital filters [19]. Galeotti et al. used a voting scheme for fusing beat annotations from ECG, BP, SV, EEG, and EOG, and this algorithm achieved a gross sensitivity of 91.08% and PPV of 86.96% on the Physionet/CinC 2014 challenge dataset [20]. Chandra et al. proposed the use of a convolutional neural network for the fusion of BP and ECG signals for heartbeat detection and achieved a gross sensitivity of 93.41% and PPV of 95.47% on the Physionet/CinC 2014 challenge [21].

ECG and PPG signals can be obtained conveniently and simultaneously using smartwatches and/or wearable sensors. Therefore it is highly beneficial to fuse ECG and PPG signals for improving detection accuracy in ambulatory conditions rather than independently interpreting features from them [13]. ECG and PPG can be more conveniently acquired compared to signals like EEG, SV, BP, EOG, etc., which is an important consideration in wearable device design.

Signals obtained from wearable devices are highly susceptible to noise due to the ambulatory nature of the recordings, and therefore the fusion algorithm needs to take into account the quality of the signals recorded prior to fusion. Methods for indicating signal quality as multiple classes or continuous values that specify the acceptability of ECG signals of diagnostic quality on wearables have been discussed in the literature [22, 23] as well as in Chap. 3. Signal quality assessment of PPG signals based on various clinical diagnostic requirements has been discussed in [24, 25]. Instantaneous signal quality indicators for heterogeneous signal inputs to use as weights for data fusion are required here. In Chap. 4, instantaneous signal quality indicators for ECG signals were developed. This chapter introduces a novel generalizable instantaneous signal quality indicator that can be applied to signals of various modalities, as long as they exhibit quasi-periodic behavior.

The novelty of the fusion methodology proposed in this chapter is the development of a transform-domain fusion algorithm for multimodal data and its application to ambulatory heartbeat detection. Unlike signal-level fusion, the proposed technique isolates the characteristics of interest in the wavelet domain. This is done by removing coefficients corresponding to noise and thus reducing its effect on the fused signal. The coefficients of interest are fused using an SQI-weighted fusion mechanism where an instantaneous SQI that is applicable to both input signals is used. Further processing is performed on the fused transform domain signal for heartbeat detection. The proposed technique is also superior to feature/ decision-level fusion, where the initial features are extracted from a noisy input. The proposed weighted fusion technique improves the reliability of heartbeat detection from ECG and PPG signals, as discussed in Sect. 5.2. The performance of the algorithm when the ECG signal is corrupted by noise is compared against the performance of the single-channel ECG beat detection algorithms made available in the WFDB toolbox [26, 27], such as GQRS, WQRS [4], and SQRS [5]. The performance of the fusion algorithm when the PPG signal is corrupted by noise is compared against the performance of the fusion algorithm discussed in this chapter, when only the PPG signal is provided as input. The results and performance comparisons are discussed in Sect. 5.3.

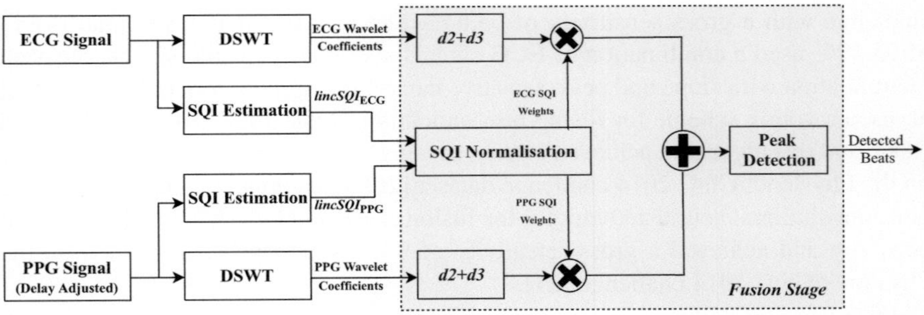

Fig. 5.1 Flow diagram of the proposed fusion algorithm. The sum of the $d2$ and $d3$ bands obtained from the discrete stationary wavelet transforms of the ECG and PPG signals fused in the ratio of the linearized SQIs

5.2 Methodology

5.2.1 Datasets and Pre-processing

Traditionally, the Physionet/CinC 2014 challenge database is used for the analysis of fusion techniques for heartbeat detection. However, it lacks simultaneous ECG and PPG data, which is essential for the use case in this chapter. Therefore, signals from the MIMIC III waveform database [27, 28] are used for analysis. MIMIC III integrates de-identified, comprehensive clinical data of patients admitted to the Beth Israel Deaconess Medical Center in Boston, Massachusetts. The dataset contains bedside vitals-monitoring signals, and subsets of the dataset have been used for various studies with regard to cardiovascular activity monitoring [29–31]. The study utilizes 27 records from the dataset, at a sampling rate of 125 Hz, of varying lengths that contain an ECG signal channel and a PPG signal channel to constitute the dataset. The PPG signals are shifted to compensate for the pulse transit time [11], such that the peaks of both signals align. Pre-recorded noise (baseline Wander, muscle artifacts, and electrode motion noise) available from the MIT-BIH noise stress test Database (NSTDB) [32] and simulated Gaussian noise, high-frequency noise, low-frequency noise, and power-line interference are added during testing. The performances of the algorithms in the presence of a combination of these noises types are also analyzed: Combination 1 consists of high-frequency noise, low-frequency noise, power-line interference, baseline wander, muscle artifacts, and electrode motion noise, and Combination 2 consists of baseline wander, muscle artifacts, and electrode motion noise.

5.2.2 Discrete Stationary Wavelet Transform

Discrete Wavelet Transform (DWT) decomposes a signal into its time-frequency components (called detailed and approximate coefficients). The DWT of a signal x is calculated by passing it through a series of high- and low-pass filters. First, the samples are passed through a low-pass filter with impulse response g and down-sampled by 2 to yield the first-level approximation coefficients (here 0–31.25 Hz) (Eq. (5.1)). The signal is also decomposed simultaneously using a high-pass filter, h, and downsampled by a factor of 2 to yield the first-level detail coefficients (Eq. (5.2)) (here 31.25–62.5 Hz, or the d1 band). The first-level approximation band is decomposed using h to obtain the second-level detail coefficients band (here 15.625–31.25 Hz, or the d2 band).

$$y_{low}[n] = (x * g) \downarrow 2 \tag{5.1}$$

$$y_{high}[n] = (x * h) \downarrow 2 \tag{5.2}$$

The Discrete Stationary Wavelet Transform (DSWT) is essentially a DWT without the downsampling process [33], with the assumed periodicity of data [33]. This can be used on ECG and PPG signals as they can be considered quasi-periodic. The advantage of DSWT is that full temporal resolution is maintained at the output, and the relevant bands can be used as features, as without the downsampling stage, the transform is translation-invariant. This is essential for temporal fusion and for subsequently detecting heartbeats. A three-level DSWT is performed on the ECG and PPG signals to capture temporal details in the 7.8125–15.625 Hz band (d3 coefficients) and 15.625–31.25 Hz band (d2 coefficients) to capture information corresponding to the QRS complex peaks. In the case of PPG signals, these wavelet bands are chosen as not much information aside from the peaks corresponding to the rising edges is desired for fusion. The information in these two bands is captured by adding the d2 and d3 coefficients of each signal, as depicted in Fig. 5.1. The Biorthogonal 6.8 wavelet (from [34]) is chosen as the mother wavelet for the ECG signal. For the PPG signal, the Symmlet 8 (from [35]) wavelet is chosen. The choice of the mother wavelet for a particular signal was based on the similarities in temporal structures of the mother wavelet and the signal.

5.2.3 Signal Quality Indicators

Signal quality indicators for signals with different temporal structures and energies would have different SQIs at different signal-to-noise ratios (SNRs). But heterogeneous signal-based weighted fusion calls for similar signal quality index characteristics for similar SNRs. Therefore, a generalized approach for SQI-based weight calculation for heterogeneous signal fusion from any m sensors is introduced in this chapter, as shown in Fig. 5.2. After the SQI calculation stage, where signal quality indicators may be derived from any of the methods described in the literature [22–25], the SQIs must be converted to a standard form such that SQI values match perfectly with the signal SNRs (low SNR should yield a low SQI value, while a high SNR should yield a high SQI value). Therefore, a mapping function for each sensor input must be derived and calibrated such that the SQI at a particular SNR for an input signal is linearly mapped to its SNR value. This SQI mapping can be called $linxSQI$, representative of the linear mapping, with the x indicating the type of SQI that is linearly mapped. Then the SQI normalization block calculates the weights as shown in Eq. (5.3), where i corresponds to the SQI of the ith information signal/sequence, and n indicates the signal sample.

$$W_i[n] = \frac{linxSQI_i[n]}{\sum_{k=1}^{m} linxSQI_k[n]}. \tag{5.3}$$

This normalization ensures that in weighted additive fusion, the weights sum to 1, preserving the signal energy of the input signals in the fused output.

Next, a signal quality indicator is developed for quasi-periodic signals with a consistent temporal structure. This SQI is then analyzed and validated using ECG and PPG signals. In comparison to the SQIs discussed in Sect. 4.3.3, which are applicable only to ECG signals, the signal quality method developed in this chapter can be applied to any quasi-periodic signal, making it suitable for the quality assessment of both ECG and PPG signals. To detect the quality of a signal, the algorithm must be capable of detecting the amount of noise contained in the signal, therefore, a good signal quality indicator must be a monotonically increasing function of the SNR, which can then be mapped as a linear function of SNR. The method works by comparing a single-cycle sequence of the signal curve with its clean signal-specific template and is therefore called curve SQI (cSQI). The SQI is calculated based on the difference between the single-cycle template and the incoming signal cycle to measure how much noise is present in the signal. Moreover, cSQI can provide instantaneous signal quality (signal quality for each sample) for a signal sequence and can be used for both ECG and PPG signals for instantaneous signal quality assessment. The SQI proposed is based on the waveform morphology of the signal of interest and tries to evaluate how similar the signal is to the expected waveform

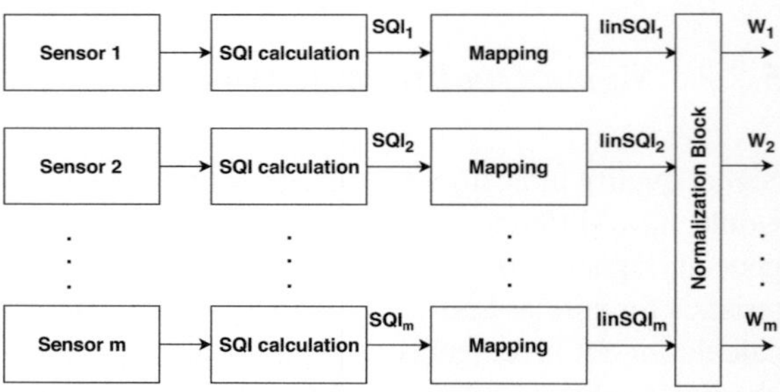

Fig. 5.2 Flow diagram for SQI-based weight calculation for fusion

shape. This SQI is then tested on both ECG and PPG signals. Since there is no widely accepted metric for evaluating signal quality indicators, their performance is assessed based on how the SQI varies as a function of the SNR.

Template Generation

The first step in carrying out signal quality analysis using cSQI is the generation of a single signal cycle curve template. This is a pre-requisite and must be done in a wearable device during the initialization stage, where the user can be prompted to stay still and clear from noise sources. Therefore, it can be assumed that the signal obtained in the initialization stage is clean. The template generation would attempt to use **N** cycles of clean data, and the time taken for the generation of the template will vary depending on the average time period of a given quasi-periodic signal. (For example: In the case of the ECG signals indicating heart rate, the knowledge that the average heart rate is 60 to 100 beats per minute is utilized. Let us consider it to be 80 bpm, therefore, 10 cycles can be obtained in approximately 7.5 s). The knowledge of this time period also helps us in fixing the template length that needs to be used for template generation. The steps involved in generating a template of length $2M + 1$ samples are shown in Algorithm 1. Once the signal for template generation is acquired, the template can be generated offline.

At this stage, it is crucial to limit **N** to a small number, as using too many cycles to create a template can overly smooth the signal template, which is not desirable.

cSQI Calculation

Once the template is generated, cSQI can be estimated. cSQI is a signal quality indicator that assigns a quality value for each signal sample. This is assigned based on the quality of the signal window centered at the signal sample of interest. cSQI at a point t_0 is defined as the inverse of the variance of the difference between the template and signal window centered at t_0 at the point of maximum correlation (the point of maximum correlation is found by circular rotation of the template, and then correlation calculation). The SQI at the next point t_1 is calculated in a similar manner by shifting the window about the center to the next sample once. This yields a signal quality level at each sample point, which can be used for weighting in the multimodal data fusion task. The steps in calculating the cSQI per sample are discussed in Algorithm 2. The block diagram of the algorithm for a circuit implementation is shown in Fig. 5.3.

cSQI for ECG Signals

ECG signals can be corrupted by a variety of noises. The most common examples of noise corrupting ECG signals are motion artifacts, electrode motion noise, muscle

Algorithm 1 Algorithm for Template Generation

Setup:

1. Signal sequence of interest $\mathbf{x}$.
2. Signal vector $\mathbf{X}_n$ of length $2M+1 \triangleq [\mathbf{x}[n-M], \cdots, \mathbf{x}[n], \cdots, \mathbf{x}[n+M]]$ where M has been decided based on the time period.
3. Let $\mathbf{X}_{n(c)}$ be the correlation Toeplitz matrix of $\mathbf{X_n}$ of dimensions $(2M+1, 4M+1)$ where
$$\mathbf{X_n}(c) = \begin{bmatrix} \mathbf{x}[n+M] & \cdots \mathbf{x}[n] & \cdots \mathbf{x}[n-M] & \cdots 0 & 0 \\ 0 & \mathbf{x}[n+M] & \cdots \mathbf{x}[n] & \cdots \mathbf{x}[n-M] & \cdots 0 \\ \cdot & \cdot & \cdot & \cdot & \cdot \\ \cdot & \cdot & \cdot & \cdot & \cdot \\ 0 & 0 & \cdots \mathbf{x}[n+M] & \cdots \mathbf{x}[n] & \cdots \mathbf{x}[n-M] \end{bmatrix},$$
where the kth column of $\mathbf{X}_{n(c)}$ is indicated by $\mathbf{X}_{n(c)}^{(k)}$ where $k \in [0, 4M]$.
4. Let $\mathbf{X_n}^{(t)}$ be the tth circular shifted version of $\mathbf{X_n}$ ie., $\mathbf{X_n}^{(t)} = [\mathbf{x}[n-M+t], \mathbf{x}[n-M+t+1], \cdots, \mathbf{x}[n+M]\mathbf{x}[n-M], \cdots, \mathbf{x}[n-M+t-1]]$.

Template Generation:

1. Identify a signal segment containing $\mathbf{N}$ consecutive fiducial features of interest (From the time period or otherwise).
2. Let $\mathbf{n_0}$ be the center of the first segment of length $2M+1$, $\mathbf{X}_{n_0}$.
3. Initialize $\mathbf{T} = \mathbf{X}_{n_0}$.
4. For $i = (M+1)$ to $(\mathbf{N}(2M+1) - M)$ in steps of $2M+1$
   a. Let $\mathbf{X}_{n_0+i\,(c)}$ be the correlation Toeplitz matrix of $\mathbf{X}_{n_0+i}$.
   b. Let $k = argmax_j\ [|\mathbf{T}.\mathbf{X}_{n_0+i(c)}^{(j)}|]$
   c. If $k \approx n_0$
      i. $\mathbf{T} = \mathbf{T} + \mathbf{X}_{n_0+i}^{(k)}$
      End

   End
5. Normalize $\mathbf{T}$ based on number of times condition 4.c was satisfied.

artifacts, baseline wander, power-line interference, and instrumentation noise due to electronic components in the acquisition system. Signal corruption in ECG signals due to leads that have fallen off, etc., can be modeled as the ECG signal being corrupted by an additive noise signal. The cSQI can be used to determine which portions of the signals are suitable for further processing (like QRS detection [36] or beat classification [37] in ECG signals) on-chip due to its low complexity [38], which is discussed in Sect. A.2 in the appendix of this book. ECG signal data for checking the suitability of cSQI for estimating ECG signal quality was taken from the MIMIC III waveform database [27, 28]. The experiment was tested on one clean ECG record and the average of 10 noise instances was used to evaluate the performance of the SQIs at different SNRs under different noise conditions. The different noise conditions considered in this analysis are electrode motion noise, muscle artifacts, baseline wander (for which the recordings were taken from the noise stress test database [27, 32]), and 50 Hz power-line interference (modeled as sinusoids and a combination of sinusoids).

Algorithm 2 Algorithm for cSQI calculation

Initialize and calculate $c[0]$**:**

1. Let $\mathbf{T}(c)$ be the Toeplitz correlation matrix of the template $\mathbf{T}$. $\mathbf{T}(c)[u]$ indicates uth row of the matrix with $u \in [0, 2M]$.
2. Signal vector $\mathbf{S}_0$ of length $2M + 1 \triangleq [\mathbf{s}[-M], \cdots, \mathbf{s}[0], \cdots, \mathbf{s}[M]]$ where M has been decided based on the time period, centered at $\mathbf{s}[0]$.
3. Lag sequence $L = [-M, -M+1, \cdots, 0, \cdots, M-1, M]$.
4. Calculate $\mathbf{A}' = \mathbf{S}_0 \times \mathbf{T}$ where $\mathbf{A}'$ is the L-R flipped result of the correlation sequence .
5. Lag $t = L[\text{argmax}(\mathbf{A})]$
6. $c[0] = (\text{var}(\mathbf{S}_0 - \mathbf{T}^{(t)}))^{-1}$ and $\mathbf{b} = s[-M]$

Loop till no signal is read: (Fast correlation calculator)
For $i = 1$ to Z in steps of 1

1. Window centered at $s[i]$, $\mathbf{S}_i$ of length $2M + 1$.
2. $\mathbf{NR} = \mathbf{s}[i + M].\mathbf{T}_{(c)}[2M]$ and $\mathbf{FR} = \mathbf{b}.\mathbf{T}_{(c)}[0]$.
3. Calculate L-R flipped correlation sequence $\mathbf{A}' = (\mathbf{A}' - \mathbf{FR})^{(1)} + \mathbf{NR}$.
4. Lag $t = L[\text{argmax}(\mathbf{A})]$
5. $c[i] = (\text{var}(\mathbf{S}_i - \mathbf{T}^{(t)}))^{-1}$ and $\mathbf{b} = s[i - M]$

End
cSQI= LPF(c), where the low pass filter could be a moving average filter to avoid rapid variations that can be attributed to edge effects of windowing.

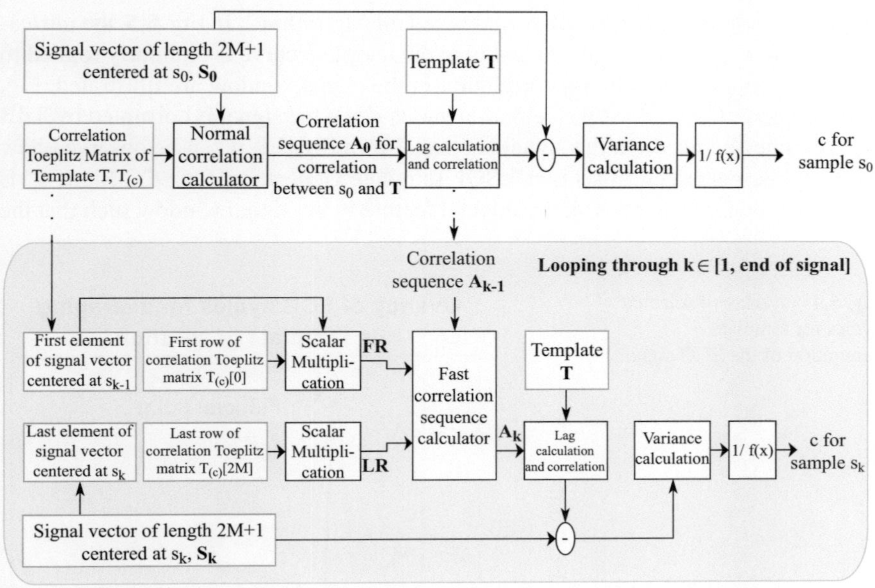

Fig. 5.3 Block diagram for implementing cSQI calculation (instantaneous c values). The boxes with green outlines show the inputs

cSQI Template Generation for ECG Signals

For generating a template for ECG signals, a clean record from the MIMIC III database was used. The first 10 cycles of the record were used for generating a template. The template length T was chosen as 70% of the average time period, so as to reduce the chances of having 2 fiducial points in one segment of data in case the time period decreases. In scenarios, when the segment completely misses a fiducial point, that segment can then be discarded due to condition 4.c in Algorithm 1. Because of this condition, all 10 cycles may not contribute to the final template. The MIMIC III database records sample signals at a sampling rate of 125 Hz, and from the time duration, the chosen template was of length 71, with $M = 35$. The samples from the 0th to the 674th sample were used for template generation, while the cSQI performance evaluation was carried out on the rest of the signal. For the ECG signals, the fiducial point was chosen at the Q wave, which is the maxima in one cycle of an ECG signal. Figure 5.4 shows the overlay of ECG cycles obtained from the first 10 cycles that are averaged to generate the template.

Calculation of cSQI for an ECG Signal

cSQI at each sample point is calculated based on Algorithm 2. In Fig. 5.5, the intricacies of how a window is selected and how the template curve is circularly rotated to match the points of maximum correlation in the signal window are illustrated. Fig. 5.5 also shows how the cSQI calculation varies when the signal is corrupted by 5 dB Gaussian noise and when the signal is clean. From Fig. 5.5, it can be observed how the template generated in section "cSQI Template Generation for ECG Signals" is circularly rotated to align with the fiducial feature in the signal window such that the

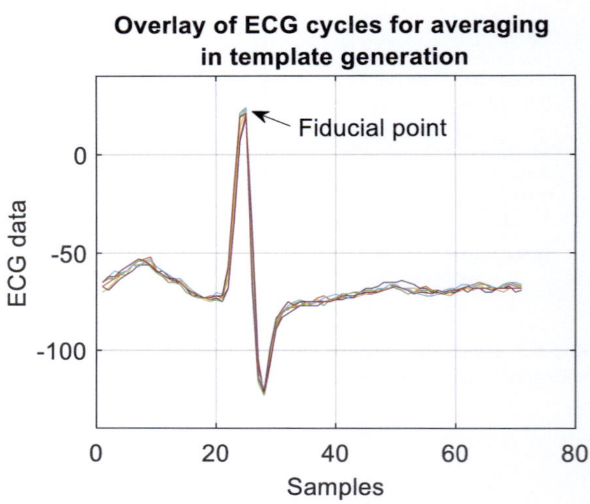

Fig. 5.4 Overlay of selected cycles for template generation of the ECG signal

difference between the corrected template and the signal window is minimal. Therefore, the difference between the signal window and the corrected template would correspond to noise (Observe in Fig. 5.5, when the signal window is noisy and the template is overlaid on it, the difference between these two signal vectors is the added noise).

Performance Evaluation of the cSQI Algorithm for ECG Signals

For clarity of understanding, the cSQI plots at each sample point for when a portion of the signal is corrupted by (a) 50 Hz power-line interference, (b) baseline wander,

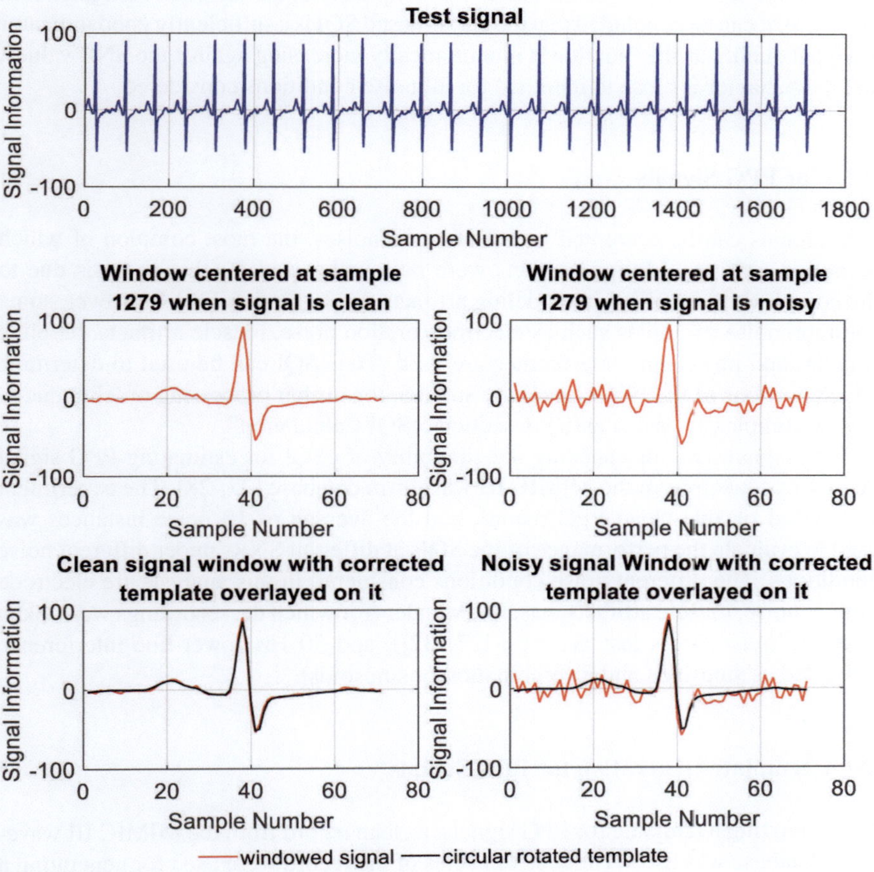

Fig. 5.5 Figure detailing how for a given ECG signal window centered at a specific point and at two different SNRs, the template is rearranged to align with the point of maximum correlation. It also shows how the cSQI changes with varying levels of noise

(c) muscle artifacts, and (d) electrode motion noise at –10 dB are shown in Fig. 5.6. Observe in the figure how cSQI values drop when the signal is noisy compared to when the signal is clean. From this figure, it can be observed that for all samples of the noisy signal, the cSQI calculated is much lower than that of when the signal is clean, which is a simple but good indicator of a good SQI. Figure 5.7a, b, c, and d show the output SQI at varying SNRs for when the ECG signal is corrupted by (a) 50 Hz power-line interference, (b) baseline wander, (c) muscle artifacts, and (d) electrode motion respectively. Here the SQI is averaged over the entire signal per noise instance per SNR. Ideally, the SQI values must vary linearly with SNR, but since this is not possible as signal quality can vary from 0 to ∞, a monotonically increasing function is a sufficiently good estimate of the signal quality. But it can be observed from Fig. 5.7a, b, c, and d that the SQI exhibits an almost linear behavior in the medium SNR ranges before saturating completely at the high SNR ranges. From the figure, it can be concluded that the curve-based SQI is a sufficiently good indicator of signal quality as the function is monotonically increasing against the SNR values, and the behavior is close to identical for all noise conditions considered.

cSQI for PPG Signals

PPG signals can be corrupted by a variety of noises, the most common of which is motion artifacts. Motion artifacts were not synthesized for this analysis due to the complexities involved in modeling artifacts due to motion [39]. However, some common noise examples such as electrode motion noise, muscle artifacts, baseline wander, and power-line interference are used. The cSQI can be used to determine which portions of the PPG signals are suitable for further processing on-chip due to its low complexity as discussed in section "cSQI Calculation".

PPG signal data for checking the suitability of cSQI for estimating PPG signal quality was taken from the MIMIC III waveform database [27, 28]. The experiment was tested on one clean PPG record, and the average of 10 noise instances was used to evaluate the performance of the SQIs at different SNRs under different noise conditions. The different noise conditions considered in this analysis are electrode motion noise, muscle artifacts, baseline wander (for which the recordings were taken from the noise stress test database [27, 32]), and 50 Hz power-line interference (modeled as sinusoids and a combination of sinusoids).

cSQI Template Generation for PPG Signals

For generating a template for PPG signals, a clean record from the MIMIC III waveform database was used. The first 10 cycles of the record were used for generating a template. The template length T was chosen as 70% of the average time period. The chosen template was of length 71, with $M = 35$. For the PPG signals, the fiducial point was chosen as the maxima in one cycle of the PPG signal. Figure 5.8 shows the

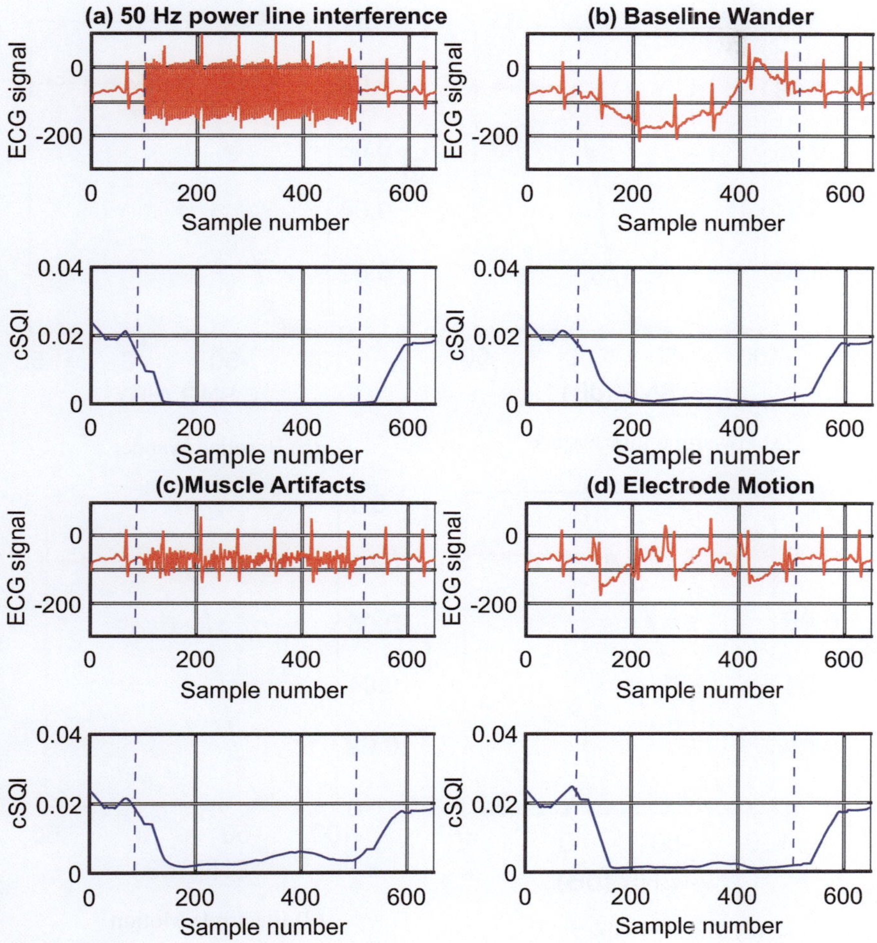

Fig. 5.6 cSQI plots at each sample point of an ECG signal when it is corrupted by **a** 50 Hz power-line interference, **b** baseline wander, **c** muscle artifacts, and **d** electrode motion at –10 dB. The vertical dotted lines delimit regions of clean signal and regions where noise is added to the clean signal

overlay of PPG cycles obtained from the first 10 cycles that are averaged to generate the template.

Calculation of cSQI for a PPG Signal

cSQI at each sample point is calculated based on the algorithm in Algorithm 2. In Fig. 5.9, the intricacies of how a window is selected and how the template curve

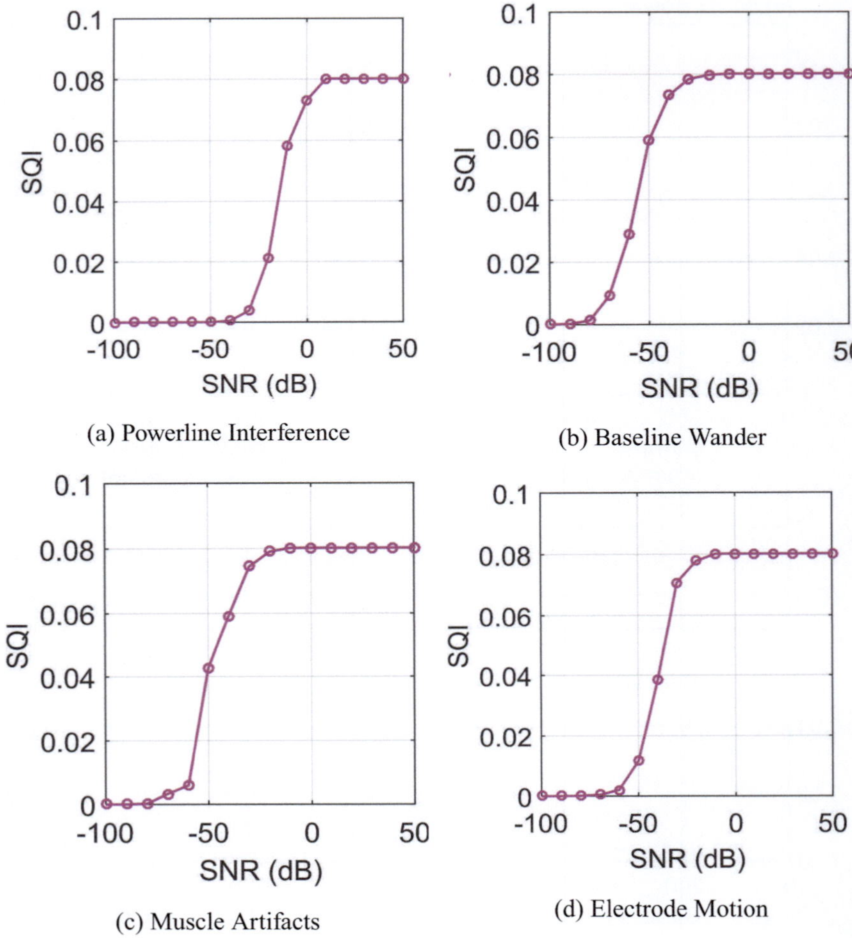

(a) Powerline Interference

(b) Baseline Wander

(c) Muscle Artifacts

(d) Electrode Motion

Fig. 5.7 cSQI versus SNR plot for when the ECG signal is corrupted by **a** 50 Hz power-line interference, **b** baseline wander, **c** muscle artifacts, and **d** electrode motion

is circularly rotated to match with the points of maximum correlation in the signal window are illustrated. Figure 5.9 also shows how the cSQI calculation varies when the PPG signal is corrupted by 5 dB Gaussian noise and when the signal is clean, and the difference in values of the SQIs is also indicated. From Fig. 5.9, it can be observed how the template generated in section "cSQI Template Generation for PPG Signals" is circularly rotated to align with the fiducial feature in the signal window such that the difference between the corrected template and the signal window is minimal. Therefore, the difference between the signal window and the corrected template would correspond to noise (observe in Fig. 5.9, when the signal window

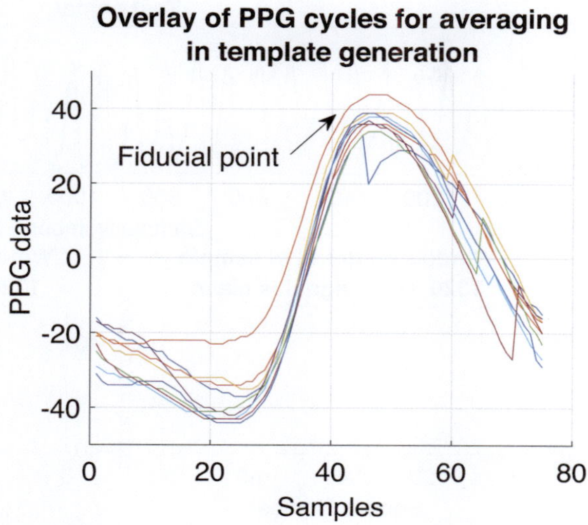

Fig. 5.8 Overlay of selected cycles for template generation of the PPG signal

is noisy and the template is overlaid on it, the difference between these two signal vectors is dominated by the added noise).

Performance Evaluation of cSQI Algorithm for PPG Signals

For clarity of understanding, the cSQI plots at each sample point for when a portion of the PPG signal is corrupted by (a) 50 Hz power-line interference, (b) baseline wander, (c) muscle artifacts, and (d) electrode motion noise at –10 dB are shown in Fig. 5.10. Observe in the figure how cSQI values drop when the signal is noisy compared to when the signal is clean. From this figure, it can be observed that for all samples of the noisy signal, the cSQI calculated is much lower than that of when the signal is clean, which is a simple but good indicator of a good SQI. Figure 5.11a, b, c, and d show the output SQI at varying SNRs for when the PPG signal is corrupted by (a) 50 Hz power-line interference, (b) Baseline Wander, (c) Muscle Artifacts, and (d) Electrode Motion respectively. Here the SQI is averaged over the entire signal per noise instance per SNR. Ideally, the SQI values must vary linearly with SNR, but since this is not possible as signal quality can vary from 0 to ∞, a monotonically increasing function is a sufficiently good estimate of the signal quality. But it can be observed from Fig. 5.11a, b, c, and d that the SQI exhibits an almost linear behavior in the medium SNR ranges before saturating completely at the high SNR ranges. From the figure, it can be concluded that curve-based SQI is a sufficiently good indicator of signal quality as the function is monotonically increasing against the SNR values, and the behavior is close to identical for all noise conditions considered.

The performance indicates the suitability of using the cSQI as a suitable instantaneous signal quality indicator for quasi-periodic signals. Further results on the

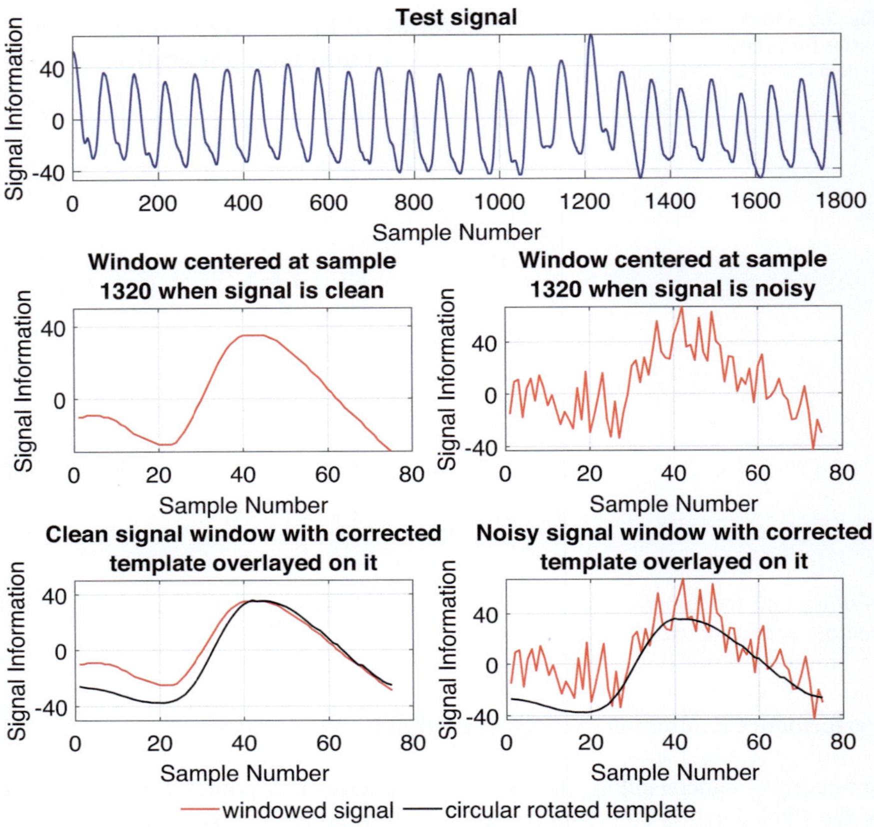

Fig. 5.9 Figure detailing how for a given PPG signal window centered at a particular point at 2 different SNRs, the template rearranges such that it is adjusted according to the point of maximum correlation and how the cSQI varies depending on noise

computational complexity of the SQI and performance on public datasets are discussed in Appendices A.2 and A.3 respectively.

cSQI for ECG and PPG Signals

The cSQI characteristics of ECG and PPG signals were obtained by corrupting a certain length of signal (which is not part of the test dataset; 8 s long at 125 Hz) with synthesized Gaussian noise at different SNRs and calculating the average cSQI for the total signal. Gaussian noise was chosen for this calibration stage by assuming that noise sequences consist of independent random variables, and the sum of these tends toward a Gaussian distribution as more variables are added, as per the central limit theorem. The cSQI characteristics of the ECG and the PPG signals obtained are as

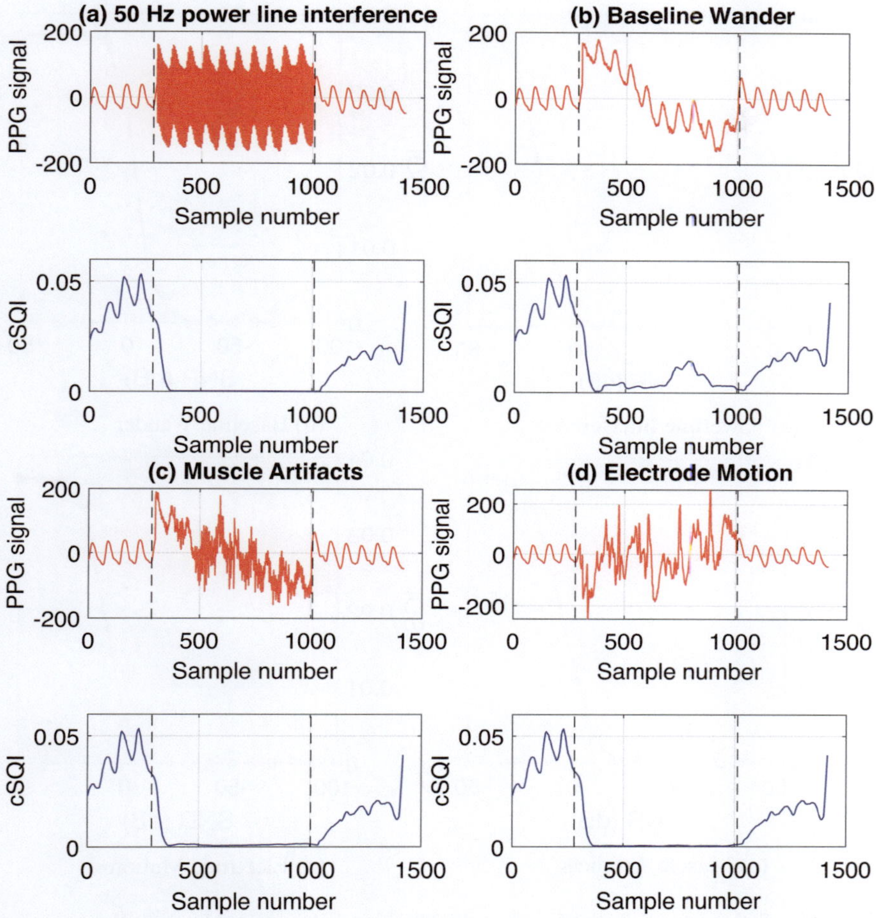

Fig. 5.10 cSQI plots at each sample point of a PPG signal when it is corrupted by **a** 50 Hz powerline interference, **b** baseline wander, **c** muscle artifacts, and **d** electrode motion at –10 dB. The vertical dotted lines delimit regions of clean signal and regions where noise is added to the clean signal

shown in Fig. 5.12a, which are dissimilar. The cSQIs were then linearly mapped to the corresponding SNRs by deriving a mapping function from the cSQI characteristics and is as shown in Fig. 5.12b. The mapping function was set to 0 for mapped cSQI values (lincSQI) less than 0 dB. In the positive SNR region, the obtained cSQIs were mapped to the corresponding SNRs to derive the corresponding lincSQIs using curve fitting methods. The mapping functions obtained were of the form $a \times \log(\text{cSQI}) + b$ where a and b need to be calculated approximately such that each mapping is as close to the desired result. The lincSQI exhibits a linear performance in the 0–25 dB region and saturates post 25 dB due to the limitation of the chosen mapping function. From

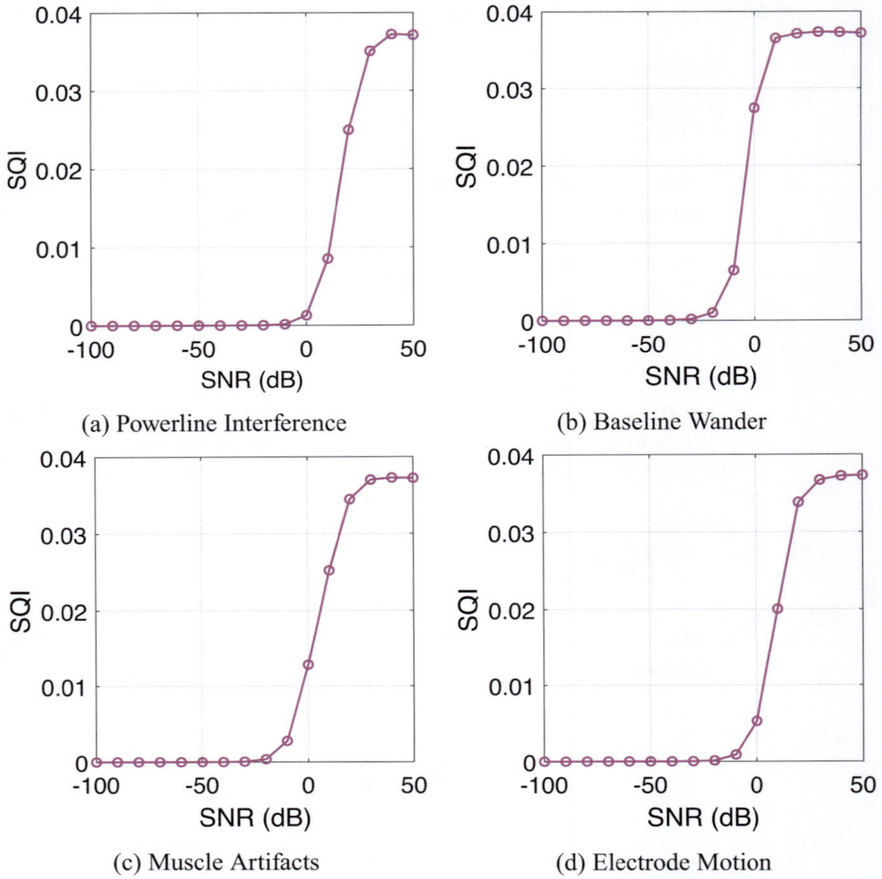

Fig. 5.11 cSQI versus SNR plot for when the PPG signal is corrupted by **a** 50 Hz power-line interference, **b** baseline wander, **c** muscle artifacts, and **d** electrode motion

Fig. 5.12b, it can be observed that the lincSQI values of ECG and PPG signals at different SNRs are comparable. The weights for the weighted additive fusion were then computed from the lincSQIs obtained as per Eq. (5.3).

5.2.4 *Data Fusion*

As discussed in Sects. 5.2.2 and 5.2.3, the ECG and PPG wavelet coefficients (with amplification of PPG signal coefficients to match the amplitude of ECG signal coefficients) and cSQIs are taken as inputs in the fusion algorithm. Initially, in the fusion stage, the SQIs were normalized as per Eq. (5.3). The $d3$ and $d2$ coefficients that

Fig. 5.12 cSQI characteristics for ECG and PPG signals, with Fig. 3a showing the raw calculated cSQIs and Fig. 3b showing the linearly mapped cSQIs

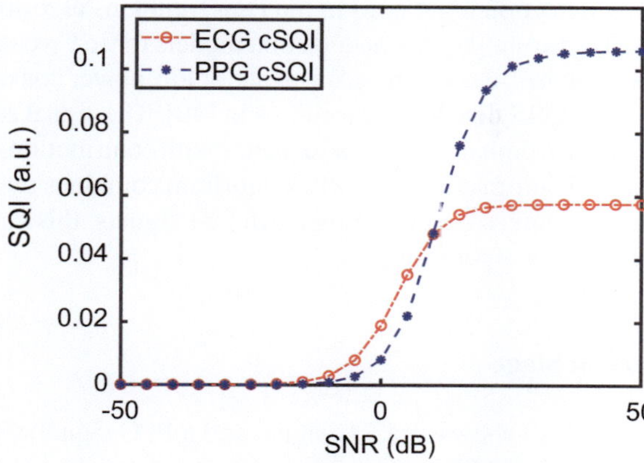

(a) cSQI of PPG and ECG signals.

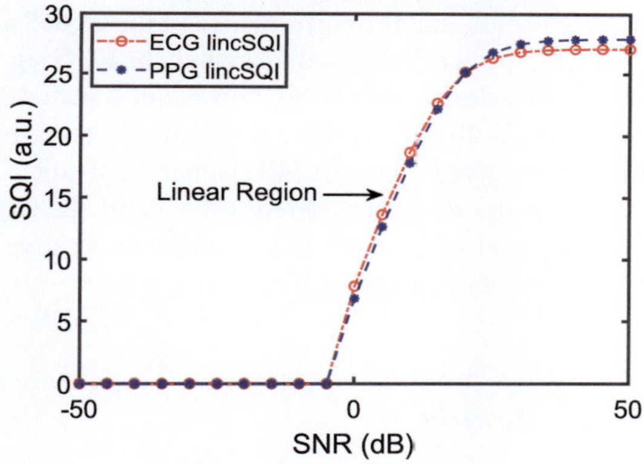

(b) lincSQIs of ECG and PPG signals.

capture the fast-moving peaks and edges in the ECG and PPG signals are added to yield the ECG and PPG wavelet transform sequences as discussed in Sect. 5.2.2. Let the wavelet transform sequences thus obtained be represented as S. A weighted additive fusion of the sequences, assuming that there are m sequences, based on the weights W_i (SQIs obtained) to obtain the fused sequence F at sample n is as shown in Eq. (5.4).

$$F[n] = \sum_{i=1}^{m} W_i[n] S_i[n] \tag{5.4}$$

The fused sequence thus obtained is then operated on by a peak detection algorithm that searches for the peaks corresponding to the QRS complex. The peak detection algorithm makes use of adaptive thresholding and search-backs for missed peaks and is derived from the peak-detection stage of the Pan-Tompkins algorithm [2]. The peak detection logic used in the Pan-Tompkins algorithm was chosen as it has been used in various heartbeat detection models for IoT wearable sensors and was found to have the best detection performance at low power consumption when compared with 8 other QRS detection algorithms in [40]. The signal sequences are combined in the transform domain to form a sequence with contributions from two multimodal signals for decision-making. Since the algorithm combines the information provided by two sensor sources observing two different signals, this can be called a complementary data fusion algorithm [41].

Fusion Stage

Figure 5.13 shows an ECG signal and a PPG signal with -10 dB SNR due to being corrupted by additive Gaussian noise, the normalized cSQI sequences, wavelet transform sequences, the weighted wavelet transform sequences, the fused signal with detected beats, and the performance of the GQRS algorithm in detecting beats. From Fig. 5.13, it can be observed that when the ECG signal is corrupted by noise, the ECG cSQI value drops, and the ECG wavelet transform sequence does not influence the fused signal, therefore, there is no influence on the final fused sequence. A similar trend is observed when the PPG signal is corrupted by noise. This is the performance desired of the algorithm. On the other hand, the GQRS algorithm performs poorly in the noisy region. Figure 5.13 is used for illustrative purposes alone, and performance analysis is not carried out in this manner.

5.2.5 *Testing*

The signals from the MIMIC III waveform database mentioned in Sect. 5.2.1 are used to test the performance of the algorithm. The performance of the data fusion algorithm is compared against the performance of three state-of-the-art single-channel ECG beat detection algorithms:

- GQRS—a peak detection algorithm provided in the WFDB toolbox, with the algorithm unpublished but exhibits promising performance as per many studies,
- WQRS—a single-channel QRS detection algorithm based on the length transform provided by the WFDB toolbox [4], and
- SQRS—a single-scan algorithm for QRS detection proposed by Englese and Zeelenberg, provided by the WFDB toolbox [5],

when the ECG signal is corrupted by noise at different SNR levels. The performance of the data fusion algorithm when the PPG signal is corrupted by noise is compared against the performance of the fusion algorithm when only the noise corrupted-PPG signal is provided as input. As the signals selected were clean and relatively free of noise, noise scenarios that are encountered in ECG or PPG signal acquisition were

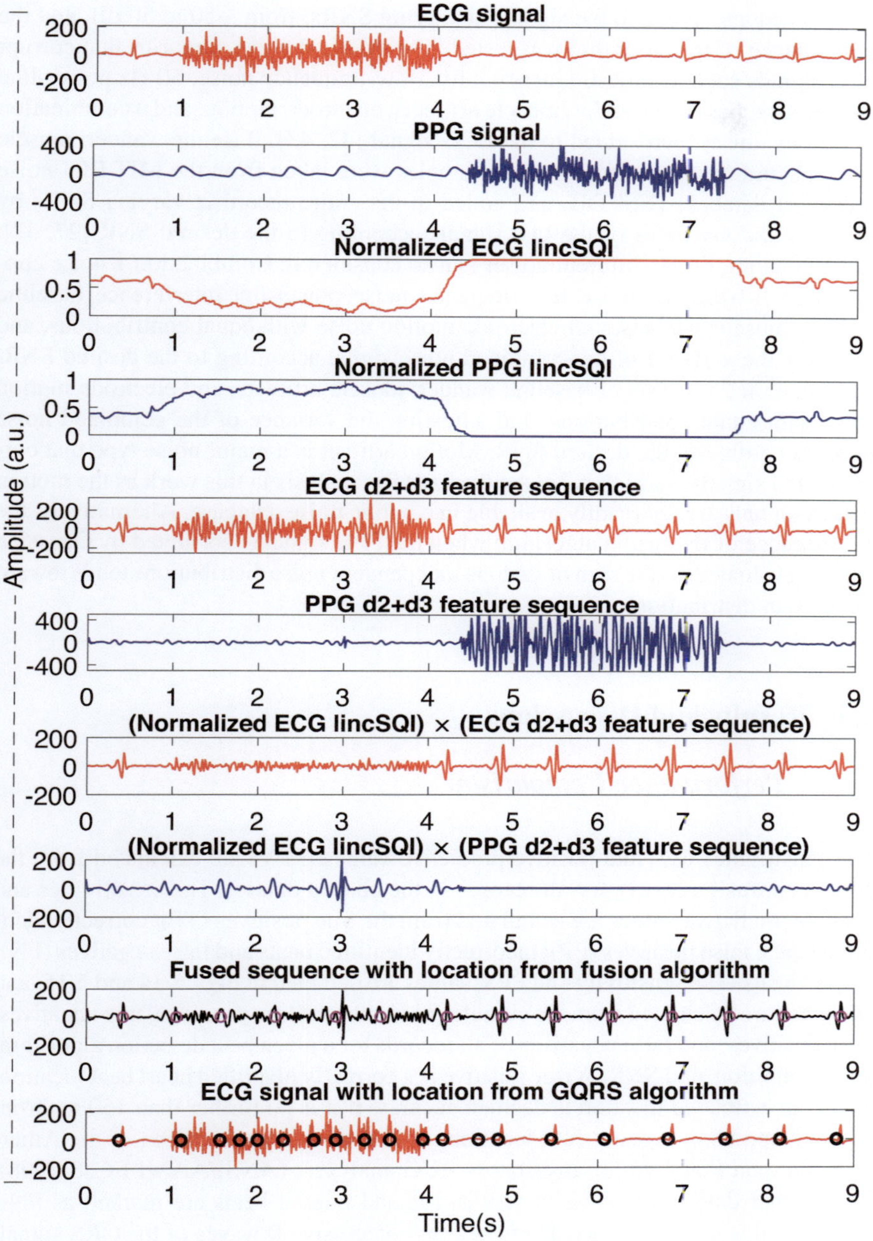

Fig. 5.13 ECG signal and PPG signal corrupted by additive Gaussian noise, resulting in SNR of −10 dB from 1 to 3 s, and 4.4 to 7.4 s respectively (for illustrative purposes), SQI sequences, wavelet $d2 + d3$ transform sequences, fused sequence with heartbeat locations from the fusion algorithm, and the ECG signal with heartbeat locations from GQRS algorithm

synthesized and added to the signals at varying SNRs, from −50 to 50 dB, and the performance of the algorithm was tested. The different noise scenarios that corrupt ECG signals such as high-frequency noise, low-frequency noise, 50 Hz power-line interference, baseline wander, muscle artifacts, electrode motion, and a combination of various noises were added to the ECG signal [42, 43]. Baseline wander, muscle artifacts, and electrode motion noise samples were taken from the MIT-BIH noise stress test database (NSTDB) and added to the entire record at varying SNRs by adjusting the variances of the two signals according to the desired SNR [27, 32]. A combination of the different noises is also considered: Combination 1 noise consists of high-frequency noise, low-frequency noise, power-line interference, baseline wander, muscle artifacts, and electrode motion noise with equal contributions, and adjusting the variance of the combined noise signal according to the desired SNR. Combination 2 consists of baseline wander, muscle artifacts, and electrode motion noise with equal contributions, and adjusting the variance of the combined noise signal according to the desired SNR. Motion artifact is a major noise type that corrupts PPG signals—which is not synthesized for analysis in this work as the motion artifact signals are not readily available in a public noise database. Alternatively, the performance of the fusion algorithm when the PPG signal is corrupted by Gaussian noise is evaluated, as the sum of various independent noise distributions tends toward a Gaussian distribution.

5.3 Results and Discussions

5.3.1 Performance Comparison

The sensitivities (Se) and positive predictive values (PPV) are calculated from the pooled confusion matrix for all records in the chosen dataset. The sensitivities and positive predictive values are calculated from the true positives (TP)/ correctly identified beats, false positives (FP)/ incorrectly identified beats and false negatives (FN)/ missed beats. The sensitivity and PPV values are depicted in Figs. 5.14 and 5.15, and all values are calculated from the cumulative number of true positives, true negatives, false positives, and false negatives of all records for a given beat detection algorithm, noise condition, and SNR. A true positive is a correctly identified heart beat location, where an estimated location is deemed accurate if it is no further than 150 ms from the corresponding ground truth location, as per the recommendation of the American National Standard for ambulatory ECG analyzers (ANSI/AAMI EC38-1998). Every other detected beat is marked as FP, and missed beats are marked as FNs. RR interval is the time elapsed between two successive R waves of the QRS signal. The difference between the groundtruth RR interval and the calculated RR intervals is also used to analyze the performance. The RR interval difference comparison was carried out by calculating the RR intervals from the beats detected by all the algorithms discussed and compared against the reference RR intervals. Beat-by-beat

Fig. 5.14 Performance comparison results of the data fusion algorithm against single-channel ECG GQRS, SQRS, and WQRS algorithm when the ECG signal is corrupted by the addition of **a** high-frequency noise, **b** low-frequency noise, **c** 50 Hz power-line interference, **d** baseline wander, **e** electrode motion, **f** muscle artifacts, **g** combination noise 1, and **h** combination noise 2 at SNRs from −50 to 50 dB

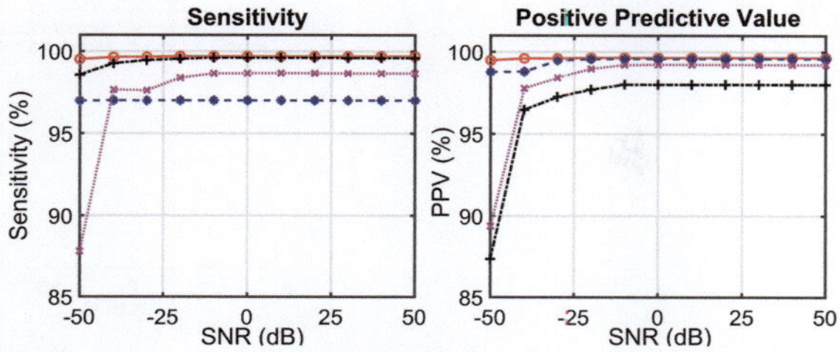

(a) High-frequency noise

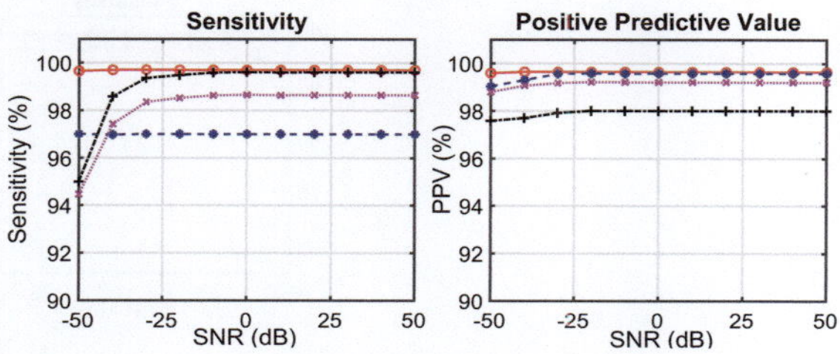

(b) Low-frequency noise

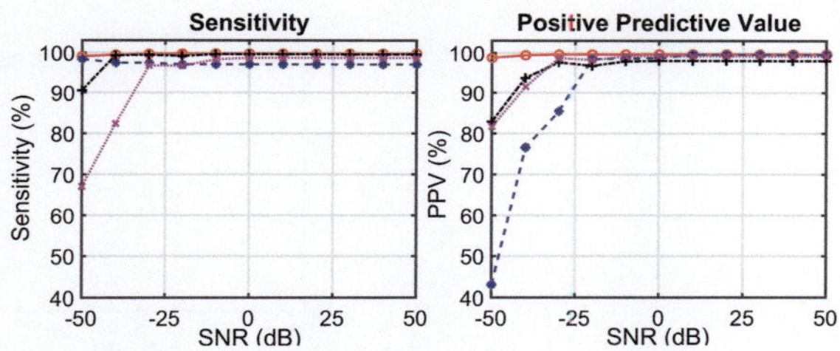

(c) 50 Hz power-line interference

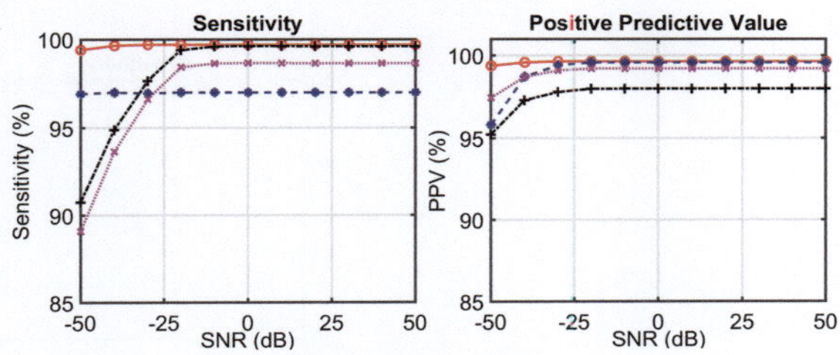

(d) Baseline wander

detection comparison is carried out based on sensitivity, positive predictive value, and F1-score, whereas the comparison in terms of RR interval calculation is based on the mean of the relative errors and 90th percentile of the relative errors to indicate the error spread.

Fig. 5.14 (continued)

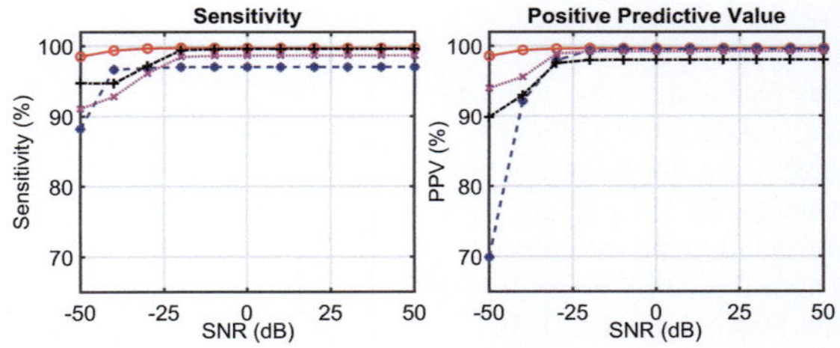

(e) Electrode motion

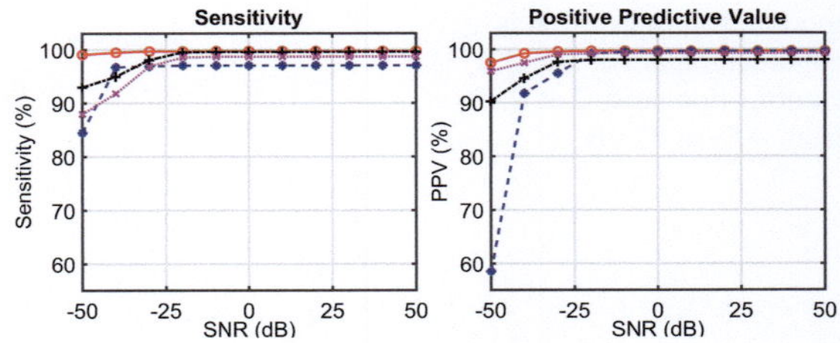

(f) Muscle artifact

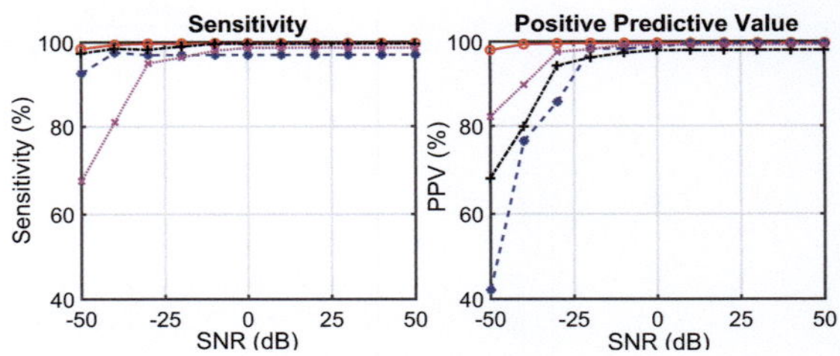

(g) Combination noise 1

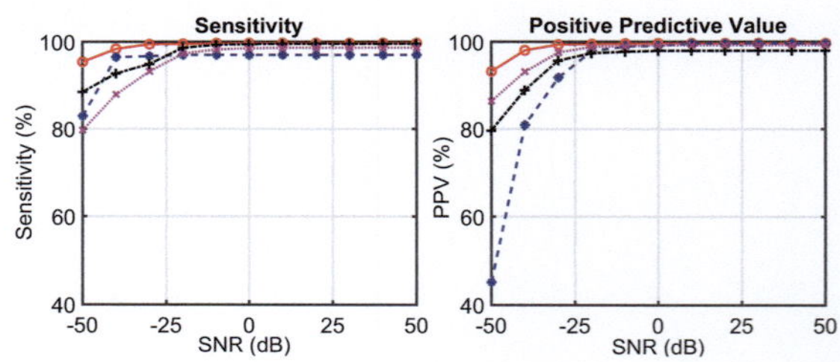

(h) Combination noise 2

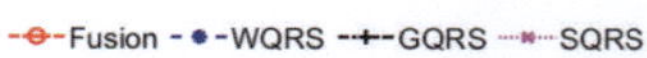

Beat-by-Beat Comparison

Sensitivity (Se), Positive Predictive Value (PPV) or precision, and F1-score (f1) are chosen as metrics for performance comparison. They are calculated as shown in Eqs. (3.9), (3.11), and (3.12).

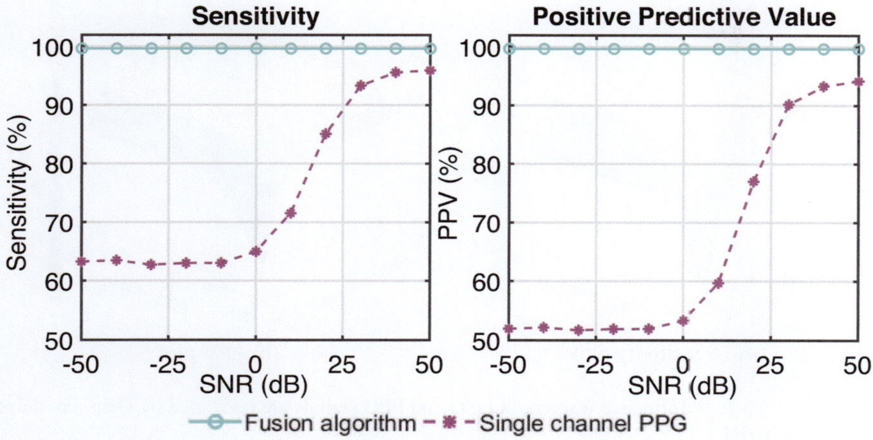

Fig. 5.15 Performance comparison results of the data fusion algorithm against single-channel PPG beat detection algorithm when the PPG signal is corrupted by the addition of Gaussian Noise at SNRs from −50 to 50 dB

From Fig. 5.14, it can be concluded that for the selected dataset, the algorithm consistently yields the same performance; i.e., a sensitivity of 99.69% and a PPV of 99.64%, regardless of the type of noise that corrupts the ECG signal or the SNR level (from −30 to 50 dB). The performance is as expected as the weights used for fusion are based on the signal quality indicators of the input signals. At SNRs lower than −30 dB, the algorithm performs slightly poorly compared to that from −30 dB to 50 dB SNRs. On investigating this per record, it was observed that random peaks in the PPG signal led to few FPs in the analysis window, which caused an actual beat to be missed by the peak detection and search algorithm and led to an increase in FNs. The increase in FPs and the resultant increase in FNs led to lower sensitivity and PPV values as per Eqs. (3.9) and (3.11). This is because of a higher weight being assigned to the PPG signal, regardless of the low signal quality of the PPG signal, due to an even lower ECG signal quality.

From Fig. 5.15, it can be observed that the single-channel PPG beat detection algorithm (fusion algorithm takes only PPG signal as input) is outperformed by the fusion algorithm in all noise scenarios considered when the PPG signal is corrupted by noise and the ECG signal is clean. It is also observed that the fusion algorithm consistently yields the same performance from −30 dB to 50 dB, i.e., a sensitivity of 99.69% and a PPV of 99.64%. The poor performance at low SNRs could be attributed to low quality and random peaks in the clean ECG signal, similar to as discussed above.

The performance of the fusion algorithm when both signals are corrupted by Gaussian noise is shown in Fig. 5.16a and b. It can be observed from the figure that above an SNR of −20 dB, the fusion algorithm exhibits a consistently high sensitivity and PPV. The performance of the fusion algorithm, when the ECG signal is corrupted

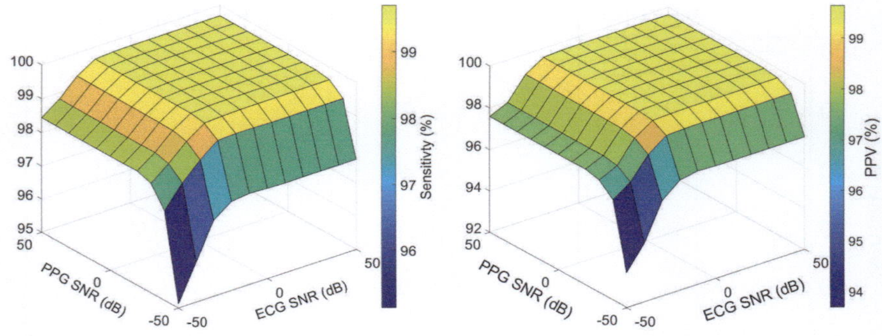

(a) Sensitivity plot (b) PPV plot

Fig. 5.16 Fusion performance when both ECG and PPG signals are corrupted by Gaussian noise at different SNRs

by Combination noise 1 at –50 dB, is found to be quite close to the performance of the fusion algorithm when the ECG signal is corrupted by Gaussian noise at –50 dB (Figs. 5.16a and b). This would theoretically imply that the combination noise has an effect very similar to Gaussian noise. However, the fusion algorithm exhibits a slightly poorer performance when the ECG signal is corrupted by combination noise 2. This could be attributed to the large contributions of muscle artifacts and electrode motion (compared to Combination 1), which are highly non-stationary in nature. This noise combination affects the ECG feature sequence amplitude more heavily than when each noise is considered separately, which leads to its contribution to the final fusion process, although the lincSQI values are small. One might expect the algorithm to exhibit very high sensitivity, PPV, and f1-score when only one channel is noisy, but from Tables 5.1 and 5.2, it can be observed that this is not the case. On investigating the results, it was observed that the algorithm missed a couple of beats at the beginning and end of a signal record due to the algorithm's specificity in marking a beat only if one whole cycle of the signal is available. However, in wearable devices for health monitoring, a continuous stream of data will be available as long as the device is powered on, and therefore this is not a concern. From the plots of sensitivity and positive predictive values in Fig. 5.16a and b, it can be observed that at −50 dB SNR of PPG signal, the performance is quite poor even when the ECG signal SNR varies from −50 to 50 dB. This suggests that optimal performance is achieved only when both signals are available with adequate signal quality, highlighting the importance of data fusion. A key limitation of this approach is that fusion performance relies on the condition that at least one channel provides reasonably clean data.

Table 5.1 Performance of the fusion algorithm, GQRS, SQRS, and WQRS algorithms for heartbeat detection when ECG signal is corrupted by different kinds of noise at −50 and 50 dB SNRs

Noise	SNR (dB)	Algorithm	Se (%)	PPV (%)	f1 (%)	E_{rel}	E_{90}
High frequency noise	−50	Fusion	99.51	99.47	99.49	0.02	0.04
		GQRS	98.56	87.35	92.62	0.20	0.45
		SQRS	87.80	89.37	88.57	0.36	0.87
		WQRS	96.98	98.76	97.86	1.05	2.32
	50	Fusion	99.69	99.64	99.66	0.01	0.02
		GQRS	99.61	98.00	98.79	0.06	0.15
		SQRS	98.64	99.21	98.92	0.12	0.31
		WQRS	97.00	99.58	98.27	0.03	0.09
Low frequency noise	−50	Fusion	99.64	99.58	99.61	0.01	0.03
		GQRS	94.99	97.58	96.27	0.30	0.91
		SQRS	94.44	98.80	96.56	2.17	2.88
		WQRS	97.00	99.03	98.00	1.05	2.38
	50	Fusion	99.69	99.64	99.66	0.01	0.02
		GQRS	99.61	98.00	98.80	0.06	0.15
		SQRS	98.64	99.21	98.92	0.12	0.31
		WQRS	97.00	99.58	98.27	0.03	0.09
50 Hz noise	−50	Fusion	99.02	98.74	98.88	0.04	0.10
		GQRS	90.53	82.88	86.54	0.38	0.89
		SQRS	67.06	81.65	73.64	3.57	5.36
		WQRS	98.33	43.06	59.89	0.85	0.71
	50	Fusion	99.69	99.64	99.66	0.01	0.02
		GQRS	99.61	98.00	98.80	0.06	0.15
		SQRS	98.64	99.21	98.92	0.12	0.31
		WQRS	97.00	99.58	98.27	0.03	0.09
Baseline wander	−50	Fusion	99.39	99.37	99.38	0.02	0.04
		GQRS	90.73	95.22	92.92	0.44	1.15
		SQRS	89.08	97.45	93.07	0.50	1.31
		WQRS	96.92	95.81	96.36	0.97	2.33
	50	Fusion	99.69	99.64	99.66	0.01	0.02
		GQRS	99.61	98.00	98.79	0.06	0.15
		SQRS	98.64	99.21	98.92	0.12	0.31
		WQRS	97.00	99.58	98.27	0.03	0.09

(continued)

Table 5.1 (continued)

Noise	SNR (dB)	Algorithm	Se (%)	PPV (%)	f1 (%)	E_{rel}	E_{90}
Electrode motion	−50	Fusion	98.49	98.55	98.52	0.02	0.04
		GQRS	94.73	89.84	92.22	0.18	0.46
		SQRS	91.08	93.98	92.51	0.22	0.61
		WQRS	88.21	69.90	77.99	0.44	0.19
	50	Fusion	99.69	99.64	99.66	0.01	0.02
		GQRS	99.61	98.00	98.80	0.06	0.15
		SQRS	98.64	99.21	98.92	0.12	0.31
		WQRS	97.00	99.58	98.27	0.03	0.09
Muscle artifact	−50	Fusion	97.71	97.45	97.58	0.03	0.10
		GQRS	92.93	90.25	91.57	0.21	0.55
		SQRS	87.92	95.90	91.74	0.36	0.91
		WQRS	84.44	58.46	69.08	0.51	0.64
	50	Fusion	99.69	99.64	99.66	0.01	0.02
		GQRS	99.61	98.00	98.80	0.06	0.15
		SQRS	98.64	99.21	98.92	0.12	0.31
		WQRS	97.00	99.58	98.27	0.03	0.09
Combination 1	−50	Fusion	98.43	98.02	98.22	0.08	0.12
		GQRS	97.43	68.08	80.15	1.04	2.07
		SQRS	67.55	82.15	74.14	0.76	1.18
		WQRS	97.00	99.58	98.27	0.03	0.09
	50	Fusion	99.69	99.64	99.66	0.01	0.02
		GQRS	99.61	98.00	98.80	0.06	0.15
		SQRS	98.64	99.21	98.92	0.12	0.31
		WQRS	97.00	99.58	98.27	0.03	0.09
Combination 2	−50	Fusion	95.43	93.23	94.32	0.21	0.55
		GQRS	88.52	79.62	83.83	0.84	1.73
		SQRS	79.67	86.42	82.91	0.47	0.93
		WQRS	82.91	45.14	58.45	1.42	2.30
	50	Fusion	99.69	99.64	99.66	0.01	0.02
		GQRS	99.61	98.00	98.80	0.06	0.15
		SQRS	98.64	99.21	98.92	0.12	0.31
		WQRS	97.00	99.58	98.27	0.03	0.09

RR Interval Comparison

The performance of each of the algorithms in different noise scenarios is compared by calculating the RR interval in beats-per-minute (bpm) from the detected beats and comparing it with the ground truth RR interval. The mean of the relative errors E_{rel} between the RR intervals estimated from the algorithms, $\widehat{RR_i}$, and the reference RR intervals RR_i is used as a metric to quantify the performance of the algorithm. E_{rel} is

Table 5.2 Performance of the fusion algorithm and single-channel PPG heartbeat detection algorithm when the PPG signal is corrupted by Gaussian of noise at −50 dB and 50 dB SNRs

Noise	SNR (dB)	Algorithm	Se (%)	PPV (%)	f1 (%)	E_{rel}	E_{90}
Gaussian Noise	−50	Fusion	99.62	99.62	99.12	0.01	0.02
		PPG alone	95.15	98.76	98.95	0.12	0.32
	50	Fusion	99.69	99.64	99.66	0.01	0.02
		PPG alone	96.75	99.58	98.14	0.07	0.18

calculated as shown in Eq. (5.5), where N is the number of RR intervals in a record and E_{rel} is calculated per record.

$$E_{\text{rel}} = \frac{1}{N} \sum_i \frac{\left|\widehat{RR_i} - RR_i\right|}{RR_i} \tag{5.5}$$

The 90th percentile of the relative errors E_{90} provides a measure of the error spread and is calculated for each record for each noise condition as shown in Eq. 5.6, where $P_{90}(\cdot)$ indicates the 90th percentile operator.

$$E_{90} = P_{90}\left(\left\{\frac{\left|\widehat{RR_i} - RR_i\right|}{RR_i}\right\}\right) \tag{5.6}$$

The E_{rel} and E_{90} values mentioned in Tables 5.1 and 5.2 are the means of all the records considered in testing performance. From the tables, it can be observed that the fusion algorithm exhibits a mean E_{rel} of 0.01 at 50 dB and not more than 0.04 at −50 dB when the ECG signal is corrupted by noise and the PPG signal is clean. The measure of error spread E_{90} in low SNR ranges in all noise scenarios is found to be as low as 0.04 and not exceeding 0.1. From Table 5.2, it can be observed that the fusion algorithm exhibits an E_{rel} no greater than 0.01 at both SNR extremes and E_{90} values not exceeding 0.02. From the sensitivities, PPVs, E_{rel}, and E_{90} values mentioned in Table 5.2, it can be concluded that the fusion algorithm with both ECG and PPG signals consistently outperforms the single-channel PPG beat detection algorithms. From the values of E_{rel} and E_{90} in Tables 5.1 and 5.2, it can be observed that the relative error and error spread are less when the PPG signal is corrupted by noise and the ECG signal is clean. This could be attributed to the fast-rising peak and magnitude of the ECG QRS complex, which makes it easier to detect the peak corresponding to a heartbeat from an ECG signal than from a PPG signal. But regardless of this observation, the fusion algorithm outperforms all the single-channel algorithms discussed.

Bland Altman Plots

Figure 5.17 shows the Bland Altman plots for when the ECG signal is corrupted by muscle artifacts at −50 dB SNR. The Bland Altman plot has the reference RR on the x-axis instead of the mean of the reference RR and calculated RR from the algorithm for ease of comparison [44]. The y-axis shows Δ, which is equal to the difference between the RR interval value calculated by the algorithm and the ground RR interval value in beats-per-minute (bpm) calculated as:

$$\Delta = \widehat{RR[n]} - RR[n]. \tag{5.7}$$

This scenario is chosen as the fusion algorithm exhibits the worst performance when the ECG signal is corrupted by muscle artifacts, as can be observed from Table 5.1. The plots are generated by calculating the inter-beat-interval to calculate the heart rate in beats per minute (bpm) from the beats detected by the fusion algorithm, along with GQRS, SQRS, and WQRS algorithms. The plots are generated from all records considered in this dataset. From Fig. 5.17, it can be observed that there is no noticeable bias visible in the mean of the differences in the case of the fusion algorithm (0.44 bpm). However, this is not so in the case of the other 3 algorithms, as there is a noticeable bias indicated by the line showing the mean of the errors (2.74 bpm, –7.17 bpm, and 39.31 bpm for GQRS, SQRS, and WQRS algorithms respectively). When the PPG signal is corrupted by Gaussian noise and the ECG signal is clean, the fusion algorithm exhibits a bias of 0.2170 bpm, while the single-channel PPG signal exhibits a bias of 19.5609 bpm. In Fig. 5.17, it can be observed that the limits of agreement (99th percentile of differences, P_{99} and 1st percentile of differences, P_1) of the fusion algorithm are at the least 20 times narrower compared to that of the ECG GQRS, SQRS, and WQRS algorithms, indicating that the errors in RR intervals in beats-per-minute are much lower in case of the fusion algorithm when compared to other single-channel, state-of-the-art ECG beat detection algorithms.

To summarize, data fusion is carried out and the decision on the point of occurrence of a beat is made in the wavelet transform space. This is possible by using the discrete stationary wavelet transform to preserve temporal information and maintaining only the frequency bands necessary for peak detection, thereby ensuring that any low or high-frequency noises do not affect the peak detection algorithm. This helps in removing the signal filtering stages and therefore reduces hardware requirements on a wearable device. The fusion algorithm exhibits a performance as shown in Tables 5.1 and 5.2 at the two extreme SNRs considered in testing the algorithm, which can be compared against single-channel beat detection algorithms. The algorithm is found to consistently outperform the single-channel state-of-the-art algorithms in terms of sensitivity and PPV in nearly all SNRs and all noise scenarios when a beat-by-beat comparison against the ground truth is carried out. The relative RR interval error and the error spread of the fusion algorithm are found to be much lower than the single-channel algorithms as seen in Tables 5.1 and 5.2. From all the quality metrics considered, it can be concluded that the transform space, temporal,

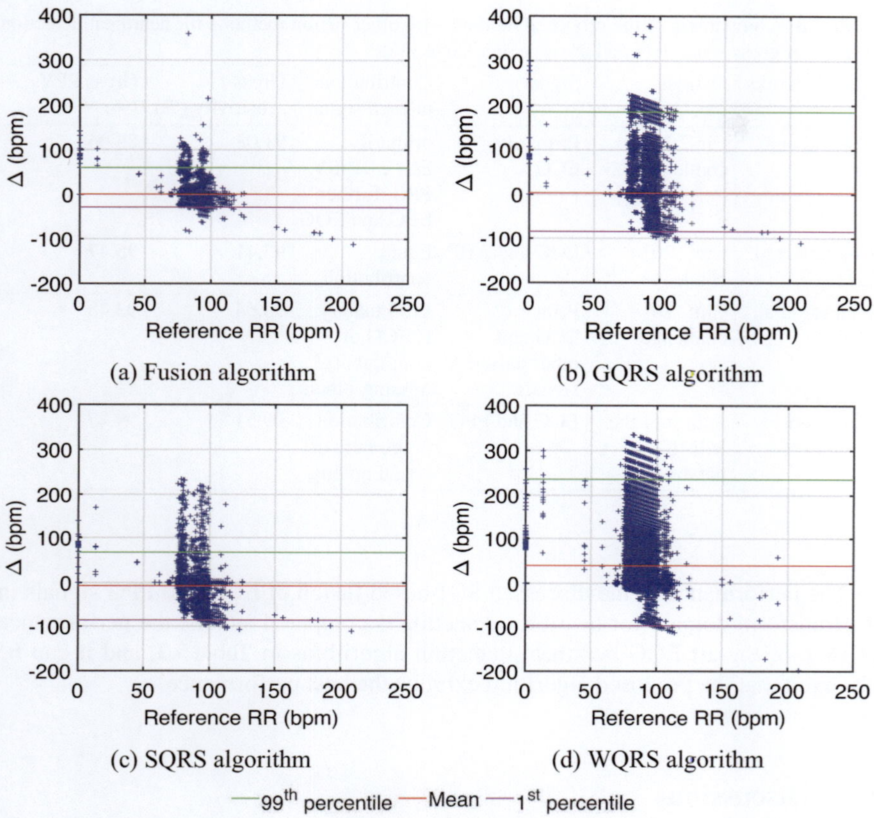

Fig. 5.17 Bland Altman plot showing the agreement between the RR interval information obtained from the **a** Fusion, **b** GQRS, **c** SQRS, and **d** WQRS algorithms and the reference RR interval for all records when the ECG signal is corrupted by muscle artifacts such that the signal SNR is −50 dB. The limits of agreements of the differences are shown through P_{99} and P_1, which are also shown in the figure to indicate where 98% of the estimates lie. The values recorded are in beats-per-minute (The y-axis limits of the plots have been adjusted for ease of comparison and therefore some outliers are not visible in the figures)

complementary, multimodal fusion algorithm presented in this chapter is a valid solution for improving accuracy in heartbeat detection algorithms in IoT wearable devices.

Table 5.3 Comparison of the proposed method with other fusion methods for heartbeat detection in terms of gross sensitivity and positive predictive value

Related works	Dataset	Signals	Contribution of each signal	Gross sensitivity (%)	Gross PPV (%)
Galeotti et al. [20]	CinC 2014 challenge	Primarily ECG	primarily: ECG, BP, SV, PPG, further EEG and EOG	91.08	86.96
Chandra et al. [21]	CinC 2014 challenge	ECG and ABP	Equal contribution	93.41	95.47
Pangerc et al. [19]	CinC 2014 challenge	Primarily ECG and other pulsatile signals	ECG mainly. If ECG of poor quality/ missing-others	95.64	93.48
This work	Subset of the MIMIC III database	ECG and PPG	Contribution in the ratio of signal quality	99.51	99.47

The performance of the discussed SQI-based fusion of ECG and PPG signals in the transform domain for heartbeat detection is compared against the performance of state-of-the-art ECG heartbeat detection algorithms in Table 5.3, and it can be observed that the proposed algorithm exhibits the best performance.

5.4 Discussions

The proposed ECG-PPG fusion algorithm significantly enhances heartbeat detection accuracy by leveraging wavelet-based signal fusion and signal quality indicator (SQI)-weighted processing. This approach ensures that only reliable signals contribute to peak detection, thereby reducing errors associated with motion artifacts, noise, and low signal amplitude. However, its performance is dependent on at least one clean input channel, meaning that in scenarios where both ECG and PPG signals are heavily corrupted, accuracy may degrade. Additionally, while fusion improves robustness, it requires greater computational resources and sensor integration, which may limit feasibility in ultra-low-power wearable devices. The peak detection step currently relies on the Pan-Tompkins algorithm, chosen for its proven effectiveness, but alternative algorithms tailored to fused signals remain unexplored. Another limitation is the use of synthetic Gaussian noise to evaluate PPG corruption, which does not fully replicate real-world motion artifacts, a dominant noise source in wearable PPG sensing.

Future improvements should focus on expanding multi-sensor fusion beyond ECG and PPG to include signals like peripheral oxygen saturation for oxygen saturation trends, and ballistocardiography for non-contact monitoring. While such multimodal fusion could further enhance noise resilience, it also increases computational complexity, requiring a careful balance between accuracy and energy efficiency. Implementing adaptive, artificial intelligence (AI)-driven fusion models that dynamically adjust fusion strategies based on sensor conditions, as discussed in the previous chapter, could improve performance across diverse user scenarios. Additionally, hardware optimization is needed to support real-time processing on wearables, potentially leveraging cloud-based analytics for long-term monitoring. Beyond cardiac monitoring, this framework could be extended to atrial fibrillation detection and sleep apnea detection, as discussed in subsequent chapters, making wearables more clinically useful.

5.5 Conclusions

This chapter presents a multimodal fusion of ECG and PPG signals for ambulatory heartbeat detection. The proposed method uses an SQI-weighted technique for the fusion of ECG and PPG signals unlike the fusion algorithms previously discussed in the literature. A generalized signal quality index-based weighted fusion for heterogeneous signals in the wavelet transform domain is proposed to ensure that only signals of reliable signal quality can contribute to decision-making. The work proposes the use of stationary wavelet transform to ensure that only relevant frequency bands contribute to decision-making and eliminates the need for a signal filtering stage while preserving temporal information that are important to detect the exact beat positions. Further steps to improve the performance of the algorithm may involve fusion for noise reduction from heterogeneous signal inputs from a wearable device. The performance of the proposed fusion algorithm was tested at different SNRs on the addition of various kinds of noise from −50 to 50 dB. The fusion algorithm yielded a sensitivity of 99.69%, a positive predictive value of 99.64%, E_{rel} of 0.01, and E_{90} of 0.02 from −30 to 50 dB SNRs for all noise scenarios considered. On comparing the performance of the proposed fusion algorithm to some state-of-the-art single-lead ECG beat detection algorithms and a single-lead PPG beat detection algorithm, the fusion algorithm yields better results in almost all noise scenarios considered. This approach is also limited by the fact that at least one information channel is reasonably clean. The proposed algorithm used two-channel inputs, but the method of SQI-weighted fusion can be extended to a multitude of multimodal signals to make the inference process robust to noise and outliers. The possibility of using a similar fusion framework for determining respiration rate using ECG and PPG signals could also be explored. The proposed fusion algorithm is an example of complementary, feature-level, and FEI-FEO fusion.

References

1. M.-C. Chambrin, Alarms in the intensive care unit: How can the number of false alarms be reduced? Crit. Care **5**(4), 184–188 (2001)
2. J. Pan, W.J. Tompkins, A real-time QRS detection algorithm. IEEE Trans. Biomed. Eng. **BME-32**, 230–236 (1985)
3. B.U. Kohler, C. Hennig, R. Orglmeister, The principles of software QRS detection. IEEE Eng. Med. Biol. Mag. **21**, 42–57 (2002)
4. W. Zong, G. Moody, D. Jiang, A robust open-source algorithm to detect onset and duration of QRS complexes. Comput. Cardiol. **2003**, 737–740 (2003)
5. W.A.H. Engelse, C. Zeelenberg, A single scan algorithm for QRS-detection and feature extraction. Comput. Cardiol. **6**, 37–42 (1979)
6. W.-L. Chin, C.-C. Chang, C.-L. Tseng, Y.-Z. Huang, T. Jiang, Bayesian real-time QRS complex detector for healthcare system. IEEE Internet Things J. **6**(3), 5540–5549 (2019)
7. C.J. Deepu, Y. Lian, A joint QRS detection and data compression scheme for wearable sensors. IEEE Trans. Biomed. Eng. **62**(1), 165–175 (2015)
8. T. Jeon, J. Yu, W. Pedrycz, Robust detection of heartbeats using association models from blood pressure and EEG signals. Biomed. Eng. Online **15**, 7 (2016)
9. A. Henriksen, M. Mikalsen, A. Woldaregay, M. Muzny, G. Hartvigsen, L. Hopstock, S. Grimsgaard, Using fitness trackers and smartwatches to measure physical activity in research: analysis of consumer wrist-worn wearables. J. Med. Internet Res. **20**, e110 (2018)
10. H. Sharma, Heart rate extraction from PPG signals using variational mode decomposition. Biocybern. Biomed. Eng. **39**(1), 75–86 (2019)
11. M. Elgendi, On the analysis of fingertip photoplethysmogram signals. Curr. Cardiol. Rev. **8**(1), 14–25 (2012)
12. A. Garde, W. Karlen, J.M. Ansermino, G.A. Dumont, Estimating respiratory and heart rates from the correntropy spectral density of the photoplethysmogram. PLoS ONE **9** (2014)
13. L.M. Feehan, J. Geldman, E.C. Sayre, C. Park, A.M. Ezzat, J.Y. Yoo, C.B. Hamilton, L.C. Li, Accuracy of Fitbit devices: Systematic review and narrative syntheses of quantitative data. JMIR Mhealth Uhealth **6**, e10527 (2018)
14. G. De Morais Borges, V. Brusamarello, Bayesian fusion of multiple sensors for reliable heart rate detection, in *2014 IEEE International Instrumentation and Measurement Technology Conference (I2MTC) Proceedings* (2014), pp. 1310–1313
15. Q. Zhang, D. Zhou, X. Zeng, Highly wearable cuff-less blood pressure and heart rate monitoring with single-arm electrocardiogram and photoplethysmogram signals. BioMed. Eng. OnLine **16**, 23 (2017)
16. V. Nathan, R. Jafari, Particle filtering and sensor fusion for robust heart rate monitoring using wearable sensors. IEEE J. Biomed. Health Inf. **22**(6), 1834–1846 (2018)
17. W. Karlen, S. Raman, J.M. Ansermino, G.A. Dumont, Multiparameter respiratory rate estimation from the photoplethysmogram. IEEE Trans. Biomed. Eng. **60**(7), 1946–1953 (2013)
18. I. Silva, B. Moody, J. Behar, A. Johnson, J. Oster, G.D. Clifford, G.B. Moody, Robust detection of heart beats in multimodal data. Physiol. Meas. **36**(8), 549–552 (2015)
19. U. Pangerc, F. Jager, Robust detection of heart beats in multimodal records using slope- and peak-sensitive band-pass filters. Physiol. Meas. **36**, 1645–1664 (2015)
20. L. Galeotti1, C.G. Scully1, J. Vicente1, L. Johannesen, D.G. Strauss, Robust algorithm to locate heart beats from multiple physiological waveforms by individual signal detector voting, Physiol. Meas. **36**(8), 1705 (2015)
21. B.S. Chandra, C.S. Sastry, S. Jana, Robust heartbeat detection from multimodal data via CNN-based generalizable information fusion. IEEE Trans. Biomed. Eng. **66**, 710–717 (2018)
22. U. Satija, B. Ramkumar, M. Sabarimalai Manikandan, Real-time signal quality-aware ECG telemetry system for IoT-based health care monitoring. IEEE Internet of Things J. **4**(3), 815–823 (2017)

23. L. Smital, C.R. Haider, M. Vitek, P. Leinveber, P. Jurak, A. Nemcova, R. Smisek, L. Marsanova, I. Provaznik, C.L. Felton, B.K. Gilbert, D.R. Holmes III., Real-time quality assessment of long-term ECG signals recorded by wearables in free-living conditions. IEEE Trans. Biomed. Eng. **67**(10), 2721–2734 (2020)
24. C. Orphanidou, *Signal Quality Assessment in Physiological Monitoring: State of the Art and Practical Considerations* (Springer International Publishing, 2018)
25. W. Karlen, K. Kobayashi, J.M. Ansermino, G.A. Dumont, Photoplethysmogram signal quality estimation using repeated gaussian filters and cross-correlation. Physiol. Meas. **33**(10), 1617–1629 (2012)
26. I. Silva, G. Moody, An open-source toolbox for analysing and processing physionet databases in MATLAB and Octave. J. Open Res. Softw. **2**, e27 (2014)
27. A.L. Goldberger, L.A. Amaral, L. Glass, J.M. Hausdorff, P.C. Ivanov, R.G. Mark, J.E. Mietus, G.B. Moody, C.K. Peng, H.E. Stanley, Physiobank, physiotoolkit and physionet: components of a new research resource for complex physiologic signals. Circulation **101**, e215–e220 (2000)
28. L.S.A. Johnson, T. Pollard, MIMIC-III, a freely accessible critical care database. Sci. Data **3** (2016)
29. E.Y. Ding, D. Albuquerque, M. Winter, S. Binici, J. Piche, S.K. Bashar, K. Chon, A.J. Walkey, D.D. McManus, Novel method of atrial fibrillation case identification and burden estimation using the MIMIC-III electronic health data set. J. Intensive Care Med. **34**(10), 851–857 (2019)
30. S.K. Bashar, Y. Noh, A.J. Walkey, D.D. McManus, K.H. Chon, VERB: VFCDM-based electrocardiogram reconstruction and beat detection algorithm. IEEE Access **7**, 13856–13866 (2019)
31. S.K. Bashar, E. Ding, A.J. Walkey, D.D. McManus, K.H. Chon, Noise detection in electrocardiogram signals for intensive care unit patients. IEEE Access **7**, 88357–88368 (2019)
32. G. Moody, W. Muldrow, R. Mark, A noise stress test for arrhythmia detectors. Comput. Cardiol. **11**, 381–384 (1984)
33. D. Sundararajan, *The Discrete Stationary Wavelet Transform* (Wiley Publication, Singapore, 2015)
34. N. Sivannarayana, D. Reddy, Biorthogonal wavelet transforms for ECG parameters estimation. Med. Eng. Phys. **21**, 167–174 (1999)
35. R. Sukanesh, R. Harikumar, Analysis of photoplethysmography (PPG) signals with motion artifacts (gaussian noise) using wavelet transforms. Int. J. Biomed. Soft Comput. Hum. Sci. Off. J. Biomed. Fuzzy Syst. Assoc. **16**(1), 135–139 (2011)
36. C. Liu, X. Zhang, L. Zhao, F. Liu, X. Chen, Y. Yao, J. Li, Signal quality assessment and lightweight QRS detection for wearable ECG smartvest system. IEEE Internet Things J. **6**(2), 1363–1374 (2019)
37. U. Satija, B. Ramkumar, M.S. Manikandan, A new automated signal quality-aware ECG beat classification method for unsupervised ECG diagnosis environments. IEEE Sens. J. **19**(1), 277–286 (2019)
38. C.J. Deepu, X.Y. Xu, D.L.T. Wong, C.H. Heng, Y. Lian, A 2.3 uw ECG-on-chip for wireless wearable sensors. IEEE Trans. Circuits Syst. II Express Briefs **65**(10), 1385–1389 (2018)
39. B. Paliakaitė, A. Petrėnas, A. Sološenko, V. Marozas, Modeling of artifacts in the wrist photoplethysmogram: application to the detection of life-threatening arrhythmias. Biomed. Signal Process. Control **66**, 102421 (2021)
40. J. Li, A. Ashraf, B. Cardiff, R.C. Panicker, Y. Lian, D. John, Low power optimisations for IoT wearable sensors based on evaluation of nine QRS detection algorithms. IEEE Open J. Circuits Syst. **1**, 115–123 (2020)
41. W. Elmenreich, *An introduction to sensor fusion* (Technische Universität Wien, Institut für Technische Informatik, Technical Report, May2001)
42. Y. Zhang, S. Wei, C.D. Maria, C. Liu, Using Lempel-Ziv complexity to assess ECG signal quality. J. Med. Biol. Eng. **36**(5), 625–634 (2016)
43. Q. Li, C. Rajagopalan, G.D. Clifford, A machine learning approach to multi-level ECG signal quality classification. Comput. Methods Programs Biomed. **117**(3), 435–447 (2014)
44. M. Cecconi, A. Rhodes, J. Poloniecki, G.D. Rocca, R.M. Grounds, Bench-to-bedside review: the importance of the precision of the reference technique in method comparison studies- with specific reference to the measurement of cardiac output. Crit. Care **13**, 201 (2009)

Chapter 6
A Multimodal and Multiresolution Fusion Using Convolutional Neural Networks

Abstract Recent advances in circuit design and sensing technology have made it increasingly feasible to collect data from multiple sensors simultaneously, enabling more accurate and comprehensive inferences in biomedical applications. This chapter presents a novel convolutional neural network (CNN) architecture designed specifically for the fusion of multimodal and multiresolution sensor data—signals acquired at differing sampling frequencies. Unlike conventional approaches that rely on signal padding or resampling to align resolutions, this model directly integrates disparate inputs, supporting high resolution temporal inference without the need for frequency normalization. The model's performance is evaluated for sleep apnea event detection, by fusing three different sensor signals, namely, electrocardiogram (ECG), peripheral oxygen saturation signal (SpO2), and abdominal movement signal, obtained from UCD St. Vincent University Hospital's sleep apnea database. Results show that fusion using all three signals yields an accuracy of 99.72% and a sensitivity of 98.98%. The chapter also introduces a selective dropout mechanism to prevent overfitting to any single high-resolution input and enhance model robustness in the presence of corrupted or missing data. Furthermore, the model exhibits incremental performance gains as more sensor inputs are added, demonstrating its scalability and generalizability. Model pruning techniques for complexity reduction are also explored, highlighting the potential for real-time deployment in resource-constrained environments.

Keywords Multiresolution data fusion · Convolutional neural networks · Sleep apnea detection · Electrocardiogram · Peripheral oxygen saturation · Abdominal movement signals · Selective dropout · Pruning

6.1 Introduction

Fusion of data obtained from multiple sensors can improve detection performance, compared to that of using data from a single sensor source [1, 2]. Fusion techniques have been demonstrated to improve the performance of a task without significant variations in the existing data acquisition setup and with minimal additional compu-

A. John et al., *Deep Learning and Signal-Processing Methods for Multisensor Data Fusion*, https://doi.org/10.1007/978-3-031-96724-5_6

tational and power consumption costs. Therefore, data fusion has become popular in the design of wearable physiological monitoring systems.

With the advances in 2-dimensional convolutional neural networks (2D-CNNs) for tasks such as object detection and image classification, a similar model for 1-dimensional (1D) time-series signals has been widely explored for biomedical applications [3–5]. Compact 1-dimensional convolutional neural networks (1D-CNNs) are less resource-intensive compared to their 2D counterparts, and the filters used in these models are equivalent to a traditional 1D time-series filter [6, 7]. Due to its efficiency and low computational requirements, 1D-CNNs are suitable for event detection and classification tasks from time-series data and are well-suited for fusion algorithms. Often, the different sensors used for physiological signal monitoring have different sampling frequencies, and therefore extraction of temporal information from fused data often requires signal padding, resampling, etc. [8].

Data fusion algorithms utilizing deep learning methods have been explored in the existing literature. Data fusion algorithms are most commonly classified based on the information abstraction level: signal-level, feature-level, or decision-level [9]. In the context of deep learning, a deep network, namely CentralNet, was proposed for the fusion of information from different sensors that could automatically balance the tradeoff between early and late fusion i.e., between the fusion of low-level versus high-level information [10]. However, the issues arising due to different sampling frequencies of time series data were not discussed. Kim and Ghosh proposed a simple convolutional fusion layer, called the latent ensemble layer, that has a structural advantage in dealing with noise, with each source being allowed to provide a different number of channels, and admitting source-specific features to survive even after the fusion procedure [11], but the issues arising due to different sampling frequencies of time series data were not analyzed. A Deep Convolutional Neural Network (DCNN), namely SensorNet, was proposed to classify multimodal and multiresolution time-series signals by generating a time-series image [8]. However, the data was resampled/ padded such that the input to the DCNN would be uniform, leading to the development of a model that does not address issues such as overfitting to signals with the larger sampling frequency. Undersampling the signal from a high sampling rate sensor can lead to the loss of important information. Oversampling/ padding the signal from a low sampling rate sensor could lead to the development of a larger model with more connections, weights, and additional computations. Therefore, traditional time series feature-based fusion algorithms are not suitable for multiresolution fusion tasks. This leads to the contributions of this chapter:

1. Development of a 1D-CNN-based fusion framework for the data-driven fusion of multi-sensor, multimodal data at different sampling frequencies for temporal inferences, without having to resort to padding or resampling, by utilizing signal-based selective dropout method that ensures that the fusion model does not overfit to the signals with higher sampling frequencies (or features from models with more samples in the input), which is applicable to biomedical IoT sensors.

2. A network development methodology that focuses on developing individual well-performing models for each sensor source prior to fusion, so that the fused model developed uses the best features from each sensor source for fusion.
3. Development of a novel multimodal sleep apnea detection technique using the proposed fusion framework that is at a higher detection resolution compared to state-of-the-art, and its performance evaluation.
4. Complexity analysis of the developed fusion models and their optimization using network pruning techniques to reduce computational costs.

The core hypothesis focuses on analyzing whether an event detection model based on 1D-CNNs can fuse multimodal and multiresolution signals without resorting to resampling of the signals, and that the fused models can improve detection accuracies compared to using 1D-CNNs for each signal individually.

The rest of the chapter is organized as follows: Section 6.2 discusses the methodology and experimental design, the fusion stage, and the method to prevent overfitting to signals with high sampling frequencies. Section 6.3 discusses the application of the proposed fusion architecture to the sleep apnea detection problem. Section 6.4 includes the performance of the fusion models developed from three different sensor sources and discussions.

6.2 1D-CNN-Based Fusion for Multi-sensor Data

This section presents the methodology for developing multimodal and multiresolution fusion models. Consider an event classification task from multi-sensor time-series data obtained from different sensor sources, where data from each source can be used to carry out the classification task with acceptable performance. At first, independent 1D-CNN models for achieving the task have to be developed based on each sensor input signal separately. Here, the hyperparameters of each model have to be tuned to design an optimal independent model based on a single sensor source data. These models are free to have any number of convolutional layers, filters, and filter sizes, such that the functions that the models learn are optimal. The number of nodes in the output layer of the model will be specified based on the type of classification. The performances of the independent models are tested on a test set to ensure that the model has acceptable performance when only data from a single sensor source is available.

6.2.1 Data-Driven Fusion Approach

Once all the independent models are able to achieve acceptable performance, the output layers of the models have to be removed, and the flatten layers have to be concatenated together. This concatenated layer should then be attached to a fully connected dense network and then to a final output layer to form the multi-sensor fusion model. The structure of the multi-sensor fusion model is shown in Fig. 6.1, and it works based on a feature-level fusion approach. The same dataset with different sensor signals is used to train the multi-sensor fusion model. All layers except the final fully connected layer are frozen during training to obtain the desired performance-equivalent to using the bottleneck features in transfer learning. The hyperparameters of the fully connected layers are then tuned to optimize the fusion model's performance using the validation set. Finally, the performance of the fusion model is evaluated on the test set.

The objective of the fusion algorithm is to enhance overall model performance. However, in certain scenarios, combining a high-performing model with a lower-performing one may result in the fused model performing worse than the individual models (catastrophic fusion). Since the initial layers are frozen, and training is carried out only on the layers after the concatenated flatten stage, the network should be able to learn the correct set of weights based on which of the signals are more reliable for improving detection accuracies. This is beneficial in scenarios where the individual models perform reasonably well, and the model performance can be improved by combining information from the features learned by the individual models (which are optimal with regard to the learning algorithm for each model), leading to a data-driven approach to fusion and improvements in the performance of the fused model.

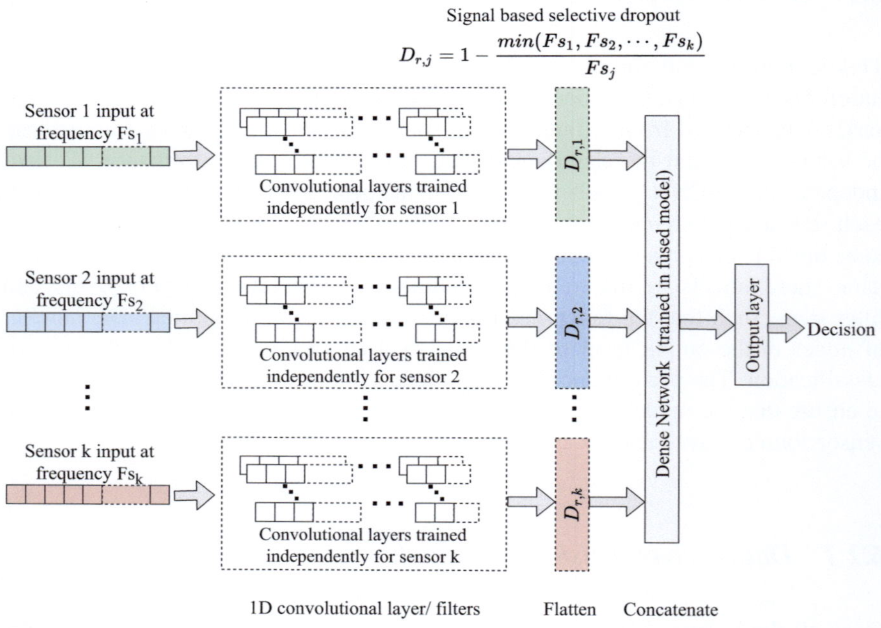

Fig. 6.1 Proposed 1D-CNN-based fusion technique for multi-sensor multiresolution data fusion

This is advantageous in wearable devices as individual sensor features in the flatten stage (as they are already optimal for the single sensor network) can be used at any point in time to carry out individual sensor source inference by using the final weights after the flatten stage (maybe due to power constraints, fusion is not preferred and the wearable device can switch between using fusion methodology or using the best performing individual sensor depending on the battery levels easily due to the shared weights prior to the flatten stage) without significant additional memory or area requirements. Moreover, the fused model training time would also be lower with this approach, with the confidence that the fused model performance would be equal to or better than the individual model performances.

6.2.2 Signal-Based Selective Dropout

Consider the scenario where some sensor sources have higher sampling rates than others, leading to more data samples per second. In this scenario, the decision-making of the fusion model will be dominated by the sensor source with a higher sampling frequency (Fs_{high}) than the sensors with lower sampling frequencies (Fs_{low}). This leads to the overfitting of the model to the data from the sensor source at the highest sampling frequency. Methods to prevent overfitting to training data, such as regularization and dropout, have been previously discussed in the literature [12]. However, methods to prevent overfitting to inputs from a single source, in the case of a multi-source input scenario, have not been explored previously. Here, enabling selective dropout during the training stage at the flatten layer for only the features from the signal source with higher sampling rates, so that the fusion model would not overfit to the higher sampling rate inputs is proposed. The advantage of the signal-based selective dropout is that during the inference stage, all the features in the flatten stage can contribute to the fusion process. On the other hand, undersampling the signal obtained at the higher sampling frequency leads to the dropped signals or features not being able to contribute at all to the model. Moreover, dropout is random for each training cycle, thereby making sure that all the features at the various locations get to contribute during the training process, which is completely different from undersampling the obtained features (which often leads to the same samples/ samples in fixed locations being dropped and not contributing to the fusion stage). Moreover, this is the first method proposed to prevent overfitting to the model with larger number of neurons in the flatten stage as optimizing individual model architectures can lead to a different number of neurons in the flatten layers. For signals with different sampling frequencies but matching network structures, the dropout rate for the model for the signal with a higher sampling frequency should be set to $1 - \frac{Fs_{\text{low}}}{Fs_{\text{high}}}$ to equalize the influence of data samples from all sensor sources. Matched network structures indicate that for two different signals with sampling rates Fs_1 and Fs_2, and the corresponding ratio of the input window lengths to the two networks being $\frac{Fs_1}{Fs_2}$, the output feature maps of each layer of the two networks should have the same ratio of $\frac{Fs_1}{Fs_2}$. In case there is a mismatch in network structure such that the ratio of the neurons in the output layer does not match the ratio $\frac{Fs_{\text{low}}}{Fs_{\text{high}}}$, the dropout ratio D_r should be set as

$$D_r = 1 - \frac{neurons_{\text{low}}}{neurons_{\text{high}}}, \tag{6.1}$$

where $neurons_{\text{low}}$ indicates the number of neurons in the flatten layer of the model for the signal with the lower sampling frequency Fs_{low}, and $neurons_{\text{high}}$ indicates the number of neurons in the flatten layer of the model for the signal with the higher sampling frequency Fs_{high}. In the scenario where the fusion of k different sensors at k different sampling frequencies need to be carried out, the dropout ratio D_r for each of the jth signal model can be calculated as

$$D_{r,j} = 1 - \frac{min(Fs_1, Fs_2, \cdots, Fs_k)}{Fs_j} \forall j \in [1, \cdots, k]. \tag{6.2}$$

Alternatively, the dropout ratio for the jth signal model in terms of the number of neurons in the flatten layer of each of the individual models can be calculated as

$$D_{r,j} = 1 - \frac{min(neurons_1, neurons_2, \cdots, neurons_k)}{neurons_j} \forall j \in [1, \cdots, k], \tag{6.3}$$

where $neurons_j$ corresponds to the number of neurons in the flatten layer of the jth signal model. This is termed signal-based selective dropout. The flow of the proposed fusion methodology is therefore:

1. Develop 1D-CNN models for each sensor input as discussed in Sect. 6.2.
2. Extract and freeze the convolutional layers from all independent models.
3. Calculate the signal-based selective dropout as per Eq. (6.2) or (6.3) and apply to the flatten layers of all the independent models.
4. Concatenate the flatten layers as discussed in Sect. 6.2.1.
5. Attach a fully connected network from the concatenated layer to the output layer.
6. Freeze the layers from independent models, and train the fully connected layers alone for optimal data-driven fusion.

6.3 Application to Sleep Apnea Detection

The abnormal pause in breathing or lowering of the breath rate during sleep is termed as sleep apnea-hypopnea syndrome and is a disorder that affects 10% of middle-aged adults [13]. There is a complete pause in breathing during an apnea event, while a hypopnea event is characterized by a drop in oxygen saturation for at least 10 s due to the reduction in airflow [14]. Overnight polysomnography is the diagnostic tool used to diagnose sleep-related disorders under the supervision of a clinician. Recording polysomnograms for clinical evaluation is costly and involves wearing sensors that are uncomfortable for the patient during sleep, like airflow thermistors at the nose. Therefore, the development of non-intrusive and automatic sleep-apnea detection methods for deployment in wearable devices like smartwatches and smart vests using physiological signals that can be easily acquired without causing discomfort to the patient is of paramount importance. However, in wearable sensing, these signals can be corrupted by electrode motion noise, power-line interference, motion artifacts, electromyographic noise, etc., leading to poor feature extraction, erroneous data interpretation, and false alarms [15]. This can lead to subsequent slowing down of response times due to alarm fatigue [16].

Electrocardiogram is a record of the electrical activity of the heart and can be obtained through non-invasive wearable devices, as discussed in previous chapters. Peripheral oxygen saturation (SpO2) is an estimation of blood oxygen levels and is usually measured with a pulse oximeter, which is non-invasive and is commonly found in smartwatches [17]. Various studies have shown that sleep apnea events can be monitored using ECG and SpO2. An ECG-based sleep apnea detection method using ECG-derived respiration (EDR) that can detect sleep apnea occurrence during a minute with an accuracy of 94.12% was proposed in [18]. Heart rate variability (HRV) features-based methods for sleep apnea detection- from ECG signals with apnea event occurrences in 30-second windows- with an accuracy of 74.85% were proposed in [19]. A SpO2-based apnea detection method that could detect apnea event occurrences in a minute with an accuracy of 85.26% was proposed in [20]. A review of the methods discussed in the literature for sleep apnea detection using deep learning was carried out in [21]. In the literature, it was observed that most studies carry out sleep-apnea detection on a minute-by-minute basis or for a window of 30 s and often use only a single sensor for inference.

Many machine learning-based methods were discussed for sleep apnea detection by combining various features like ECG-derived respiration (EDR) and HRV derived from ECG signals [18, 22–25]. Convolutional neural networks have been used to generate features that can be used for sleep apnea detection in [26]. CNNs were used in combination with long short-term memory networks for sleep apnea detection from ECG signals [26–29]. Deep learning methods for apnea detection from SpO2 signals have been discussed in the literature [20]. Fusion methods that combine multiple sensor signals for the detection of apnea events are also studied. In [30], fuzzy structural algorithms were utilized to identify and characterize apnea and hypopnea episodes using respiratory airflow and SpO2. Fusion methods based on combinations of features derived from ECG signal and SpO2 for a time segment with a resolution of 1 minute were proposed for apnea detection in [14]. A 2D-CNN-based method for sleep apnea detection using SpO2, oronasal airflow, ribcage, and abdominal movement was proposed in [31]. A method to fuse ECG and SpO2 signals through a combination of CNNs and LSTM was proposed in [29]. These studies and their corresponding performance in sleep apnea detection were instrumental in determining which complementary signals should be used for fusion to enhance accuracy in cases where any of the signals are corrupted or lost.

A method for sleep apnea detection with a very high-resolution of one second is proposed in this chapter. The advantage of a high-resolution model is that apnea events can be detected quickly, and the patient can be alerted quickly as loss of oxygen for an extended period of time can lead to potential brain damage [32, 33], and faster detection of apnea events could help to reduce the impact. Additionally, depending on the power constraints of a wearable device (remaining battery power), the resolution can be adjusted in real-time by carrying out predictions once every few seconds. Since the model is trained to infer events on a per-second basis, resolution can be reduced purely by skipping the signal windows to reduce power consumption, and high-resolution processing can be resumed without a large wait time when needed. Apnea detection with a resolution of 1 s is carried out by considering a single signal window of 11 s and employing a 1D-CNN for apnea detection. An 11-second window is used as an apnea event is classified as sleep apnea if the patient does not breathe for at least 10 s. To achieve this, overlapping windows of a duration of 11 s and an overlap of 10 s are generated. A window is assigned the label (apnea/non-apnea) depending on whether the 2nd 1 s signal in that window is apneic or non-apneic.

6.3.1 Dataset and Models

For this work, the sleep apnea database from UCD St. Vincent's University Hospital, Dublin, is used [34]. The database contains polysomnogram data from 25 patients with annotations from a sleep expert. Signals recorded were: electroencephalogram, electrooculogram, submental electromyogram, oro-nasal airflow, ribcage movements, abdomen movements, snoring, body position, as well as the signals of interest in this study- ECG signals at 128 Hz (Fs_{high}) and SpO2 (from finger pulse oximeter) at 8 Hz (Fs_{low}). The ECG and SpO2 signals are suitable for apnea detection as they can be easily incorporated into wearable devices. The labels provided by the sleep experts were used to mark sleep seconds as apneic or non-apneic. The dataset is split into training, validation, and test sets in the ratio of 8:1:1. All the patient records, except the ones without any apnea events (ucddb008, ucddb011, ucddb013, and ucddb018), were used. The class imbalance problem was addressed by oversampling the minority class. It is to be noted that the oversampling of the minority class (apnea events) signal examples is different from signal resampling to correct for resolution differences as discussed in the introduction to this chapter. The oversampling of the minority classes increases the number of examples used to train the 1D-CNN model, while signal resampling increases or decreases the signal segment length of the data fed into the model.

6.3.2 ECG Model

ECG is a record of the electrical activity of the heart. Sleep apnea detection methods based on ECG signals, EDR, and HRV have been proposed and could detect sleep apnea occurrence during a minute [14, 18, 23, 24], and occurrences in a 30-second window [19]. It was found that most studies on sleep-apnea detection using deep learning methods have a resolution of 1 min or 30 s i.e., inferences are made on a minute-by-minute basis or for a window of 30 s [25, 28, 35, 36]. The highest resolution for sleep apnea detection from ECG signals (every 10 s) was studied by Urtnasan et al. [35].

The ECG signal-based model uses a 1D-CNN that has 1408 (128 samples per second) nodes in the input layer. The samples are standard normalized prior to the input layer and the input layer is followed by a batch-normalization stage. The 1D-CNN uses 3 convolution layers with 3, 50, and 30 filters of size 100, 10, and 30, respectively. The stride of the first convolution layer of 3 filters was set to 2, while the stride was maintained as 1 for the other two convolution layers. The flow diagram of the neural network is shown in Fig. 6.2. In the model, all pooling layers use the maximum pooling method, and all activations use the ReLU activation function. The fully connected layers implement weight dropout with a probability of 0.25 during training. The model optimization used binary cross-entropy as the loss function with the Adam optimizer. Adam can be looked at as a combination of the RMSprop that uses the squared gradients to scale the learning rate individually for each parameter (2nd moment) and by using a moving average of the gradient instead of the gradient itself, like in stochastic gradient descent with momentum. The advantage of the Adam optimizer is that it foregoes the need for a learning rate scheduler and is robust to any reasonably bounded learning rate for the problem at hand. The learning rate was fixed at 0.001, and the softmax activation function was applied to the output layer.

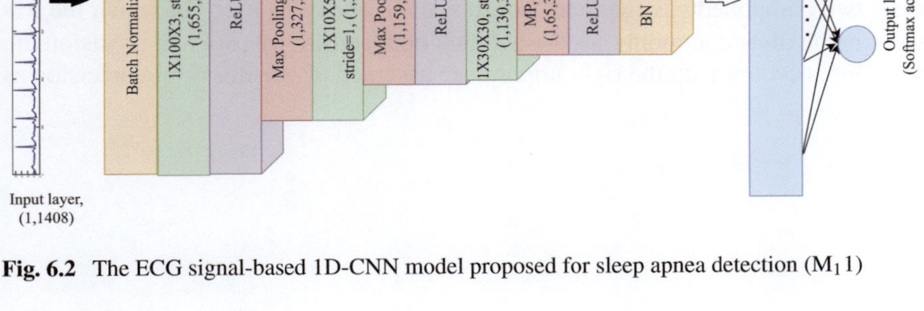

Fig. 6.2 The ECG signal-based 1D-CNN model proposed for sleep apnea detection ($M_1$1)

Table 6.1 ECG signal model parameters for sleep apnea detection

Layer	Kernel		Output shape	# Parameters
	Size	# kernels		
Batch normalization	–	–	1408,1	4
Convolution	100	3	655,3	303
RELU Activation	–	–	655,3	0
Maxpooling	–	–	327,3	0
Convolution	10	50	318,50	1550
Maxpooling	–	–	159,50	0
RELU Activation	–	–	159,50	0
Convolution	30	30	130,30	45030
Maxpooling	–	–	65,30	0
RELU Activation	–	–	65,30	0
Batch normalization	–	–	65,30	120
Flatten	–	–	1950	0
Dropout	–	–	1950	0
Dense, Softmax	–	–	2	3902

The model considered has a total of 50,909 parameters. This model is referred to as $M_1 1$ (1 in the subscript indicates a single sensor model). The parameters of the model are shown in Table 6.1.

The model $M_1 1$ was trained over the full training set and simultaneously validated on the validation set for each epoch, where a true positive stands for an accurately detected apnea event. Validation callback was used to determine the best weights during training based on which set of weights exhibited the highest validation accuracy. The model performance is evaluated based on the Sensitivity (Se), Specificity (Sp), and Accuracy (Ac) that are calculated as shown in Eqs. (3.9), (3.10), and (3.8).

However, due to the huge class imbalance with 50368 test set examples of non-apnea events and just 1368 apnea event samples, missing a few apnea events (FNs) leads to very low sensitivity values, although the overall accuracy doesn't get affected much due to a large number of true negatives. Therefore, a metric that can capture these two components is required. For this, the F_β-score is used- of which the F1-score is the most commonly used- such that both sensitivity (Se) and precision are taken into account with the right importance assigned to sensitivity and precision as

Table 6.2 Comparison of the performance of the proposed CNN model for apnea detection with state-of-the-art ECG record-based sleep apnea detection algorithms

Article	Urtnasan et al. [35]	Wang et al. [36]	Dey et al. [28]	Chen et al. [25]	This work
Dataset	Samsung medical center dataset	UCD apnea database	Physionet Apnea ECG database	Physionet Apnea ECG database and UCD apnea database	UCD apnea database
Resolution	10 s	1 min	1 min	Variable resolution upwards of 30 s	1 s
Method	CNN, 6 layers	CNN, 2 layers	CNN, 3 layers	Support vector machines with selected features	1D-CNN $M_1 1$
Accuracy	96.00%	71.80%	98.91%	92.87%	99.56%
Sensitivity	96.00%	26.6%	97.82%	97.41%	96.05%

is required. The precision/ Positive Predictive Value (PPV) is calculated as shown in Eq. (3.11). The F_β- score is calculated from Se and PPV as shown in Eq. (6.4).

$$F_\beta - score(\%) = \frac{(1+\beta^2).\text{PPV.Se}}{\beta^2.\text{PPV} + \text{Se}} \times 100 \tag{6.4}$$

In this case, correctly identifying true positives (sensitivity) is weighted β times over reducing the number of false positives or false negatives (precision). Since there are approximately 36 times more non-apnea events than apnea events, and β is set to $\sqrt{36} = 6$, and therefore this metric is called the F6-score (F6).

The model $M_1 1$ exhibited an F6-score of 95.82%, an accuracy of 99.56%, a specificity of 99.66%, and a sensitivity of 96.05% on the test set consisting of data from all the patients.

To ensure that the obtained ECG signal-based model is well-performing, it is compared with a few state-of-the-art algorithms that use ECG signals, as shown in Table 6.2. It can be observed from Table 6.2 that the proposed model $M_1 1$ exhibits the best performance in terms of accuracy and sensitivity, at the highest resolution of 1 s. It should be noted that 1D-CNN models are computationally expensive. Therefore, methods to reduce the computational complexity of the ECG-based sleep apnea detection model through pruning [37] and binarization [38] is discussed in Appendix B.1 in the appendix of this book.

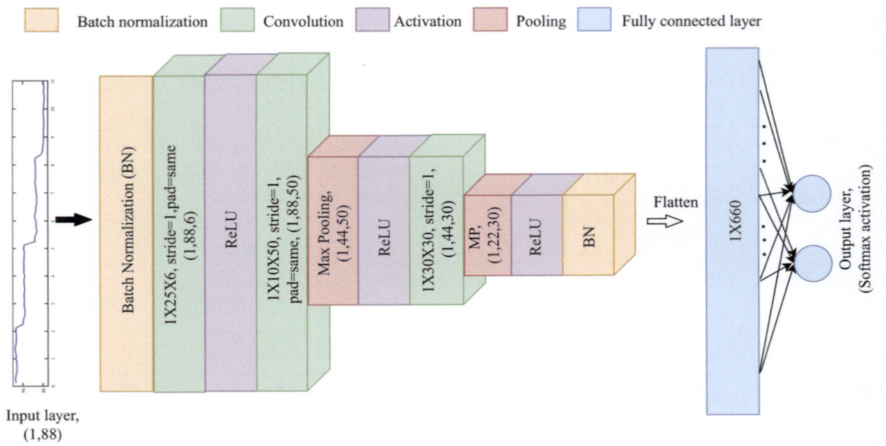

Fig. 6.3 The SpO2 signal-based 1D-CNN model proposed for sleep apnea detection ($M_1 2$)

6.3.3 SpO2 Model

Methods for analyzing obstructive sleep apnea events from pulse oximeter data were reviewed in [39]. It was found that most studies on sleep-apnea detection using deep learning methods have a resolution of 1 minute, i.e., inferences are made on a minute-by-minute basis [14, 20, 22]. Moreover, SpO2-based sleep apnea detection algorithms are often accompanied by other signals to improve inferences [14, 31]. The highest resolution of sleep apnea detection from SpO2 signals (in combination with other signals) was explored in [31] with a resolution of 1 s and with a performance accuracy of 79.61%.

The 1D-CNN SpO2 model has 88 (8 samples per second) nodes in the input layer and the input layer is followed by a batch-normalization stage. The 1D-CNN uses three convolution layers with 6, 50, and 30 filters of size 25, 10, and 30, respectively. The strides of all the convolutional layers were maintained as 1, and zero padding was used. The flow diagram of the neural network is shown in Fig. 6.3. In the model, all pooling layers use the maximum pooling method, and all activations use the ReLU activation function. The fully connected layers implement weight dropout with a probability of 0.25 during training. The model optimization used binary cross-entropy as the loss function with the Adam optimizer, and the softmax activation function was applied to the output layer. The model considered has a total of 27,182 parameters. This model is referred to as $M_1 2$ (1 in the subscript indicates a single sensor model). The parameters of the model are as shown in Table 6.3.

Model $M_1 2$ was trained over the full training set and simultaneously validated on the validation set for each epoch. Validation callback was used to determine the best weights during training based on which set of weights exhibited the highest

Table 6.3 SpO2 signal model parameters for sleep apnea detection

Layer	Kernel		Output shape	# Parameters
	Size	# kernels		
Batch normalization	–	–	88,1	4
Convolution	25	6	88,6	156
RELU Activation	–	–	88,6	0
Convolution	–	–	88,50	3050
Maxpooling	10	50	44,50	0
RELU Activation	–	–	44,50	0
Convolution	–	–	44,30	22530
Maxpooling	15	30	22,30	0
RELU Activation	–	–	22,30	0
Batch normalization	–	–	22,30	120
Flatten	–	–	660	0
Dropout	–	–	660	0
Dense, Softmax	–	–	2	1322

validation accuracy. $M_1 2$ exhibited an F6-score of 82.86%, an accuracy of 97.08%, a specificity of 97.42%, and a sensitivity of 84.65% on the test set.

To ensure that the obtained SpO2 signal-based model is well-performing, the model performance is compared against the performances of a few state-of-the-art algorithms that use SpO2 as shown in Table 6.4. It can be observed from Table 6.4 that the proposed model $M_1 2$ exhibits the best performance in terms of accuracy and comparable performance in terms of sensitivity, at the highest resolution of 1 s. It should be noted that 1D-CNN models are computationally expensive. Therefore, methods to reduce the computational complexity of the ECG-based sleep apnea detection model through pruning [37] and binarization [38] is discussed in Appendix B.2 in the appendix of this book.

For the individual signal models, the ECG and SpO2 models discussed above are used. Since the number of neurons in the flatten layer of the ECG model was 1950 and that for the SpO2 model was 660, the dropout rate for the features extracted from the ECG model in the flatten layer for the fusion model was fixed at 0.67. The flattened layers from the two models are concatenated and batch-normalized prior to the output layer, which uses a softmax activation.

Table 6.4 Comparison of the performance of the proposed CNN model for apnea detection with state-of-the-art SpO2 record-based sleep apnea detection algorithms

Article	Mostafa et al. [20]	Almazayadeh et al. [22]	Xie et al. [14]	Cen et al. [31]	This work
Dataset	Physionet Apnea ECG database and UCD apnea database	Physionet Apnea ECG database	UCD apnea database	UCD apnea database	UCD apnea database
Signal	SpO2	SpO2	SpO2 and ECG	SpO2, oronasal airflow, ribcage and abdominal movement	SpO2
Resolution	1 min	1 min	1 min	1 s	1 s
Method	3 layer deep belief network	Multilayer neural network	Bagging with RepTree	2D CNN	1D-CNN, $M_1 2$
Accuracy	97.64 %	93.30 %	84.40 %	79.61%	97.08 %
Sensitivity	78.75 %	87.50 %	79.75 %	–	84.65 %

6.4 Results

The total number of parameters from the fused model using ECG and SpO2 signals is 88,529, out of which 78,087 parameters are from the convolution layers and are frozen. The model was trained using the Adam optimizer. The model was trained over the full training set and simultaneously validated on the validation set for each epoch, where a true positive stands for an accurately detected apnea event. Validation callback was used to determine the best weights during training based on which set of weights exhibited the highest validation accuracy.

The ECG signal-based model ($M_1 1$) was found to exhibit an F6-score of 95.82%, a sensitivity of 96.05%, and a specificity of 99.75%. The SpO2 signal-based model ($M_1 2$) was found to exhibit an F6-score of 82.86%, a sensitivity of 84.65%, and a specificity of 97.42%. The performance of the SpO2 model is not as good as the ECG signal-based model, however, the overall performance of the fused model should be better than that of the individual models. The fused model ($M_2 1$ (the 2 in the subscript indicates a two-sensor model)) was found to exhibit an F6-score of 97.20%, a sensitivity of 97.44%, and a specificity of 99.68%. On comparing the performance of the fusion model with the SpO2 alone model, there is a significant performance improvement. However, when comparing the performance of the fusion model with the ECG signal model, there is only a slight improvement in overall performance. The performance of the fusion model ($M_2 1$) is compared against the performance of a fusion model without the dropout layer ($\tilde{M}_2 1$) applied to the sensor model with the higher sampling frequency/ higher number of neurons in the flatten layer as discussed in Sect. 6.2.2. In the case where the possibility of overfitting to the ECG signal-based model is ignored, the ECG+Spo2 fused model ($\tilde{M}_2 1$) exhibited an F6-score of only 96.86%, which is poorer than the model that accounted for the possibility of overfitting to the ECG based model ($M_2 1$), thereby proving that the selective dropout method is suitable to prevent overfitting. However, this is not evident in the case where both signals are clean. Overfitting to the sensor source with a higher sampling rate becomes more pronounced in the case where the signal sources are corrupted by noise, which is a major challenge in wearable health monitoring. Severe motion artifacts, electrode motion noise, and electromyographic noise can affect data obtained from wearable devices. The main advantage of fusion methods becomes evident when any of the sensor sources are corrupted by noise, which is discussed in the next subsection.

6.4.1 Performance Evaluation with Noisy Data

To analyze the performance of fusion models with noisy data, simulated scenarios with noisy signal windows are employed, where all the samples in a noisy signal window are corrupted by noise. This is achieved by adding –20 dB white Gaussian noise to the signal segments. The approach of using white Gaussian noise is to capture all noise frequencies, although, during sleep, the low-frequency noises would be dominant. However, as per the central limit theorem, the sum of multiple independent distributions tends to a Gaussian distribution, and therefore white Gaussian noise is added to the signals to study the performance in noisy scenarios. For the training and validation set, all the signal samples in a window in 20% of the ECG signal set are noisy while the corresponding SpO2 window set are clean, all the signal samples in a window in 20% of the SpO2 signal set are noisy while the corresponding ECG window set are clean, and all the signal samples in a window in 20% of both SpO2 and ECG signal set are noisy for both apnea and non-apnea events. For the test set, all the signal samples in a window in 11.42% of the ECG window set are noisy while the corresponding SpO2 window set are clean, all the signal samples in a window in 11.42% of the SpO2 window set are noisy while corresponding ECG window set are clean, and all the signal samples in a window in 11.42% of both SpO2 and ECG window set are noisy for both apnea and non-apnea events.

The individual models for ECG and SpO2 were retrained using this dataset with simulated noise and fused as discussed in Sect. 6.3.1. The ECG signal-based model $M_{1,n}1$ (the 'n' in the subscript indicates that the model is generated for a dataset containing noisy samples) was found to exhibit an F6-score of 74.80%, a sensitivity of 74.85%, and a specificity of 99.24%, and the SpO2 signal-based model ($M_{1,n}2$) was found to exhibit an F6-score of 72.22%, a sensitivity of 77.85%, and a specificity of 91.56%. The fused model with selective dropout ($M_{2,n}1$) was found to exhibit an F6-score of 82.3%, a sensitivity of 85.01%, and a specificity of 96.33%. From the results, it can be observed that the F6-score and sensitivity are significantly higher for the fusion model compared to the ECG and SpO2 models. A fused model without the selective dropout stage ($\tilde{M}_{2,n}1$) was found to exhibit an F6-score of 81.55%, a sensitivity of 83.41%, and a specificity of 97.26%. From the performance of the ECG and SpO2 models as well as the fused model without selective dropout, it can be seen that the models are highly biased toward the majority class, even with minority class oversampling. These observations indicate that when the signals are corrupted by noise, important features that can be learned on clean signals are not learned by the model, and in such scenarios, the fusion algorithm proposed in this chapter will be useful in tackling this issue.

6.4.2 Generalization to Multiple Sensors

To expand and generalize the proposed method, the fusion model is generalized to include a third sensor source. Here, the abdominal movement signals are considered, as the sensors used to obtain abdominal movement signals can be easily incorporated in a wearable smart vest. Abdominal movements are found to be highly reliable parameters for the sleep apnea detection problem [40–42]. Lin et al. proposed an adaptive nonharmonic model to model thoracic and abdominal movement signals to design features for sleep apnea events [40]. In [41], wavelet domain features from abdominal, chest, and nasal way signals are used to detect obstructive sleep apnea events using ensembles. In [42], a measuring module integrating abdominal and thoracic triaxial accelerometers, SpO2, and an ECG sensor was devised, and a long short-term memory recurrent neural network model was proposed to classify four types of sleep breathing patterns. The abdomen movement signals recorded using strain gauges from the UCD St. Vincent's University Hospital's sleep apnea database [34] are used here. The dataset is split into training, validation, and test sets in the ratio of 8:1:1. The class imbalance problem was addressed by oversampling the minority class. A 1D-CNN model for the abdominal signal was developed, which is discussed next.

1D-CNN Model Based on Abdominal Signal

The 1D-CNN model developed for abdominal movement signals for sleep apnea detection is as shown in Fig. 6.4, and the model parameters are detailed in Table 6.5. The model optimization used binary cross-entropy as the loss function with the Adam optimizer, and the training procedure was similar to that used to generate the ECG- and SpO2-based models. The weights and biases were selected during training by using a validation callback method that stored the best parameters that exhibited the highest accuracy on the validation set. The abdominal movement-based single sensor model is referred to as $M_1 3$.

To ensure that the obtained abdominal signal-based model is well-performing, the model performance is compared against a few state-of-the-art algorithms that

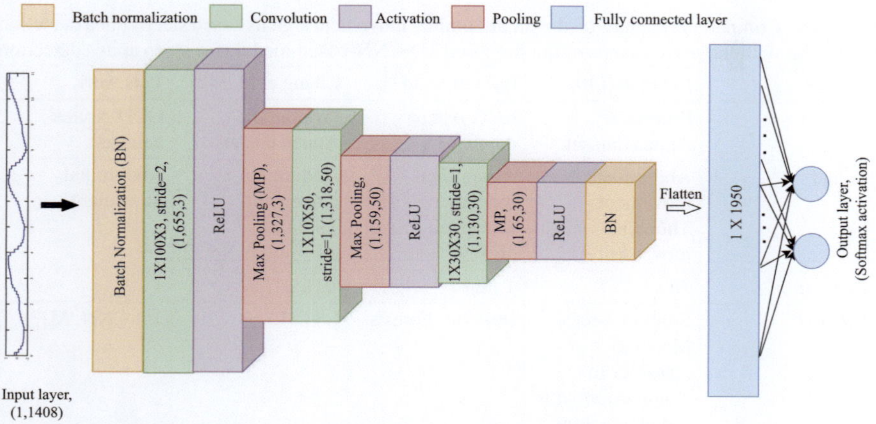

Fig. 6.4 Independent 1D-CNN fusion model for sleep apnea detection from abdominal movements ($M_1$3)

Table 6.5 The model parameters for the abdomen movement signal-based model for sleep apnea detection

Layer	Kernel		Output shape	# Parameters
	Size	# kernels		
Batch normalization	–	–	1408,1	4
Convolution	100	3	655,3	303
RELU Activation	–	–	655,3	0
Maxpooling	–	–	327,3	0
Convolution	10	50	318,50	1550
Maxpooling	–	–	159,50	0
RELU Activation	–	–	159,50	0
Convolution	30	30	130,30	45030
Maxpooling	–	–	65,30	0
RELU Activation	–	–	65,30	0
Batch normalization	–	–	65,30	120
Flatten	–	–	1950	0
Dropout	–	–	1950	0
Dense, Softmax	–	–	2	3902

Table 6.6 Comparison of state-of-the-art abdominal movement signals based sleep apnea detection algorithms with the performance of the proposed 1D-CNN-based models for sleep apnea detection

Article	Lin et al. [40]	Avcı et al. [41]	Chang et al. [42]	This work
Dataset	Proprietary Apnea dataset	ECG Apnea database	Proprietary Apnea dataset	UCD Apnea database
Signal	Abdominal movement, Thoracic movement	Abdominal movement, chest movement, nasal airflow[a]	Abdominal movement, Thoracic movement, SpO2	Abdominal movement
Resolution	30 s	1 min	10 s	1 s
Method	Support Vector Machine classifier in conjunction with a state machine	Random Forests	LSTM	1D-CNN, $M_1 3$
Accuracy	87.8%	92.75%	92.3%	99.52%
Sensitivity	–	–	–	98.76%

[a]Avc et al. carry out sleep apnea detection on different signals, but the accuracy depicted in this table is for only abdominal movement signal

Table 6.7 Performance of the single signal models for sleep apnea detection

Model	F6 (%)	Se (%)	Sp (%)	Ac (%)
$M_1 1$ (ECG)	95.82	96.05	99.75	99.56
$M_1 2$ (SpO2)	82.86	84.65	97.42	97.08
$M_1 3$ (Abdo)	98.34	98.76	99.54	99.52

use abdomen movement signals for sleep apnea detection, as shown in Table 6.6. It can be observed from Table 6.6 that the proposed model $M_1 3$ exhibits the best performance in terms of accuracy at the highest resolution of 1 s.

The results of the three independent models are summarized in Table 6.7. The model $M_1 3$ exhibited an F6-score of 98.34%, a sensitivity of 98.76%, and a specificity of 99.54% on the test set consisting of data from all the patients.

6.4.3 *Fusion Algorithm with Abdominal Signal*

The 1D-CNN model based on abdominal signal is fused with the SpO2 signal-based model (SpO2+Abdo $\rightarrow$ $M_2 2$) and the ECG signal-based model (ECG+Abdo $\rightarrow$ $M_2 3$). Since the number of samples used in the abdominal signal-based model is equal to the number of samples used in the ECG signal model, sampling frequency-based selective dropout is not required. However, when the abdominal signal-based model is fused with the SpO2 signal-based model, the possible overfitting to the

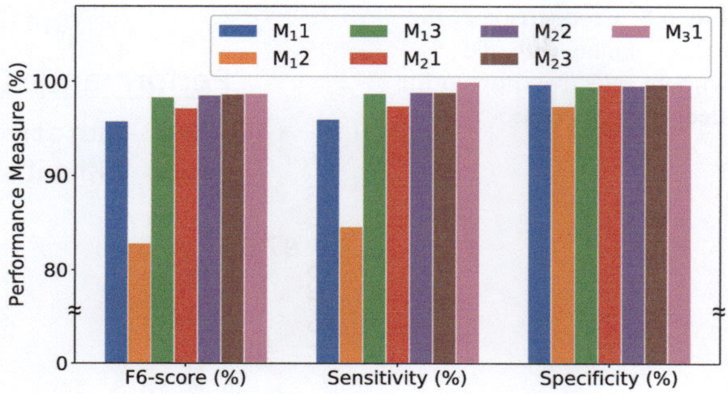

Fig. 6.5 Performance of the single signal models and fusion models employing selective dropout for sleep apnea detection

abdominal signal model features in the fusion model needs to be accounted for, and therefore signal features-based selective dropout needs to be carried out. For the abdominal signal-based model, the number of neurons in the flatten layer is 1950, the number of neurons in the flatten layer of the SpO2 signal-based model is 660, and therefore the selective dropout ratio is set to 0.67. The performance of the fusion models in terms of F6-score, sensitivity, and specificity is as shown in Fig. 6.5, along with the three new fusion models, namely SpO2+Abdo ($M_2 2$), ECG+Abdo ($M_2 3$), and ECG+SpO2+Abdo ($M_3 1$). On comparing with the ECG+SpO2 model ($M_2 1$), it can be seen that the sensitivity is higher when abdominal signal data is included (in models $M_2 2$, $M_2 3$, and $M_3 1$), and with a higher F6-score for when ECG and abdominal models are fused ($M_2 3$). The results demonstrate that increasing the number of sensor sources generally improves the performance of the fusion models. The efficacy of the selective dropout method to prevent overfitting to the sensor source with a larger sampling rate is also analyzed. In the case of the SpO2+Abdo model with selective dropout ($M_2 2$), an F6-score of 98.57% is exhibited, while for a fused SpO2+Abdo model without selective dropout ($\tilde{M}_2 2$), the F6-score was just slightly lower at 98.56%. However, in the case of the ECG+SpO2+Abdo model, the difference is slightly more pronounced. For the ECG+SpO2+Abdo model with selective dropout ($M_3 1$), an F6-score of 98.75% is exhibited, while for a fused ECG+SpO2+Abdo model without selective dropout ($\tilde{M}_3 1$), the F6-score was lower at 98.65% as shown in Fig. 6.6. In the case of model $M_2 3$, since both ECG and abdominal movement signals have the same number of samples, selective dropout is not employed. The results of the fusion algorithms with and without selective dropout are discussed in Table 6.8. The performance of the fusion model with noisy input data is analyzed next.

Performance of Fusion with Three Inputs and Noisy Data

For the training and validation set, a few windows in the dataset have been corrupted by noise in the following manner:

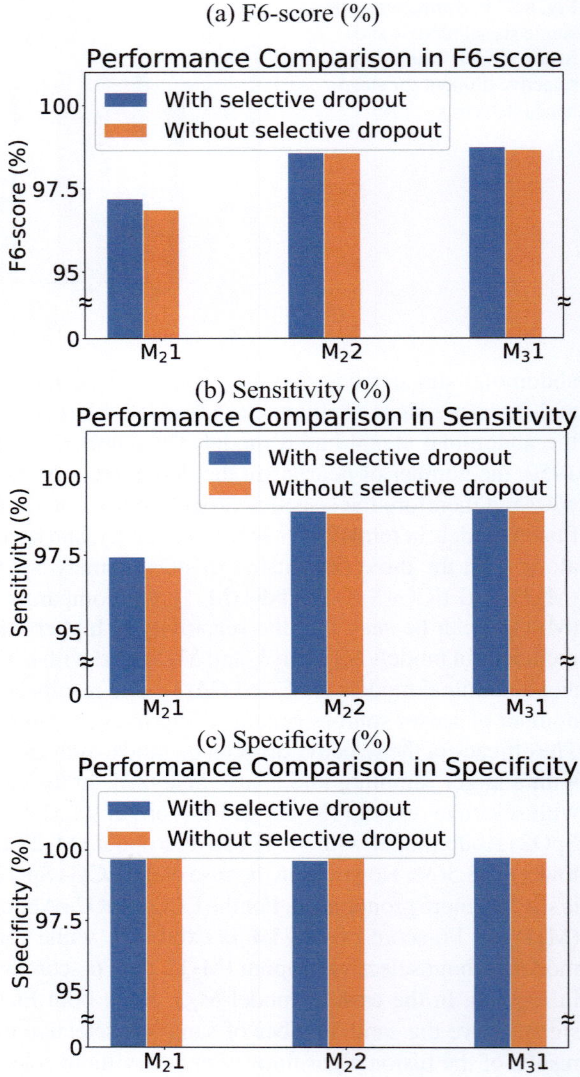

Fig. 6.6 Comparison of the fusion models with and without selective dropout for sleep apnea detection

1. All the signal samples in a window in 10% of the ECG window set are noisy while corresponding SpO2 and abdomen movement signal window sets are clean, all the signal samples in a window in 10% of the SpO2 window sets are noisy while corresponding ECG and abdomen window sets are clean, and all the signal samples in a window in 10% of the abdomen movement window set is noisy while corresponding ECG and SpO2 window sets are clean.
2. All the signal samples in a window in 10% of the ECG and SpO2 window sets are noisy while the corresponding abdomen window set is clean, all the signal

Table 6.8 Performance of the fusion models with and without selective dropout for sleep apnea detection

Model	F6 (%)	Se (%)	Sp (%)	Ac (%)
With selective dropout				
$M_2 1$ (ECG+SpO2)	97.20	97.44	99.68	99.62
$M_2 2$ (Abdo+SpO2)	98.57	98.90	99.63	99.61
$M_2 3$ (ECG+Abdo)	98.69	98.90	99.76	99.73
$M_3 1$ (ECG+SpO2+Abdo)	98.75	98.98	99.74	99.72
Without selective dropout				
$\tilde{M}_2 1$ (ECG+SpO2)	96.87	97.08	99.70	99.63
$\tilde{M}_2 2$ (Abdo+SpO2)	98.56	98.83	99.70	99.68
$\tilde{M}_2 3$ (ECG+Abdo)[a]	98.69	98.90	99.76	99.73
$\tilde{M}_3 1$ (ECG+SpO2+Abdo)	98.65	98.90	99.71	99.70

[a]The individual ECG and Abdo models have the same number of neurons in the flatten layer, and therefore selective dropout ratio for both would be 0. Hence, $M_2 3$ and $\tilde{M}_2 3$ exhibit the same performance

samples in a window in 10% of the SpO2 and abdomen window sets are noisy while corresponding ECG window set is clean, and all the signal samples in a window in 10% of the ECG and abdomen movement window sets are noisy while corresponding SpO2 window set is clean.

3. 10% of the ECG signal, SpO2 signal, and abdomen signal window sets are noisy.

For the test stage, the few samples in the test set have been corrupted by adding noise to the samples in a similar manner as discussed for the training and validation set:

1. 5.7% of the ECG signal windows are noisy while corresponding SpO2 and abdomen signals are clean, 5.7% of the SpO2 signal windows are noisy while corresponding ECG and abdomen signals are clean, and 5.7% of the abdomen movement signal windows are noisy while corresponding ECG and SpO2 signals are clean.
2. 5.7% of the ECG signal and SpO2 signal windows are noisy while corresponding abdomen signals are clean, 5.7% of the SpO2 signal windows and abdomen signal windows are noisy while corresponding ECG signals are clean, and 5.7% of the ECG signal windows and abdomen movement signal windows are noisy while corresponding SpO2 signals are clean.
3. 5.7% of the ECG signal, SpO2 signal, and abdomen signal windows are noisy.

The individual models were retrained using this dataset with added noise and fused as discussed in Sect. 6.3.1. The performances of the independent models are summarized in Table 6.9. The performance of the fusion models in terms of F6-score, sensitivity, and specificity is as shown in Fig. 6.7. From the results, it can be observed

Table 6.9 Performance of the single signal models when the dataset contains noisy samples for sleep apnea detection

Model	F6 (%)	Se (%)	Sp (%)	Ac (%)
$M_{1,n}1$ (ECG)	74.80	74.85	99.24	98.60
$M_{1,n}2$ (SpO2)	72.22	77.85	91.56	91.20
$M_{1,n}3$ (Abdo)	83.05	88.45	93.15	93.03

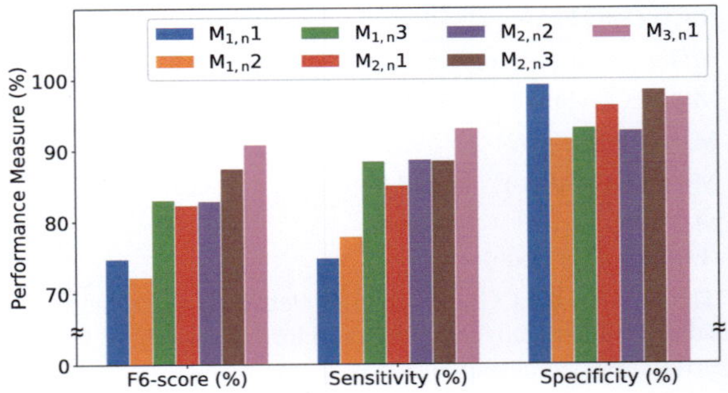

Fig. 6.7 Performance of the single signal models and fusion models with selective dropout in the presence of noise for sleep apnea detection

that F6-score is higher for the fusion models compared to the individual models. From the performance of the ($M_{1,n}1$) and ($M_{1,n}2$) models, it can be seen that the models are highly biased toward the majority class, even with minority class oversampling. These observations indicate that when the signals are corrupted by noise, important features that can be learned on clean signals are not learned by the model. The performance of the fusion model is also dependent on the quality of the individual models, like in the case of model $M_{2,n}2$, the performance is poorer with an F6-score of 82.91%, compared to the single sensor model $M_{1,n}3$ with an F6-score of 83.05%, as when fusing with the SpO2 model with just an F6-score of 72.22%, reduced the specificity of the fused model.

The performance of the fused model $M_{2,n}2$ could have been improved upon by changing the training routine, reducing the regularization coefficients, etc., but consistency across training routines across all models was maintained for comparison, and these results are reported here. From the results, it can be observed that in the noisy scenario, the three-sensor model exhibits a significant improvement in F6-score and sensitivity compared to the single sensor or two-sensor fusion models, which was not as evident in the non-noisy scenario. From the results, it is noticeable that the fusion model with ECG+SpO2 model ($M_{2,n}1$ and M_21) does not outperform the Abdo model ($M_{1,n}3$ and M_13), leading to the conclusion that abdominal movement needs to be incorporated into the fusion algorithms for best performance, but the Abdo movement recordings is not a standard signal acquired from wearable devices. However, the ECG and SpO2 signals can be easily incorporated into the current wearable devices ecosystem, and the ECG+SpO2 (M_21) fusion algorithm is well suited for wearable devices as they exhibit significant performance improvement over the ECG alone ($M_1 1$) or SpO2 alone ($M_1 2$) model for both clean signal and noisy signal scenarios. However, in the scenario where abdominal movement signals can be obtained, the Abdo+ECG model ($M_2 3$) can be used to improve performance over the single sensor models, and the larger ECG+SpO2+Abdo model ($M_3 1$) can be used when the power constraints are not pressing. The efficacy of the selective dropout method to prevent overfitting to the sensor source with the larger sampling rate is also analyzed. In the case of the $M_{2,n}2$ model, an F6-score of 82.91% is exhibited, while for the $M_{2,n}2$ model without selective dropout ($\tilde{M}_{2,n}2$), the F6-score was lower at 81.86% as shown in Fig. 6.8. For the model $M_{3,n}1$, an F6-score of 89.69% is exhibited, while for the fused model without selective dropout ($\tilde{M}_{3,n}1$), the F6-score was lower at 89.25%. The results of the fusion algorithms with and without selective dropout when the dataset contains noisy samples are as discussed in Table 6.10. The studies with and without selective dropout indicate that using selective dropout improves the model sensitivity with no additional computational cost during inference, as selective dropout is applied only during training to prevent overfitting to the model that provides more features at the flatten stage of the fusion algorithm.

6.4.4 Complexity Analysis

The computational complexities of the three networks were calculated in terms of the number of multiplications and additions required for each second. The computations involved in the convolution layers were calculated based on a simple filtering calculation count and without assuming a fast Fourier transform approach [43]. For the maxpooling layers, the operation of selecting the maximum is approximated to an addition operation. For the dense layers, the calculations are carried out as discussed in the multi-layer perceptron classifier in Chap. 3. Detection of sleep apnea events with ECG signal using $M_1 1$ requires 6534116 multiplications and 6546647 additions, and with SpO2 signal using $M_1 2$ requires 1270016 multiplications and 1272876 additions. The fusion model $M_2 1$ requires 7809352 multiplications and 7824743 additions. Detection of sleep apnea events with abdomen movement signals using $M_1 3$ requires 6534116 multiplications and 6546647 additions like the ECG-based model $M_1 1$ due to the same network structure. The fusion model $M_2 2$ that fuses the SpO2 and abdomen movement signal requires 7809352 multiplications and 7824743 additions. The fusion model $M_2 3$ that fuses the ECG and abdomen movement signal requires 13076032 multiplications and 13101094 additions. The fusion model $M_3 1$ that fuses all the three signals requires 14347368 multiplications and 14375290 additions. The total energy consumption during prediction is estimated by assuming that the energy required for a 16-bit multiplication accumulation (MAC) operation is 0.39 pJ [44, 45], and for a 16-bit adder is around 20 fJ [46] in 28nm FD-SOI technology. The energy consumption during prediction using $M_1 1$ is found to be 2.55 μJ, prediction using $M_1 2$ is found to be 0.50 μJ, and with $M_1 3$ is found to be 2.55 μJ. The total energy consumption during prediction using the fusion models $M_2 1$ is found to

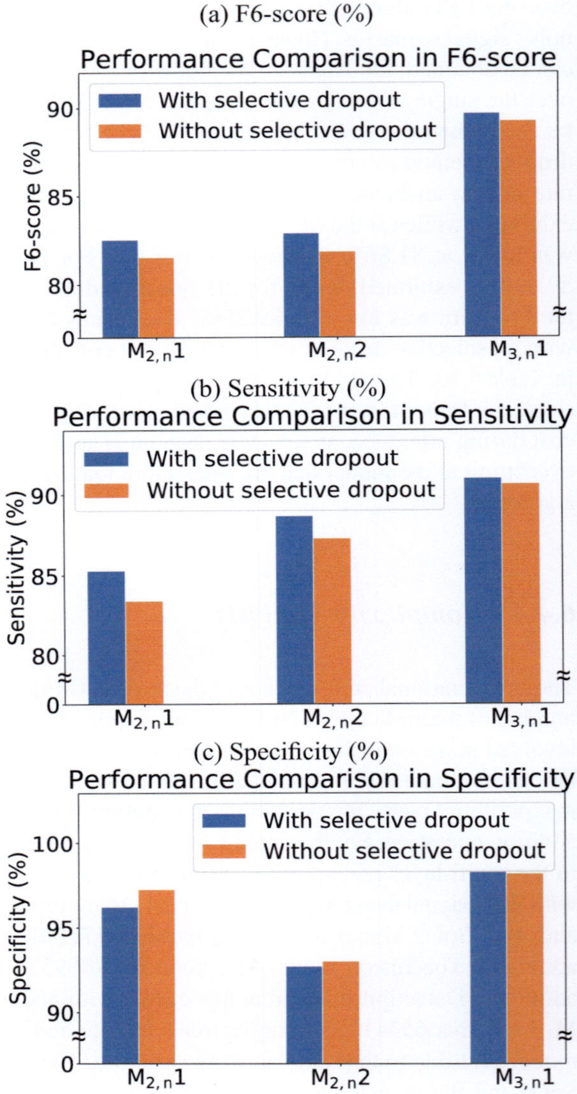

Fig. 6.8 Comparison of the fusion models with and without selective dropout when the dataset contains noisy signal samples for sleep apnea detection

be 3.05 μJ, prediction using $M_2 2$ is found to be 3.05 μJ, and with $M_2 3$ is found to be 5.10 μJ. The energy consumption for prediction using the three-sensor fusion model $M_3 1$ is 5.61 μJ. The complexity in terms of multiplications and additions and the corresponding energy consumption are discussed in Table 6.11. The corresponding models trained without selective dropout and the models generated for the dataset with noisy samples require the same number of computations as discussed above due to the same network structure. From performance and computational complex-

Table 6.10 Performance of the fusion models with and without selective dropout when the dataset contains noisy samples for sleep apnea detection

Model	F6 (%)	Se (%)	Sp (%)	Ac (%)
With selective dropout				
$M_{2,n}1$ (ECG+SpO2)	82.34	85.01	96.33	96.04
$M_{2,n}2$ (Abdo+SpO2)	82.91	88.67	92.72	92.61
$M_{2,n}3$ (ECG+Abdo)	87.47	88.52	98.48	98.22
$M_{3,n}1$ (ECG+SpO2+Abdo)	90.85	93.05	97.37	97.25
Without selective dropout				
$\tilde{M}_{2,n}1$ (ECG+SpO2)	81.55	83.41	97.26	96.89
$\tilde{M}_{2,n}2$ (Abdo+SpO2)	81.86	87.28	93.00	92.85
$\tilde{M}_{2,n}3$ (ECG+Abdo)[a]	87.47	88.52	98.48	98.22
$\tilde{M}_{3,n}1$ (ECG+SpO2+Abdo)	89.25	90.64	98.18	97.98

[a]The individual ECG and Abdo models have the same number of neurons in the flatten layer, and therefore selective dropout ratio for both would be 0. Hence, $M_{2,n}3$ and $\tilde{M}_{2,n}3$ exhibits the same performance

ity analysis, abdominal signal-based models are found to be most suitable for the task. However, for ease of integration into the current generation of wearables, the ECG+SpO2 model is ideal. For the dataset used in the clean signal scenario, the fusion model that combines all three signals does not significantly outperform the ECG+SpO2 model or the ECG+Abdo model. However, in the scenario where signals are noisy, the fusion model that combines all three signals significantly outperforms the 2-signal fusion models, indicating that as the number of sensors used in fusion increases, the reliability of inferences also increases. As discussed in the previous section the ECG+SpO2 ($M_2 1$) fusion algorithm is well suited for integration into current wearable devices ecosystems. However, in the scenario where abdominal movement signals can be obtained, the Abdo+SpO2 model ($M_2 2$) which is the least computationally intensive of the abdominal movement-based fusion models can be used. But the Abdo+SpO2 ($M_2 2$) model does not perform as well as the more computationally intensive ECG+Abdo model ($M_2 3$), and the larger ECG+SpO2+Abdo model ($M_3 1$) can be used when the power constraints are not pressing. It can also be observed that the addition of the SpO2 signal to the ECG+Abdo ($M_2 3$) fusion does not lead to a very large increase in energy requirements due to the SpO2 ($M_1 2$) alone model being very small. The computational complexity of the fusion models can be reduced by investigating model pruning, which is discussed next.

Model Pruning

Pruning is the systematic removal of parameters from an existing network for reducing resource requirements at prediction time [47]. In this work, magnitude-based

Table 6.11 Computational complexities of the sleep apnea detection models in terms of the number of multiplications and additions required, and the estimated energy consumption

Models	Multiplications	Additions	Energy (μJ)
$M_1 1$	6534116	6546647	2.55
$M_1 2$	1270016	1272876	0.50
$M_1 3$	6534116	6546647	2.55
$M_2 1$	7809352	7824743	3.05
$M_2 2$	7809352	7824743	3.05
$M_2 3$	13076032	13101094	5.10
$M_3 1$	14347368	14375290	5.61

weight pruning is used, which gradually zeros out model weights during the training process [37]. This enables compression of the model and is suitable for deployment in resource-constrained wearable devices. The pruning of the fused models to reduce the computational complexity that results from the additional sensor sources used in fusion is considered here, and the individual models are also pruned for the purpose of comparison. For this, the models are pruned with sparsity varying from 50 to 80% (aggressive pruning) using the weight pruning method. The performances of the pruned models of models $M_1 1$, $M_1 2$, and $M_1 3$ are shown in Fig. 6.9. The performance of $M_2 1$, $M_2 2$, $M_2 3$, and $M_3 1$ pruned models are as shown in Fig. 6.10a–d respectively, with model sparsity varying from 50 to 80%. From the figures, it can be observed that for models $M_2 3$, and $M_3 1$, as sparsity increases, the F6-score, and sensitivity drops while accuracy and specificity stay above approximately 97%. However, for models $M_2 1$ and $M_2 2$, the F6-score and sensitivity behave unpredictably. The F6-score and sensitivity increase when sparsity is increased from 50% to 60% and drops again after sparsity is increased to 70%. It can be observed that both $M_2 1$ and $M_2 2$ used the SpO2 signal, and this trend is similar to that observed for the individually sparsified SpO2 model as shown in Fig. 6.9, even though the overall performance drops (in terms of accuracy) with an increase in sparsity. In this case, it appears to be that the weights that make the network highly specific are pruned away with an increase in sparsity, making the network more sensitive. Therefore, this trend is the contribution of the SpO2-based signal model. A similar trend is observed in the model $M_3 1$ that fuses all three signals, but with F6-score and sensitivity increasing when sparsity is increased from 60 to 70% and drops again at a sparsity of 80%. The computational complexity and energy consumption of the pruned models are discussed in Table 6.12.

From the performance figures and model complexities, it can be concluded that an increase in model sparsity may lead to a drop in performance of the fused model, but the pruned fusion model performance can outperform individual model performances at lower computational complexities, which is promising. However, in this work, magnitude-based aggressive pruning (sparsity above 50%) alone is discussed, and therefore, the performances of the pruned models are not as compelling. Hence,

Fig. 6.9 Performance of the pruned models for sleep apnea detection with sparsity varying from 50 to 80% for the **a** ECG signal model, **b** SpO2 signal model, and **c** abdomen movement signal model

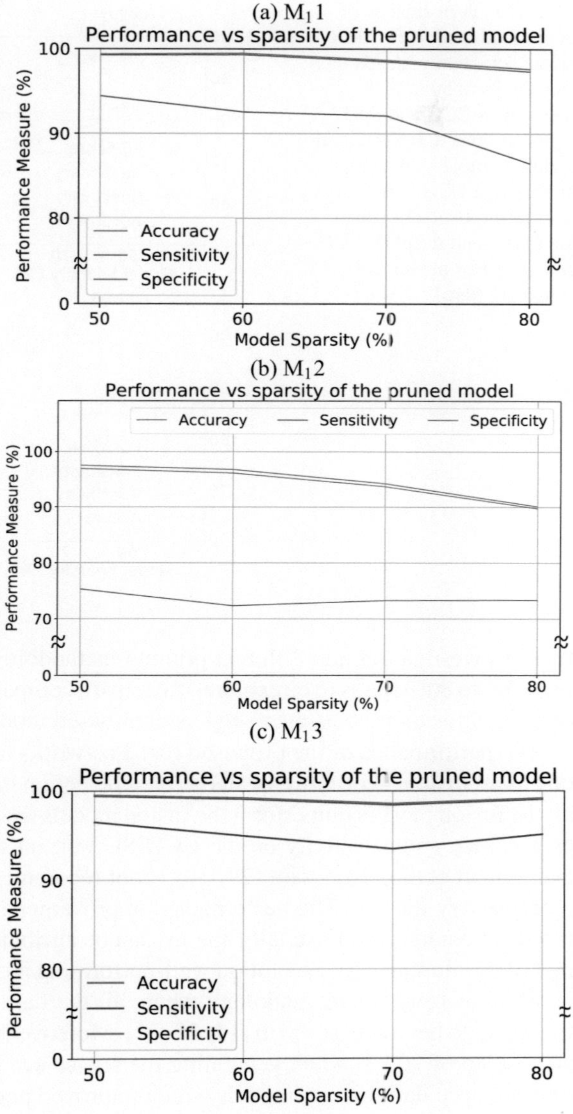

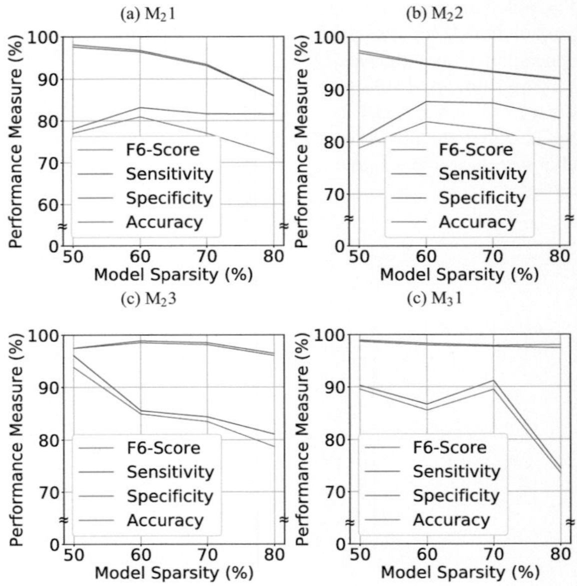

Fig. 6.10 Performance of the pruned models for sleep apnea detection with sparsity varying from 50 to 80% when the **a** ECG and SpO2 signal are fused, **b** SpO2 and abdomen movement signal are fused, **c** ECG and abdomen movement signal are fused, and **d** SpO2, ECG, and abdomen movement signal are fused

further investigation into different pruning methodologies at sparsity levels varying from 5% to 80% needs to be explored to derive a compelling argument for the benefits of pruned fusion models over single signal-based models.

The performance of the proposed fused networks is compared with that of state-of-the-art fusion algorithms for sleep apnea detection in Table 6.13. It can be seen that all the fusion models outperform the fusion models discussed in [14, 29–31] in terms of accuracy and sensitivity on the UCD St. Vincent's sleep apnea database. However, direct comparison with [29, 30] is not recommended due to the studies using a proprietary dataset. The performance improvements observed with the proposed model in comparison to state-of-the-art can be attributed to the proposed methodology of developing high-resolution well-performing individual sensor models, which is different from the methodology where all the data is fed to a single model and trained together as proposed in [14, 29–31]. Moreover, the windowing approach with an overlap of 90.9% when generating the signal windows for training helped with training deep data-hungry models which improved performance. The pruned fusion models are also found to outperform the state-of-the-art methods.

6.5 Discussions

The proposed multimodal multiresolution data fusion framework introduces a 1D-CNN-based feature-level fusion approach that enables direct integration of sensor signals with different sampling rates without requiring resampling or padding. This

Table 6.12 Computational complexity in terms of multiplication and addition operations and the energy consumption of the sparsified sleep apnea detection models during prediction

Models	Multiplications	Additions	Energy (μJ)
$M_1$1 (50%)	3270416	3289663	1.28
$M_1$1 (60%)	2617676	2630207	1.03
$M_1$1 (70%)	1964936	1977467	0.77
$M_1$1 (80%)	1312232	1324763	0.52
$M_1$2 (50%)	636020	638880	0.25
$M_1$2 (60%)	508904	511764	0.20
$M_1$2 (70%)	382316	385176	0.15
$M_1$2 (80%)	255200	258060	0.10
$M_1$3 (50%)	3270416	3289663	1.28
$M_1$3 (60%)	2617676	2630207	1.03
$M_1$3 (70%)	1964936	1977467	0.77
$M_1$3 (80%)	1312232	1324763	0.52
$M_2$1 (50%)	3911656	3933763	1.53
$M_2$1 (60%)	3131800	3147191	1.22
$M_2$1 (70%)	2352472	2367863	0.92
$M_2$1 (80%)	1572616	1587553	0.61
$M_2$2 (50%)	3911656	3933763	1.53
$M_2$2 (60%)	3131800	3147191	1.22
$M_2$2 (70%)	2352472	2367863	0.92
$M_2$2 (80%)	1572616	1587553	0.61
$M_2$3 (50%)	6548632	6587126	2.55
$M_2$3 (60%)	5243152	5268214	2.04
$M_2$3 (70%)	3937672	3962734	1.54
$M_2$3 (80%)	2632192	2656354	1.03
$M_3$1 (50%)	7185972	7227326	2.80
$M_3$1 (60%)	5753376	5781298	2.25
$M_3$1 (70%)	4321308	4349230	1.68
$M_3$1 (80%)	2888712	2915734	1.13

is particularly advantageous for wearable IoT sensing, where different physiological signals operate at varying frequencies. The introduction of selective dropout ensures that the fusion model does not overfit to high-resolution signals, effectively balancing the contribution of each sensor modality. The application to sleep apnea detection demonstrates the model's capability to integrate ECG, SpO2, and abdominal movement signals, leading to improved detection accuracy compared to single-sensor methods. This deep learning-driven fusion strategy provides a robust alternative to conventional signal processing techniques, particularly in real-world scenarios where individual sensor data may be unreliable. However, one notable limitation of this

Table 6.13 Comparison of state-of-the-art fusion-based sleep apnea detection algorithms with the performance of the proposed 1D-CNN-based fusion models for apnea detection

Article	Otero et al. [30]	Bernardini et al. [29]	Xie et al. [14]	Cen et al. [31]	This work			
Dataset	Proprietary polysomnogram data	Proprietary stroke unit database	UCD St. Vincent's database	UCD St. Vincent's database	UCD St. Vincent's database			
Resolution	2 min	3 min	1 min	1 s	1 s			
Method	Fuzzy structural algorithm	CNN+ LSTM	Bagging with RepTree	2D-CNN	1D-CNN fusion			
Signal	Respiratory airflow and SpO2	ECG and SpO2	ECG and SpO2	SpO2, oronasal airflow, ribcage, and Abdo	ECG and SpO2	SpO2 and Abdo	ECG and Abdo	ECG, SpO2, and Abdo
Accuracy (%)	–	81.50	84.40	79.61	99.62	99.61	98.69	99.71
Sensitivity (%)	96.00	67.20	79.75	–	97.44	98.90	98.90	98.98

approach is that it does not incorporate signal quality indicators (SQI) to assess the reliability of individual sensor inputs. In the fusion methods discussed in the previous chapters, SQI-based weighting mechanisms dynamically adjust sensor contributions based on real-time quality assessments, ensuring that degraded signals have minimal influence on the final output. In contrast, the approach presented in this chapter treats all input signals equally, irrespective of their quality. As a result, if multiple sensor sources are simultaneously corrupted, the model lacks an explicit mechanism to reduce the influence of unreliable signals. This could lead to misclassifications in cases of severe noise contamination. This limitation is addressed in the next chapter. Additionally, the computational complexity of fusing multiple signals poses challenges for real-time, low-power implementation in wearable devices. While pruning techniques help reduce model size and energy consumption, further optimization, such as quantization or neuromorphic computing is necessary to make this approach feasible for always-on monitoring.

6.6 Conclusions

In this work, a 1D-CNN model for the fusion of multimodal and multiresolution signals is proposed to capture temporal information, without having to resort to resampling of the individual signals. Here, whether the fusion model can improve accuracy over that of using 1D-CNNs for each signal individually is explored. An experimental study in which selective dropout of the features obtained from the sensor with a larger sampling rate is used to prevent the fusion model from overfitting to features from the sensor with the larger sampling rate is also proposed. In the proposed architecture, the 1D-CNN network learns the optimal parameters for fusion-a data-driven approach instead of a heuristic method for fusion. This modeling procedure is applied to the sleep apnea detection problem using ECG and SpO2 signals with the signals sampled at different sampling rates. Prior literature in the field indicates that these two signals can detect apnea events with reasonably good accuracy. The performance of the fusion model and individual signal models were analyzed over a common test set to confirm the hypothesis that the fusion model would perform better. The advantage of including the selective dropout in the training process over models that do not use selective dropout is also analyzed. The fusion model is then extended to include a third sensor source to exhibit the generalizability to multi-sensor fusion problems using abdomen movement signals for sleep apnea detection. The performance of the three signal fusion models and various combinations of two signal fusion models were analyzed to exhibit the merits of the fusion algorithm. In the case of the model that fuses ECG and SpO2 signals, the F6-score was found to be 1.6% higher on the test set compared to the ECG signal-based model and 18.03% higher than the F6-score of the SpO2-based model. In the case of the model that fuses ECG and abdomen movement signal, the F6-score was found to be 2.99% higher on the test set compared to the ECG signal-based model and 0.35% higher than the F6-score of the abdomen movement-based model. Pruning methods were also analyzed to compensate for the increase in computational complexity arising from additional sensor sources. It was observed that although an increase in model sparsity may lead to a drop in performance of the fused model, the pruned fusion model performance can outperform individual model performances at lower computational complexities-indicating that pruned fusion models are a promising direction for sleep apnea detection in wearable devices. The performances of the fusion models were compared against other fusion models in the literature, and the proposed models outperformed the state-of-the-art models discussed in the literature. The model complexities were also analyzed, and it was observed that the fusion model that combines all three sensor sources does not significantly outperform models that combine two sensor sources in the scenario where the signal samples are noise-free. However, in scenarios with noisy signals, the fusion model that integrates all three signals significantly outperforms the two-signal fusion models. This suggests that as the number of sensors used in fusion increases, the reliability of inferences improves, highlighting the potential of data fusion for sleep apnea detection in wearable devices. The proposed fusion algorithm is an example of complementary, feature-level, and FEI-DEO fusion.

References

1. B.V. Dasarathy, Sensor fusion potential exploitation-innovative architectures and illustrative applications. Proc. IEEE **85**, 24–38 (1997)
2. D.L. Hall, J. Llinas, An introduction to multisensor data fusion. Proc. IEEE **85**, 6–23 (1997)
3. J. Acharya, A. Basu, Deep neural network for respiratory sound classification in wearable devices enabled by patient specific model tuning. IEEE Trans. Biomed. Circuits Syst. **14**(3), 535–544 (2020)
4. J. Zhang, Y. Wu, A new method for automatic sleep stage classification. IEEE Trans. Biomed. Circuits Syst. **11**(5), 1097–1110 (2017)
5. M. Zanghieri, S. Benatti, A. Burrello, V. Kartsch, F. Conti, L. Benini, Robust real-time embedded EMG recognition framework using temporal convolutional networks on a multicore IoT processor. IEEE Trans. Biomed. Circuits Syst. **14**(2), 244–256 (2020)
6. D. Biswas, L. Everson, M. Liu, M. Panwar, B.-E. Verhoef, S. Patki, C.H. Kim, A. Acharyya, C. Hoof, M. Konijnenburg, N. Helleputte, CorNET: deep learning framework for ppg-based heart rate estimation and biometric identification in ambulant environment. IEEE Trans. Biomed. Circuits Syst. **13**(2), 282–291 (2019)
7. M. Singha Roy, B. Roy, R. Gupta, K. Das Sharma, On-device reliability assessment and prediction of missing photoplethysmographic data using deep neural networks. IEEE Trans. Biomed. Circuits Syst. **14**(6), 1323–1332 (2020)
8. A. Jafari, A. Ganesan, C.S.K. Thalisetty, V. Sivasubramanian, T. Oates, T. Mohsenin, SensorNet: a scalable and low-power deep convolutional neural network for multimodal data classification. IEEE Trans. Circuits Syst. I Regul. Pap. **66**(1), 274–287 (2019)
9. F. Castanedo, A review of data fusion techniques. Sci. World J. (2013)
10. V. Vielzeuf, A. Lechervy, S. Pateux, F. Jurie, Multilevel sensor fusion with deep learning. IEEE Sens. Lett. **3**(1), 1–4 (2019)
11. T. Kim, J. Ghosh, On single source robustness in deep fusion models, in *Advances in Neural Information Processing Systems 32: Annual Conference on Neural Information Processing Systems* (2019), pp. 4815–4826
12. N. Srivastava, G. Hinton, A. Krizhevsky, I. Sutskever, R. Salakhutdinov, Dropout: A simple way to prevent neural networks from overfitting. J. Mach. Learn. Res. **15**, 1929–1958 (2014)
13. P.E. Peppard, T. Young, J.H. Barnet, M. Palta, E.W. Hagen, K.M. Hla, Increased prevalence of sleep-disordered breathing in adults. Am. J. Epidemiol. **177**, 1006–1014 (2013)
14. Y. Xie, J. Li, T. Zhu, C. Liu, Continuous-valued annotations aggregation for heart rate detection. IEEE Access **7**, 37664–37671 (2019)
15. U. Satija, B. Ramkumar, M. Sabarimalai Manikandan, Real-time signal quality-aware ECG telemetry system for IoT-based health care monitoring. IEEE Internet of Things J. **4**(3), 815–823 (2017)
16. M.-C. Chambrin, Alarms in the intensive care unit: How can the number of false alarms be reduced? Crit. Care **5**(4), 184–188 (2001)
17. D.L.T. Wong, J. Yu, Y. Li, C.J. Deepu, D.H. Ngo, C. Zhou, S.R. Singh, A. Koh, R. Hong, B. Veeravalli, M. Motani, K.C. Chua, Y. Lian, C.-H. Heng, An integrated wearable wireless vital signs biosensor for continuous inpatient monitoring. IEEE Sens. J. **20**(1), 448–462 (2020)
18. R. Atri, M. Mohebbi, Obstructive sleep apnea detection using spectrum and bispectrum analysis of single-lead ECG signal. Physiol. Meas. **36**(9), 1963–1980 (2015)
19. B. Sulistyo, N. Surantha, S.M. Isa, Sleep apnea identification using HRV features of ECG signals. Int. J. Electric. Comput. Eng. (IJECE) **8**, 3940–3948 (2018)
20. S.S. Mostafa, F. Mendonça, F. Morgado-Dias, A. Ravelo-GarcíÂa, SpO2 based sleep apnea detection using deep learning, in *2017 IEEE 21st International Conference on Intelligent Engineering Systems (INES)* (2017), pp. 091–096
21. S.S. Mostafa, F. Mendonça, A.G. Ravelo-García, F. Morgado-Dias, A systematic review of detecting sleep apnea using deep learning. Sensors **19**(22), 4934 (2019)

22. L. Almazaydeh, K. Elleithy, M. Faezipour, Detection of obstructive sleep apnea through ECG signal features, in *2012 IEEE International Conference on Electro/Information Technology* (2012), pp. 1–6
23. C. Varon, A. Caicedo, D. Testelmans, B. Buyse, S. Huffel, A novel algorithm for the automatic detection of sleep apnea from single-lead ECG. IEEE Trans. Biomed. Eng. **62**(9), 2269–2278 (2015)
24. H.D. Nguyen, B.A. Wilkins, Q. Cheng, B.A. Benjamin, An online sleep apnea detection method based on recurrence quantification analysis. IEEE J. Biomed. Health Inform. **18**(4), 1285–1293 (2014)
25. L. Chen, X. Zhang, C. Song, An automatic screening approach for obstructive sleep apnea diagnosis based on single-lead electrocardiogram. IEEE Trans. Autom. Sci. Eng. **12**(1), 106–115 (2015)
26. E. Urtnasan, Y.-J. Kim, J.-U. Park, E.-Y. Joo, K.-J. Lee, Deep learning approaches for automatic detection of sleep apnea events from an electrocardiogram. Comput. Methods Programs Biomed. **180**, 105001 (2019)
27. X. Liang, X. Qiao, Y. Li, Obstructive sleep apnea detection using combination of CNN and LSTM techniques, in *2019 IEEE 8th Joint International Information Technology and Artificial Intelligence Conference (ITAIC)* (2019), pp. 1733–1736
28. D. Dey, S. Chaudhuri, S. Munshi, Obstructive sleep apnoea detection using convolutional neural network based deep learning framework. Biomed. Eng. Lett. **8**, 95–100 (2017)
29. A. Bernardini, A. Brunello, G.L. Gigli, A. Montanari, N. Saccomanno, AIOSA: An approach to the automatic identification of obstructive sleep apnea events based on deep learning. Artif. Intell. Med. **118**, 102133 (2021)
30. A. Otero, P. Felix, M.R. Alvarez, C. Zamarron, Fuzzy structural algorithms to identify and characterize apnea and hypopnea episodes, in *2008 30th Annual International Conference of the IEEE Engineering in Medicine and Biology Society* (2008), pp. 5242–5245
31. L. Cen, Z. L. Yu, T. Kluge, W. Ser, Automatic system for obstructive sleep apnea events detection using convolutional neural network, in *2018 40th Annual International Conference of the IEEE Engineering in Medicine and Biology Society (EMBC)* (2018), pp. 3975–3978
32. P.M. Macey, Damage to the hippocampus in obstructive sleep apnea: a link no longer missing. Sleep **42** (2019). (zsy266)
33. M.J. Morrell, M.L. Jackson, G.L. Twigg, R. Ghiassi, D.W. McRobbie, R.A. Quest, H. Pardoe, G.S. Pell, D.F. Abbott, P.D. Rochford, G.D. Jackson, R.J. Pierce, F.J. O'Donoghue, D.R. Corfield, Changes in brain morphology in patients with obstructive sleep apnoea. Thorax **65**(10), 908–914 (2010)
34. Physionet, St. Vincents University Hospital/University College Dublin sleep apnea database (2011)
35. E. Urtnasan, J.-U. Park, E.-Y. Joo, K.-J. Lee, Automated detection of obstructive sleep apnea events from a single-lead electrocardiogram using a convolutional neural network. J. Med. Syst. **42**, 1–8 (2018)
36. T. Wang, C. Lu, G. Shen, F. Hong, Sleep apnea detection from a single-lead ECG signal with automatic feature-extraction through a modified LeNet-5 convolutional neural network. PeerJ **7**, e7731 (2019)
37. D. Blalock, J.J.G. Ortiz, J. Frankle, J. Guttag, What is the state of neural network pruning? (2020)
38. M. Courbariaux, I. Hubara, D. Soudry, R. El-Yaniv, Y. Bengio, Binarized neural networks: training deep neural networks with weights and activations constrained to +1 or -1 (2016)
39. P.I. Terrill, A review of approaches for analysing obstructive sleep apnoea-related patterns in pulse oximetry data. Respirology **25**(5), 475–485 (2020)
40. Y. Lin, H. Wu, C. Hsu, P. Huang, Y. Huang, Y. Lo, Sleep apnea detection based on thoracic and abdominal movement signals of wearable piezoelectric bands. IEEE J. Biomed. Health Inform. **21**(6), 1533–1545 (2017)
41. C. Avcı, A. Akbaş, Sleep apnea classification based on respiration signals by using ensemble methods. IOS Press **26**, S1703–S1710 (2015)

42. H.-C. Chang, H.-T. Wu, P.-C. Huang, H.-P. Ma, Y.-L. Lo, Y.-H. Huang, Portable sleep apnea syndrome screening and event detection using long short-term memory recurrent neural network. Sensors **20**(21), 6067 (2020)
43. K. Abdelouahab, M. Pelcat, J. Serot, F. Berry, Accelerating CNN inference on FPGAs: a survey (2018)
44. H. Reyserhove, N. Reynders, W. Dehaene, Ultra-low voltage datapath blocks in 28nm UTBB FD-SOI, in *2014 IEEE Asian Solid-State Circuits Conference (A-SSCC)* (2014), pp. 49–52
45. R. Taco, I. Levi, M. Lanuzza, A. Fish, An 88-fJ/40-MHz [0.4 V]-0.61-pJ/1-GHz [0.9 V] dual-mode logic 8 × 8 bit multiplier accumulator with a self-adjustment mechanism in 28-nm FD-SOI. IEEE J. Solid-State Circuits **54**, 560–568 (2019)
46. R. Taco, I. Levi, M. Lanuzza, A. Fish, Evaluation of dual mode logic in 28nm FD-SOI technology, in *2017 IEEE International Symposium on Circuits and Systems (ISCAS)* (2017), pp. 1–4
47. G. Castellano, A. Fanelli, M. Pelillo, An iterative pruning algorithm for feedforward neural networks. IEEE Trans. Neural Netw. **8**(3), 519–531 (1997)

Chapter 7
A Multi-level and Multimodal Data Fusion Algorithm Using Neural Networks

Abstract Data fusion—the integration of signals from multiple sensors in real time—has become a rapidly evolving field with significant implications for biomedical signal processing. This chapter introduces a novel convolutional neural network (CNN)-based methodology for fusing multimodal, multi-sensor data in a fully automated and data-driven manner. Unlike traditional fusion approaches that require manual specification of the fusion level, the proposed model learns to determine the optimal level of information abstraction autonomously, integrating both feature extraction and fusion within a unified architecture. A key innovation of this approach is the inclusion of signal quality indicators (SQIs) as additional input streams, allowing the model to dynamically adjust the fusion process based on the reliability of each sensor input. The method is validated through the task of atrial fibrillation detection using multimodal data—electrocardiogram (ECG) and photoplethysmogram (PPG) signals—sourced from the MIMIC III database. The model, trained using an average loss function, achieves impressive performance metrics with an accuracy of 99.33% and sensitivity of 99.74%. Its robustness under noisy conditions further underscores the effectiveness of the SQI-informed, multi-level fusion strategy. This chapter highlights the potential of intelligent, quality-aware fusion techniques to advance the reliability and efficiency of health monitoring systems.

Keywords Multi-level data fusion · Convolutional neural networks · Atrial fibrillation detection · Signal quality indicators · Electrocardiogram · Photoplethsymogram · Central loss · Average loss · Continuous signal quality indicators · Signal-quality aware fusion

7.1 Introduction

Multi-sensor data fusion combines data from multiple sensors to enhance the accuracy and reliability of health inferences in biomedical devices. A method of classifying data fusion algorithms is based on the information abstraction level at which fusion occurs: signal-level, feature-level, or decision-level [1, 2]. Signal-level fusion is used to create one or more signals with the same form as the input signal but of

A. John et al., *Deep Learning and Signal-Processing Methods for Multisensor Data Fusion*, https://doi.org/10.1007/978-3-031-96724-5_7

higher quality, which involves combining multiple signals at the signal level. Signal-level fusion requires a centralized processing approach and is typically used for redundancy to improve robustness [3]. However, direct signal-level fusion requires data from homogeneous measurement systems. For example, multiple channels of electrocardiogram (ECG) signals can be fused at the signal level without compromising the fused output, as discussed in Chap. 4. Fusion can also be carried out after the feature extraction phase. When it is necessary to combine disparate data sources of different modalities, like ECG signals and accelerometer data, feature-level fusion is advantageous, especially when raw signals are noisy, as discussed in Chap. 5. This is commonly known as multimodal fusion [4] and often requires a transformation into a common representational format. However, the performance of multimodal feature-level fusion heavily depends on the identification of the most relevant features for the fusion task. Fusion can also be carried out closer to the output stage, known as decision-level fusion. When the signals from the different measurement sources exhibit varying modalities, decision-level fusion is commonly employed [5]. Decision fusion aids in communication bandwidth savings, but the decision fusion algorithms must account for the reliability of each of the decision nodes during fusion, as poor decisions from a few measurement systems can lead to inaccurate outputs. Therefore, early-stage fusion involves integrating data closer to the raw signal level, where input signals are combined before significant processing or feature extraction, such as the addition of signals of the same modality prior to processing. On the other hand, late-stage fusion occurs at a higher level of abstraction, closer to the decision-making stage, where each signal is processed independently, and their results or features are merged afterwards. This allows for more refined, signal-specific processing but risks losing potential interactions between signals that could have been captured earlier. The information abstraction level at which fusion should be carried out is a decision that the designer makes based on empirical evidence and prior knowledge [6]. This strategy, therefore, is not ideal for identifying the optimal information abstraction level for a given task, as it can vary based on the experiences of different designers.

In the context of fusion using deep-learning approaches, self-learning or automatic identification of the information abstraction stage at which fusion should occur through a data-driven approach allows for greater flexibility, as relevant information in the early stages is either fused or propagated to the later stages. Depending on whether fusion needs to occur early on in the information processing pipeline, further later down the line, or if fusion needs to happen at multiple stages, the network is capable of learning this in an optimal manner. To automatically balance the trade-off between early and late fusion, or between the fusion of low-level (signal-level) and high-level (decision-level) information, Vielzuf et al. developed a deep-learning-based fusion network termed CentralNet [7]. However, a method for autonomous identification of the fusion stage for multimodal fusion was never studied. In this chapter, a fusion approach that is data-driven is proposed, where the information abstraction level at which fusion should occur is learned by the network. Moreover, most deep learning-based fusion algorithms for physiological signals do not account for the signal quality of the input signals, even when the training data is corrupted by noise. This is often done under the assumption that the neural network is capable of generalizing and placing greater importance on the clean signals. However, this may not always be the case, and therefore, signal quality indicators (SQIs) need to be considered during fusion, especially in the case of wearable devices for health monitoring. It is expected that SQIs lead to performance improvements, due to

1. Signal-quality-aware fusion: SQIs serve as weights in the fusion process, prioritizing higher-quality signals and reducing the impact of noisy inputs.
2. Noise robustness: The model dynamically adjusts to varying noise levels, improving reliability under real-world conditions.
3. Enhanced learning: SQIs help the model focus on meaningful features, improving its ability to generalize across datasets.

This leads to the contributions of this chapter:

1. A novel 1-dimensional CNN-based self-learning fusion methodology for multimodal multi-sensor data, in which the network autonomously determines the optimal information abstraction level for fusion, eliminating the need for manual selection;
2. A robust and adaptable model development framework that integrates SQIs directly into the input stream, enabling dynamic adjustments to signal variability and ensuring reliable fusion, even under noisy conditions; and
3. Demonstration of the proposed fusion framework through a novel multimodal atrial fibrillation detection system that leverages the fusion of electrocardiogram (ECG) and photoplethysmogram (PPG) signals to achieve state-of-the-art accuracy and resilience in real-world scenarios.

This chapter examines whether a 1D-CNN-based event detection model can fuse multimodal signals with the proper trade-off between early and late fusion phases while also taking input signal quality into account. The main goal of the investigation is to determine whether employing self-learning fusion networks can improve detection accuracy compared to using 1D-CNN-based networks for event detection from each measurement system separately.

7.2 Atrial Fibrillation Detection and State-of-the-Art

Atrial fibrillation (AF), which affects 1–2% of the general population, is characterized by rapid and erratic atrial activation that impairs atrial function and is the most common type of arrhythmia [8, 9]. Due to its short duration and sometimes asymptomatic nature, AF is hard to detect [10]. ECG recordings are commonly used to detect atrial fibrillation in clinical settings. Various rule-based algorithms for detecting atrial fibrillation from RR intervals obtained from ECG signals have been proposed in the literature [11–13]. Features derived from P wave absence, heart rate irregularity, and atrial activity were used to train a neural network to detect atrial fibrillation in [14]. Convolutional neural networks were used for atrial fibrillation detection in [15]. An ensemble of convolutional neural networks (two VGG nets) and a long short-term memory network was used to detect atrial fibrillation from the spectrogram and Poincare plots of ECG segments [16]. In [17], multiscale CNNs were proposed for atrial fibrillation detection from short single-lead ECG recordings. Recurrent neural networks (RNNs) were proposed for atrial fibrillation detection in [18].

PPG signal recordings-based atrial fibrillation detection is being explored these days due to the ease of acquisition from wearable devices. Beat-to-beat interval features derived from PPG signals acquired from smartphones were used for atrial fibrillation detection through thresholding rules in [19]. Support Vector Machines were used to distinguish between atrial fibrillation and normal sinus rhythms using features extracted from PPG signals [20]. A CNN combined with a long short-term memory network was proposed for atrial fibrillation detection, using the time-frequency chromatograph of PPG signal segments in [21].

Although there are quite a few works focusing on atrial fibrillation detection from wearable sensors, the literature on multi-sensor fusion for atrial fibrillation detection is sparse. In [22], a Markov model-based classifier that used inter-pulse interval features from pulses detected from the PPG signal was proposed to detect atrial fibrillation while using wrist accelerometer data to reject noisy pulses. In [23], a similar approach was used to distinguish between atrial fibrillation, atrial flutter, and normal sinus rhythm using other waveform features extracted from the PPG signal, as well as the accelerometer data. A few machine learning-based approaches for atrial fibrillation detection are also discussed in the literature. A hybrid classifier that leveraged a machine learning classifier as a statistical rule-based classifier for atrial fibrillation detection was proposed in [9], where the inter-beat intervals obtained from the PPG signal and related features were used for classification, along with PPG SQIs, where motion detection from accelerometer data was used to eliminate noisy inter-beat intervals from the decision classifier. An elastic net logic was used to detect atrial fibrillation using a combined feature vector obtained from a CNN network (whose input was 8-channel PPG recordings) as well as features obtained from 8-channel PPG recordings and their SQIs in [24].

The rest of the chapter is structured as follows: Section 7.3 explains the fusion architecture and network development method. The suggested fusion network is discussed in Sect. 7.4 in relation to the atrial fibrillation detection task. Section 7.5 details the loss functions used for performance evaluation, the performance evaluation of the fusion models created from the two separate measurement systems, and additional comments.

7.3 1D-CNN-Based Multi-level Fusion for Multi-sensor Data

This section presents the methodology for developing multimodal and multi-level fusion models through a data-driven approach. Consider an event classification task from multi-sensor time-series data obtained from different sensor sources, where data from each source can be used to carry out the classification task with acceptable performance. At first, independent streams of 1D-CNNs need to be used to extract the relevant features for the task prior to fusion for each sensor input signal separately as discussed in Chap. 6. This method would ensure that the individual sensor feature maps would not affect the training of the CNN streams of the other sensors. However, this method is not suitable in scenarios involving multi-level fusion of signals corrupted by noise, where the performance of individual sensors should influence other sensor signal processing streams. The interaction between the sensor processing streams allows the network to learn to place more importance on the clean sensor sources over the noisy sensor sources. Therefore, in the proposed multi-level fusion model, the individual streams, signal quality indicator streams, and the fusion network are trained together in a single step, rather than separately, allowing the signal features and SQI features to interact during the training process.

7.3.1 Multi-level Fusion Architecture

The goal of multi-level fusion is to fuse information from two sensor sources S_0 and S_1 that use neural networks M_0 and M_1 to provide decisions D_0 and D_1 respectively, to obtain an improved inference or decision $D_{0,1}$. In data fusion, a decision needs to be made on the information abstraction level at which fusion needs to happen: early fusion (closer to the signal level stage), late fusion (closer to the decision stage), or somewhere in between. Here, neural networks that sequentially process data to generate feature maps after each layer for each sensor source separately for fusion are considered. For the fusion of information from two sensor sources that have their individual neural network models for decision-making, fusion can be carried out after any of the individual layers. However, this work proposes a data-driven approach that utilizes a separate central network to determine the optimal stage for fusion, enhancing the CentralNet architecture introduced in [7] for signal-quality-based multi-level fusion. In CentralNet, features from individual neural networks, each corresponding to a different sensor source, are combined by feeding each layer a weighted sum of the layers from the respective single-sensor networks along with its own preceding layers. This is illustrated in Fig. 7.1. Such fusion layers can be defined by the following Eq. (7.1) [7]:

$$h_{C_{i+1}} = \alpha_{C_i} h_{C_i} + \sum_{k=1}^{N-1} \alpha_{S_i^k} h_{S_i^k} \tag{7.1}$$

where N is the number of individual signal sources being fused, α_{C_i} and $\alpha_{S_i^k}$ are trainable scalar weights, $h_{S_i^k}$ is the hidden representation (or feature maps) of each modality at layer i, and h_{C_i} is the central hidden representation. The resulting representation $h_{C_{i+1}}$ is then fed to the next layer. However, it can be observed that the

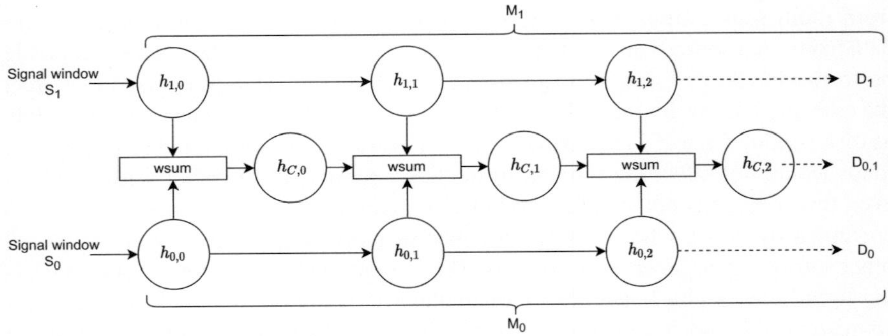

Fig. 7.1 Multi-level fusion architecture as proposed in [7]

learning doesn't take into account the quality of the signals in this method. When noisy signals are used for training and validation, it is advantageous to provide the signal quality indicators of the signals as input to the neural network to aid in the fusion process.

7.3.2 *Multi-level Data-Driven Fusion Approach with Signal Quality Indicators*

Although the fusion approach discussed in [7] addresses the issue of the fusion level, it does not account for the quality of the signals acquired from different measurement sources. High noise levels render signals unsuitable for diagnostic applications. Often, model generation and testing are conducted under sterile or ideal conditions, and the unpredictable nature of noise corrupting the signals makes it difficult for fusion models to generalize. SQIs represent the level of uncertainty in the sensor models and are useful in preventing poor performance of fusion algorithms. In neural networks, where training may use noisy data due to the limited availability of large datasets with high-quality diagnostic signals, SQIs can enhance the fusion process. Therefore, a multi-level data fusion approach incorporating SQIs is proposed here. Consider the fusion of $N+1$ multi-sensor inputs $S_k \forall k \in [0, \cdots, N]$, with their corresponding SQIs $Q_k \forall k \in [0, \cdots, N]$. Let the models for the signal inputs be $M_k \forall k \in [0, \cdots, N]$ and for the corresponding SQIs be $Ms_k \forall k \in [0, \cdots, N]$. The fusion process occurs in the central network, which integrates the signal quality indicators into the fusion process. The equations for the individual signal streams, SQI streams, and integration of SQIs in the fusion process are discussed below. The initial input to the central network is termed $h_{C_{init}}^{-1}$, which is defined as

$$h^0_{C_{init}} = \sum_{k=0}^{N} \lambda_k h^{-1}_{C_k}, \tag{7.2}$$

where

$$h^{-1}_{C_k} = tanh[conv[S_k]], \tag{7.3}$$

and λ_k are trainable weights. $tanh[.]$ and $conv[.]$ represent the hyperbolic tangent activation function and a convolution layer with padding is used to maintain the output length the same as that of the input sequence. Here, $h^{-1}_{C_{init}}$ is initialized as shown, but it can be set to 0 as suggested in [7]. To obtain the first hidden feature set, $h^{-1}_{C_{init}}$ is passed through the first computational block $Block_{C_0}$ of the central network (a $Block$ corresponds to a computational unit that can involve convolutional layers, pooling layers, and activations):

$$h^0_C = Block_{C_0}[h^0_{C_{init}}]. \tag{7.4}$$

Once the initial features of the central network are computed, the first set of fused features is calculated as

$$h^1_{C_{init}} = \alpha_{C_0} h^0_C + (1 - \alpha_{C_0}) \sum_{k=0}^{N} g^0_{k,j} h^0_{k,j}, \tag{7.5}$$

where α_{C^0} is a trainable parameter, j indicates that the multiplications are carried out element-wise (the blocks across the individual signal networks and the central network must have a uniform structure to be able to carry out these operations in an element-wise fashion), $h^0_{M_{k,j}}$ corresponds to the hidden features obtained from the first computational block of network M_k, which is $Block_{M_{0,k}}$ (the neural network model for signal S_k), as shown here:

$$h^0_k = Block_{M_{0,k}}[S_k]. \tag{7.6}$$

The term $g^0_{k,j}$ is obtained after a normalization stage as shown:

$$g^0_{k,j} = \frac{g^0_{k,j,init}}{\sum_{k=0}^{N} g^0_{k,j,init}}, \tag{7.7}$$

and $g^0_{k,j,init}$ corresponds to the hidden features obtained from the first computational block of network Ms_k (the neural network model for the SQI signal Q_k), as shown here:

$$g^0_{k,j,init} = Block_{Ms_{0,k}}[Q_k]. \tag{7.8}$$

Here, j indicates that the multiplications are carried out element-wise. After Eq. (7.5), the final hidden features of the central network are then calculated as

$$h_C^1 = Block_{C_1}[h_{C,init}^1]. \tag{7.9}$$

It should be noted that for Eq. 7.5 to perform as expected, the outputs from $Block_{M_{0,k}} \forall k \in [0, \cdots, N]$, $Block_{Ms0,k} \forall k \in [0, \cdots, N]$, and $Block_{C_0}$ must produce outputs of the same dimensions for sample-wise fusion. Further to this, the following computations are carried out for every subsequent layer denoted by i, and T is the number of the final computational block/layer. The following equations are generalized versions of the network equations discussed above, where i indicates the layer number.

$$g_{k,j,init}^i = Block_{Ms_{i,k}}[g_{k,j,init}^{i-1}] \;\; \forall i \in [1, \cdots, T]. \tag{7.10}$$

$$g_{k,j}^i = \frac{g_{k,j,init}^i}{\sum_{k=0}^N g_{k,j,init}^i} \;\; \forall i \in [1, \cdots, T]. \tag{7.11}$$

$$h_k^i = Block_{M_{i,k}}[h_k^{i-1}] \;\; \forall i \in [1, \cdots, T]. \tag{7.12}$$

$$h_{C_{init}}^i = \alpha_{C_{i-1}} h_C^{i-1} + (1 - \alpha_{C_{i-1}}) \sum_{k=0}^{N} g_{k,j}^{i-1} h_{k,j}^{i-1} \;\; \forall i \in [1, \cdots, T]. \tag{7.13}$$

$$h_C^i = Block_{C_i}[h_{C,init}^i] \;\; \forall i \in [0, \cdots, T]. \tag{7.14}$$

The flow diagram of the proposed fusion architecture with SQIs with two sensors, or $N + 1 = 2$ is as shown in Fig. 7.2. Here, the two-sensor model is shown, as for the atrial fibrillation detection problem, only two sensor inputs are used. The generalized flow diagram of the fusion architecture with $N + 1$ number of sensors is depicted in Appendix C of the appendix of this book. It can be seen from the architecture that the central network can self-learn the fusion approach to determine whether the fusion process should occur at an early stage, late stage, or a combination of these. The SQI processing pipeline aids in this. The proposed model dynamically selects between early and late fusion based on input characteristics to optimize performance. Early fusion, which combines raw features, is advantageous when the input signals (like ECG and PPG) are clean, so that it captures inter-modality correlations effectively. In contrast, late fusion, which combines high-level features extracted independently, is better suited for noisy inputs, as it isolates relevant features from each modality before fusion. Since it is not always possible to know the quality of input data beforehand in biomedical applications, the model must self-determine the best fusion methodology.

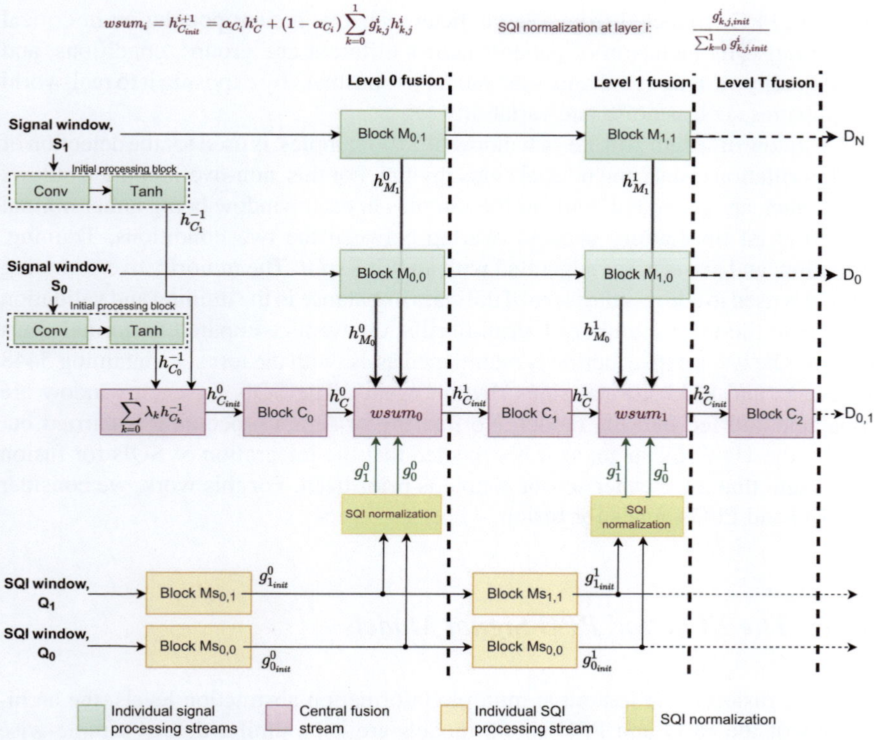

Fig. 7.2 Proposed multi-level fusion architecture of two sensors with SQIs as priors

7.4 Application to Atrial Fibrillation Detection

7.4.1 Dataset

In the literature, the MIT-BIH atrial fibrillation database [25] is commonly used to assess the performance of atrial fibrillation classifiers. However, given that this database only contains ECG recordings, the performance of the proposed fusion models cannot be analyzed using this dataset. Other works that have proposed fusion-based methods for AF classification have utilized various proprietary datasets. Therefore, for this work, the MIMIC III waveform database matched subset is utilized [26], which is a subset of the MIMIC III Waveform Database [27]. The database comprises anonymized waveforms measured from critical care patients at the Beth Israel Deaconess Medical Center in Boston. These waveforms include ECG, PPG, and arterial blood pressure (ABP) signals sampled at 125 Hz, as well as other signals such as respiration rate. These signals are not annotated with atrial fibrillation events; therefore, we make use of the publicly available annotations provided in [28]. The MIMIC III dataset provides a diverse range of multimodal physiological signals, including

ECG and PPG, collected from a large, heterogeneous patient population in critical care settings. Its inclusion of patients across different age groups, conditions, and signal quality variations enhances the model's robustness by exposing it to real-world complexities such as noise and variability.

A window of length 20 s, i.e., a window of 2500 samples, is used for the detection of atrial fibrillation events and normal sinus rhythm. For this, non-overlapping windows of 20 s each are generated, with all the samples in each window being either normal or with atrial fibrillation, with no overlap between the two conditions. Training, validation, and test sets are generated with an 8:1:1 split. The minority oversampling method is used to address the issue of the class imbalance in the training and validation sets due to the lower number of atrial fibrillation events compared to normal sinus rhythm. The test set distribution is maintained as-is, with the test set containing 5448 AF events and 8511 normal sinus rhythm events. The SQIs for each window are calculated and fed into the model. No filtering or signal processing is carried out prior to the 1D-CNN stage, as it is expected that the integration of SQIs for fusion will ensure that the cleaner sensor source is prioritized. For this work, we consider the ECG and PPG signals for fusion.

7.4.2 The ECG and PPG Signal Models

Since the fusion is carried out at multiple information abstraction levels, the architectures of the ECG and PPG signal models are kept similar due to sample-wise operations being carried out for fusion with the SQIs. If the model architectures need to be different, a transform layer can be added to manage the issue of dimensionality. However, this adds another layer of information abstraction, thereby preventing any successful attempts at early-stage fusion. The ECG signal neural network and the PPG signal network share the same structure and are shown in Table 6.1. The ECG neural network architecture was initially designed with four convolutional layers, with the stride of the first layer set to 2 to reduce dimensionality in the final output layer. Leaky ReLU was selected as the activation function for all intermediate layers, enabling negative outputs to contribute to the model's training. A softmax activation function was employed at the output layer to facilitate binary classification. Hyperparameter optimization was carried out through random search, and the final structure and configuration of the layers were refined based on the model's performance on the validation set. Once the ECG network architecture was established, the PPG network was designed to have the exact same structure to enable sample-wise operations in the central network (Table 7.1).

Table 7.1 ECG and PPG signal network architecture and parameters for atrial fibrillation detection

Block	Layer	Kernel			Output shape
		Size	# Kernels	Stride	
$Block_{M_{0,k}}$	Batch Normalization	–	–	–	2500, 1
	Convolution	30	5	2	1236, 5
	Leaky ReLU activations	–	–	–	1236, 5
$Block_{M_{1,k}}$	Maxpooling	2		2	618, 5
	Dropout	–	–	–	618, 5
	Convolution	25	5	1	594, 5
	Maxpooling	2		2	297, 5
	Leaky ReLU activations	–	–	–	297, 5
$Block_{M_{2,k}}$	Dropout	–	–	–	297, 5
	Convolution	20	5	1	278, 5
	Dropout	–	–	-	278, 5
	Convolution	15	5	1	264, 5
	Maxpooling	2		2	132, 5
	Leaky ReLU activations	–	–	–	132, 5
$Block_{M_{3,k}}$	Batch normalization	–	–	–	132, 5
	Flatten	–	–	–	660
$Block_{M_{4,k}}$	Dropout	–	–	–	660
	Dense, Softmax	–	–	–	2

7.4.3 Signal Quality Indicators for ECG and PPG Signals

Methods for indicating signal quality as multiple classes or continuous values that specify the acceptability of ECG signals of diagnostic quality on wearables have been discussed in the literature [29, 30] as well as in Chap. 3. Signal quality assessment of PPG signals based on various clinical diagnostic requirements has been discussed in [31, 32] as well as in Chap. 5. Here, instantaneous signal quality indicators for heterogeneous signal inputs are required to use as weights for data fusion, and in this work, the instantaneous signal quality of the physiological signals that are acquired using independent sensors is estimated using the curve SQI (cSQI) technique proposed in Sect. 5.2.3 is used. To refresh, the cSQI calculation process is as follows:

1. Template Initialization: A clean, signal-specific single-cycle template is generated during initialization by averaging 10 clean cycles identified using fiducial features.

2. Signal Alignment: Incoming signal cycles are aligned with the template using cross-correlation and circular shifting.
3. Quality computation: The inverse of the variance of the difference between the aligned cycle and the template is calculated as the SQI.
4. Instantaneous Quality Assessment: SQIs are computed for each sample in the signal sequence, providing real-time quality information.

Moreover, cSQI can provide instantaneous signal quality (signal quality for each sample) for a signal sequence and can be used for both ECG and PPG signals for instantaneous signal quality assessment. The cSQIs are not subjected to low pass filtering as discussed in Algorithm 2 in Chap. 5, as in this work, the aim is to avoid any signal processing or filtering stages on the signals prior to being fed into the network. The ECG and PPG cSQIs for when the signals are clean and are noisy (corrupted by Gaussian noise at –10 dB) are illustrated in Fig. 5.6. From the figure, it can be seen that the SQIs drop when the corresponding signals are noisy. The ECG and PPG SQI network architectures have a similar structure to that of the ECG and PPG signal networks (Table 7.2). The intermediate feature maps of the SQIs are used in the weighted fusion of the intermediate features generated from the signal networks (Fig. 7.3).

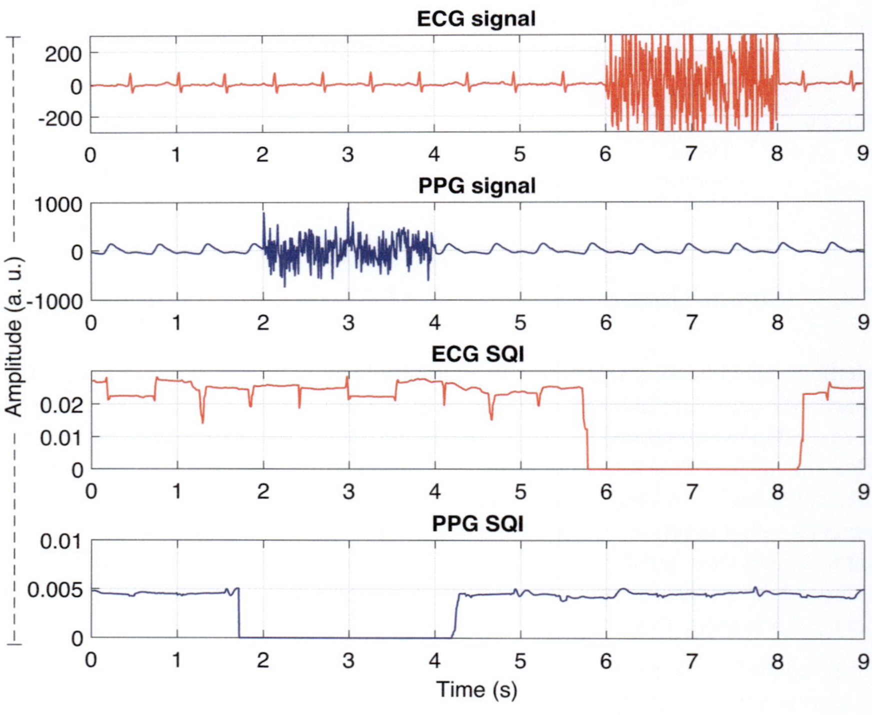

Fig. 7.3 ECG and PPG signals with their corresponding cSQIs

Table 7.2 ECG and PPG signal SQI network architecture and parameters

Block	Layer	Kernel			Output shape
		Size	# Kernels	Stride	
$Block_{M_{S0,k}}$	Batch Normalization	–	–	–	2500, 1
	Convolution	30	5	2	1236, 5
	Leaky ReLU activations	–	–	–	1236, 5
$Block_{M_{S1,k}}$	Maxpooling	2		2	618, 5
	Dropout	–	–	–	618, 5
	Convolution	25	5	1	594, 5
	Maxpooling	2		2	297, 5
	Leaky ReLU activations	–	–	–	297, 5
$Block_{M_{S2,k}}$	Dropout	–	–	–	297, 5
	Convolution	20	5	1	278, 5
	Dropout	–	–	–	278, 5
	Convolution	15	5	1	264, 5
	Maxpooling	2		2	132, 5
	Leaky ReLU activations	–	–	–	132, 5
$Block_{M_{S3,k}}$	Batch normalization	–	–	–	132, 5
	Flatten	–	–	–	660
$Block_{M_{S4,k}}$	Dropout	–	-	–	660
	Dense, Softmax	–	–	–	2

7.4.4 Fusion Network Architecture

The fusion network, or the central network, fuses the intermediate feature maps obtained from the individual signal networks along with the SQIs. In the case of two-signal fusion, as is the case here in Eq. 7.2, which generates the inputs to the central network, can be modified as

$$h_{C_{init}}^{-1} = \lambda h_{C_0}^{-1} + (1 - \lambda) h_{C_1}^{-1}, \tag{7.15}$$

so that the sum of the weights of the two components adds up to 1.

The fusion network architecture is shown in Table 7.3, where the intermediate layers depicted through equations indicate the fusion stages at which the feature maps of the individual signal networks and the corresponding SQI feature maps interact.

Table 7.3 Central network architecture and parameters of the multi-level fusion model for atrial fibrillation detection

Block	Layer	Kernel			Output shape
		Size	# Kernels	Stride	
$h_{C_0}^{-1}$	Convolution	10	1	1	2500, 1
	Tanh activations	–	–	–	2500, 1
$h_{C_1}^{-1}$	Convolution	10	1	1	2500, 1
	Tanh activation	–	–	–	2500, 1
Equation (7.15)					
$Block_{C_0}$	Batch normalization	–	–	–	2500, 1
	Convolution	30	5	2	1236, 5
	Sigmoid activations	–	–	–	1236, 5
Equation (7.13) with $i = 1$					
$Block_{C_1}$	Maxpooling	2		2	618, 5
	Dropout	–	–	–	618, 5
	Convolution	25	5	1	594, 5
	Maxpooling	2		2	297, 5
	Sigmoid activations	–	–	–	297, 5
Equation (7.13) with $i = 2$					
$Block_{C_2}$	Dropout	–	–	–	297, 5
	Convolution	20	5	1	278, 5
	Dropout	–	–	–	278, 5
	Convolution	15	5	1	264, 5
	Maxpooling	2		2	132, 5
	Sigmoid activations	–	–	–	132, 5
Equation (7.13) with $i = 3$					
$Block_{C_3}$	Batch normalization	–	–	–	132, 5
	Flatten	–	–	–	660
Equation (7.13) with $i = 4$					
$Block_{C_4}$	Dropout	–	–	–	660
	Dense, Softmax	–	–	–	2

It can be observed that fusion algorithms are comparatively more expensive than using single-channel-based methods, especially because additional SQI calculations also need to be carried out. However, various methods to reduce neural network complexity through pruning [33] and binarization [34] can be employed to reduce computational costs.

7.5 Results

The model with the four separate signal-specific streams (ECG, PPG, ECG SQI, PPG SQI) and the central fusion stream was trained together. The binary cross-entropy loss function was used for training, with optimization using the Adam optimizer. The training process was carried out such that a true positive represented an accurately detected atrial fibrillation event. The training batch size was set to 32 and the models were trained for a maximum of 25 epochs. The network was trained with a validation callback such that for each epoch, the performance on the validation set was used to retain the optimal kernel parameters that produced the minimum validation loss throughout the training process. The individual streams employed the leaky ReLU activation function, while the central fusion network employed the sigmoid activation function. The performance of the network was evaluated in terms of sensitivity (Se), specificity (Sp), and accuracy (Ac), which are calculated as shown in Eqs. (3.9), (3.10), and (3.8), where a correctly detected atrial fibrillation event is considered a true positive (TP), a correctly detected normal sinus rhythm is treated as a true negative (TN). The false positives (FP) and false negatives (FN) are defined as inverse of the above.

Due to the lower number of AF events in the test set, very low sensitivity values (due to missed atrial fibrillation events) may result in high accuracy values due to the model accurately identifying normal sinus rhythms, which are significantly more frequent compared to atrial fibrillation events. Therefore, a metric that can capture the effect of this class imbalance in the performance evaluation is required. To account for both sensitivity (Se) and precision (also called positive predictive value (PPV), which quantifies the number of TPs among all detected positives), the F1-score is used in this situation. The equation for precision (PPV) is given in Eq. (3.11). The sensitivity and precision are used to calculate the F1-score, as shown in Eq. 3.12. For the model training and evaluation, the performances of the individual signal networks trained separately were studied. The ECG signal model exhibited an accuracy of 95.96%, a sensitivity of 99.10%, a specificity of 93.95%, and an F1-score of 95.04% on the test set. The PPG signal model exhibited an accuracy of 86.71%, a sensitivity of 99.10%, a specificity of 78.78%, and an F1-score of 85.34%. For the fusion model, two different kinds of loss functions were considered for training, which is discussed next.

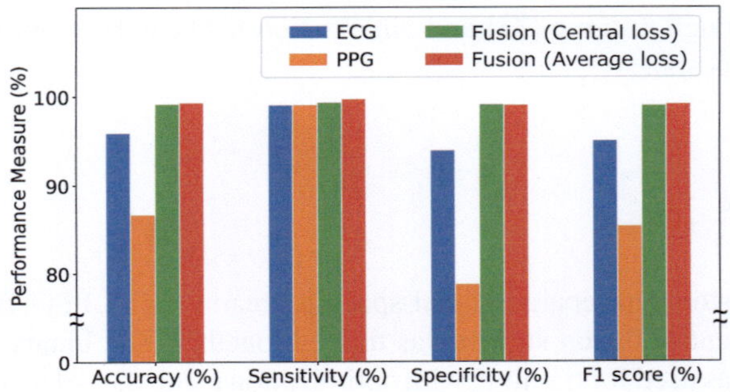

Fig. 7.4 Bar chart illustrating the performance of the single-signal ECG and PPG networks, along with the fusion network for AF detection, trained using two different loss functions (central loss and average loss) under clean signal conditions. The legend: ECG–ECG signal-only model, PPG–PPG signal-only model, Fusion (Central loss)–Fusion model trained with central loss, Fusion (Average loss)–fusion model trained with average loss

7.5.1 *Loss Functions*

Two different loss functions were evaluated for the fusion model. The first, known as the central loss, is based solely on the decisions of the central network, with this loss propagated to train the individual signal and SQI networks. The second, referred to as the average loss, is calculated as the mean of the losses from the individual signal models and the central fusion network. For a fusion network, minimizing the central loss alone may be sufficient to obtain a good fusion model. However, for wearable devices, minimizing the losses of individual networks is advantageous, as well-performing signal networks can function independently when the fusion algorithm needs to be deactivated due to low battery or other power constraints. Therefore, in this context, an average loss function is employed to minimize a global loss, calculated as the mean of the losses from both the individual networks and the central fusion network. The fused model that used the binary cross-entropy loss of the fusion network exhibited an accuracy of 99.20%, a sensitivity of 99.36%, a specificity of 99.11%, and an F1-score of 98.99%. The fused model that used the averages of the binary cross-entropy losses of all three networks exhibited an accuracy of 99.33%, a sensitivity of 99.74%, a specificity of 99.06%, and an F1-score of 99.14%. The performances of the individual ECG and PPG signal models, as well as the fused models, are depicted in Fig. 7.4 and in Table 7.4.

On comparing the performance of the fused models, there is a significant performance improvement compared to the individual models. The performance of the two different kinds of loss functions is also compared. It can be observed that although the accuracy is higher for the loss function that takes into account the decisions of the individual streams, it is not significantly higher than that of the model that takes into account only the loss function associated with the central network. However, since the performances are quite good in both scenarios, no conclusion on which is

Table 7.4 Performance of the single-signal ECG and PPG networks and the fusion network for atrial fibrillation detection using two different loss functions (central loss and average loss) under clean signal conditions

Model	Ac (%)	Se (%)	Sp (%)	F1 (%)
ECG	95.96	99.10	93.95	95.04
PPG	86.71	99.10	78.78	85.34
Fusion (central loss)	99.20	99.36	99.11	98.99
Fusion (average loss)	99.33	99.74	99.06	99.14

the better loss function can be derived, although intuitively, better performance of the individual streams should lead to better performance in the fusion network.

The fusion of different sensors is most suitable in scenarios when the individual sensors are corrupted by noise. Severe motion artifacts, electrode contact noise, and electromyographic noise can affect data obtained from wearable devices, and therefore obtaining accurate inferences is a major challenge in wearable devices. The SQI streams are especially useful in the scenarios where noisy data is also used for training. Therefore, the performance of the SQI-based multi-level fusion model is analyzed when the data is corrupted by noise in the next subsection.

7.5.2 Performance Evaluation with Noisy Data

The performance of the models when noisy data are utilized as inputs is assessed using simulated situations with noisy input signal windows. To accomplish this, noise is used to corrupt each sample within a noisy signal frame. White Gaussian noise is added to the signal segments, resulting in an output signal-to-noise ratio of –20 dB. White Gaussian noise is used in this scenario to capture noise across all frequencies. The training, validation, and test sets are each partitioned into quarters with:

1. 25% of the signal windows containing ECG signals that are corrupted by noise alongside clean PPG signals,
2. 25% of the signal windows containing PPG signals that are corrupted by noise alongside clean ECG signals,
3. 25% of the signal windows in which both signals are corrupted by noise, and
4. 25% of the signal windows in which both signals are clean.

The individual ECG and PPG signal-based atrial fibrillation detection models were retrained using the dataset with simulated noise. The ECG-based model achieved an accuracy of 84.40%, a sensitivity of 77.07%, a specificity of 89.08%, and an F1-score of 79.41%. The PPG-based model demonstrated an accuracy of 81.33%, a sensitivity of 91.56%, a specificity of 74.79%, and an F1-score of 79.29%. From the

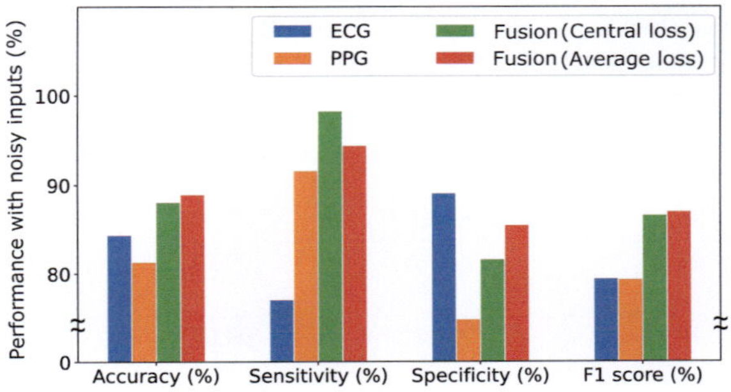

Fig. 7.5 Bar chart illustrating the performance of the single-signal ECG and PPG networks, along with the fusion network for AF detection, trained using two different loss functions (central loss and average loss) in the noisy signal scenario. The legend: ECG–ECG signal-only model, PPG–PPG signal-only model, Fusion (Central loss)–Fusion model trained with central loss, Fusion (Average loss)–fusion model trained with average loss

Table 7.5 Performance of the single-signal ECG and PPG networks and the fusion network using two different loss functions (central loss and average loss) for atrial fibrillation detection in the noisy signal scenario

Model	Ac (%)	Se (%)	Sp (%)	F1 (%)
ECG	84.40	77.07	89.08	79.41
PPG	81.33	91.56	74.79	79.29
Fusion (central loss)	88.09	98.24	81.60	86.56
Fusion (average loss)	88.94	94.38	85.45	86.95

results, it can be observed that the accuracy and sensitivity are significantly higher for the fusion model compared to ECG and PPG models. The fused model with the average loss was found to exhibit an accuracy of 88.94%, a sensitivity of 94.38%, a specificity of 85.45%, and an F1-score of 86.95%. The performances of the ECG and PPG signal models, along with the fusion model using the two different losses, are depicted in Fig. 7.5 and Table 7.5. From the performance of the ECG and PPG models as well as the fused models, it can be seen that the models are highly biased toward naming most events as atrial fibrillation events when there are noisy signals being used during the training process. These observations indicate that when the signals are corrupted by noise, important features that can be learned on clean signals are not learned by the model, and in such scenarios, the fusion algorithm proposed in this chapter will be useful in tackling this issue. It can also be observed that the average loss function performs better than the central loss function in terms of the F1-score, but the model trained with the central loss function is more accurate.

7.5.3 *Performance Evaluation Without SQI Inputs for Both Clean and Noisy Inputs*

The models were designed based on the assumption that SQI streams are particularly beneficial in scenarios where noisy data is included in training. This approach was taken with the expectation that the deep learning models would exhibit behavior similar to the signal processing models developed in Chaps. 4 and 5. To verify this assumption, the fusion model is trained without SQI inputs to assess how much the model's performance deteriorates when SQI streams are not fed into the model. For this experiment, the model described in Sect. 7.4.4 is utilized. However, instead of providing the input SQI streams for each training example, a random sequence is generated and maintained as a constant for all data samples, which is then fed into the model in place of the SQI streams. This setup is equivalent to using a constant input for each data sample in the database. The performance of these SQI-free fusion models, with both the central loss and average loss, is assessed and compared to the previously discussed fusion models for both clean and noisy signal inputs, as shown in Table 7.6. From Table 7.6, it can be seen that the integration of SQIs, even in the case of clean signals, improves performance in terms of accuracy and F1-score for both the loss functions considered. This may occur because the raw data is not entirely noise-free. When noisy signals are used as inputs for training and evaluation, the fusion models that utilize SQIs outperform the models that do not incorporate SQIs, for both central and average losses, in terms of accuracy and F1-score.

It can be observed from Table 7.6 that in the clean signal scenario, the models that use the SQI input streams outperform the models that do not. It can be observed that in both the clean signal and noisy signal scenarios, the usage of the SQI input streams improves the performance of the models. In the noisy signal scenarios,

Table 7.6 Performances of the fusion architecture for AF detection, with and without SQI inputs for both clean and noisy signal inputs

Metric	With SQI		Without SQI	
	Central loss (%)	Average loss (%)	Central loss (%)	Average loss (%)
Clean signal inputs				
Accuracy	99.20	99.33	98.47	99.24
Sensitivity	99.36	99.74	99.91	98.95
Specificity	99.11	99.06	97.54	99.42
F1-score	98.94	99.14	98.07	99.03
Noisy signal inputs				
Accuracy	88.09	88.94	87.43	85.68
Sensitivity	98.24	94.38	93.77	97.28
Specificity	81.60	85.45	84.65	78.25
F1-score	86.56	86.95	86.39	84.13

higher sensitivities are observed when the SQI streams are used as inputs. It can be observed that in the clean signal scenario without SQI inputs, the average loss outperforms the central loss model, but in the noisy signal input scenario, the central loss is more beneficial in the scenario where the SQIs are not used as inputs. It should be noted that even the model that uses the constant random sequence as SQI inputs still benefits from the structure of the proposed fusion network. Therefore, to observe more significant performance improvements due to the use of SQIs, the fusion network should not have any of the processing networks used for the SQIs. The results here indicate that it is especially beneficial to provide SQI streams as inputs to the model when noisy signal inputs may be present during training and inference, thereby verifying the importance of multi-level fusion models that account for the signal quality indicators of the input signals.

The performance of the proposed multi-level fusion model are compared with those of state-of-the-art fusion algorithms for AF detection in Table 7.7. From the table, it is evident that both fusion models proposed in this work outperform the fusion models discussed in [9, 22, 24]. However, the use of proprietary datasets in these works indicates that a direct comparison is not advisable. The performance improvements observed with the proposed model in comparison to state-of-the-art can be attributed to the proposed methodology of developing a fusion model that takes into account the SQIs of the signals considered, while also using a data-driven approach to select the information abstraction stage at which fusion happens.

Table 7.7 Comparison of the proposed 1D-CNN-based signal-quality-aware multi-level fusion model for detecting atrial fibrillation against state-of-the-art fusion-based atrial fibrillation detection algorithms

Article	Bonomi et al. [22]	Zhu et al. [35]	Shashikumar et al. [24]	This work	
Dataset	Proprietary dataset	Proprietary dataset	Proprietary dataset	MIMIC III subset	
Method	Markov model based on PPG pulse peaks	Machine learning and statistical decision model	CNN + hand engineered features	CNN with SQIs	
Signal	PPG, Accelerometer data	PPG, Accelerometer data	8 lead PPG	ECG and PPG	
Accuracy	98%	–	91.8%	Central loss	Average loss
				99.20%	99.33%
Sensitivity	97%	87.8%	89%	99.36%	99.74%

7.6 Discussions

The proposed fusion framework introduces a self-learning multi-level fusion approach using 1D-CNNs for atrial fibrillation (AF) detection, integrating ECG and PPG signals. Unlike traditional fusion methods, which rely on predefined fusion levels, this model automatically learns the optimal information abstraction level for combining multimodal signals. This flexibility is essential in biomedical applications where signal characteristics can vary due to noise, sensor placement, and patient-specific variations. Additionally, the incorporation of signal quality indicators (SQIs) directly into the fusion process ensures that low-quality signals contribute less to the decision-making process, improving robustness. The results demonstrate that this approach outperforms single-sensor models, highlighting the benefits of multimodal data fusion in wearable healthcare applications.

Looking ahead, real-world validation with diverse patient populations is crucial to assess model robustness in uncontrolled environments, especially with wearable devices. Another promising direction is real-time implementation on wearable devices, which would require balancing accuracy, power efficiency, and latency. Furthermore, integrating explainability techniques to interpret how the network prioritizes different fusion levels could enhance clinical adoption and trust in AI-driven healthcare solutions. By addressing these challenges, this work contributes to advancing multimodal AI-powered health monitoring, paving the way for more reliable and adaptive wearable healthcare systems.

7.7 Conclusions

In this work, a multi-level fusion architecture for 1D time-series signals is proposed to capture temporal information like event detection, while also using the signal quality indicators of the individual signals to aid in the training and inference stage. Here, an analysis on whether the fusion model can improve accuracy over that of using CNN models for each signal individually is carried out. In the proposed architecture, the fusion network learns the optimal parameters for fusion—a data-driven approach instead of a heuristic method for fusion, where the information abstraction level at which fusion happens is learned by the network. This method allows for more flexibility as relevant information in the early stages is fused and propagated to the later stages if required, when fusion at later stages also occurs. This fusion model is applied to the atrial fibrillation detection problem using ECG and PPG signals. Prior literature in the field indicates that these two signals can detect AF events with reasonably good accuracy. To aid in the fusion process, especially for wearable sensor signal processing, the individual signal quality indicators of the ECG and PPG signals are fed into the fusion network. The performance of the fusion model and individual signal models were analyzed over a common test set to confirm the hypothesis that the fusion model would perform better. The performance of the fusion model was observed when only the fusion network loss was considered for training, versus when the average of the losses from both the fusion network and the individual networks was used for training. It was found that using the average loss showed a performance improvement of 0.13% compared to using just the central loss, indicating that well-performing individual models improve the performance of the fusion network in the case of clean signal inputs. In the case of the model that uses the average loss, the accuracy was found to be 3.51% higher on the test set compared to the ECG signal-based model and 14.55% higher than the accuracy of the PPG signal-based model. The performance of the fusion models, when the signals were corrupted by noise, was also analyzed, and it was observed that the model that used the average loss-based training was better than the central loss-based training. The average loss-based model outperformed the ECG signal model by 5.38% and the PPG signal model by 9.36% in terms of accuracy. The performance of the fusion architecture without SQI input streams was also analyzed to demonstrate the benefits of providing SQI streams as inputs to the fusion models. It was observed that in the noisy signal scenario, the model that uses the SQI information for fusion outperforms the model that does not take into account the input signal SQI by 0.19% when the central loss is considered, and by 3.35% when the average loss is considered, in terms of accuracy. The performances of the proposed fusion models were compared against other fusion models in the literature, and the proposed fusion model outperforms the state-of-the-art models discussed in the literature. The proposed fusion algorithm is an example of complementary and multi-level fusion.

References

1. A. González Bermúdez, D. Carramiñana, A.M. Bernardos, L. Bergesio, J.A. Besada, A fusion architecture to deliver multipurpose mobile health services. Comput. Biol. Med. **173**, 108344 (2024)
2. F. Castanedo, A review of data fusion techniques. Sci. World J. (2013)
3. R. Gravina, P. Alinia, H. Ghasemzadeh, G. Fortino, Multi-sensor fusion in body sensor networks: state-of-the-art and research challenges. Inf. Fusion **35**, 68–80 (2017)
4. K. Tan, W. Huang, X. Liu, J. Hu, S. Dong, A multi-modal fusion framework based on multi-task correlation learning for cancer prognosis prediction. Artif. Intell. Med. **126**, 102260 (2022)
5. R.C. Luo, M.G. Kay, A tutorial on multisensor integration and fusion, in *[Proceedings] IECON'90: 16th Annual Conference of IEEE Industrial Electronics Society*, vol. 1 (Nov. 1990), pp. 707–722
6. E. Besler, Y.C. Wang, A.V. Sahakian, Early and late fusion machine learning on multi-frequency electrical impedance data to improve radiofrequency ablation monitoring. IEEE J. Biomed. Health Inf. **24**(8), 2359–2367 (2020)
7. V. Vielzeuf, A. Lechervy, S. Pateux, F. Jurie, Multilevel sensor fusion with deep learning. IEEE Sens. Lett. **3**(1), 1–4 (2019)
8. J. Pellman, F. Sheikh, Atrial fibrillation: mechanisms, therapeutics, and future directions. Compr. Physiol. **5**, 649–665 (2015)
9. L. Zhu, V. Nathan, J. Kuang, J. Kim, R. Avram, J. Olgin, J. Gao, Atrial fibrillation detection and atrial fibrillation burden estimation via wearables. IEEE J. Biomed. Health Inf. **26**(5), 2063–2074 (2022)

10. X. Wu, Y. Zheng, C.-H. Chu, Z. He, Extracting deep features from short ECG signals for early atrial fibrillation detection. Artif. Intell. Med. **109**, 101896 (2020)
11. K. Tahsin, M.F. Hossain, M.A. Rahman, Computer aided atrial fibrillation detection from the statistical attributes of ECG signal, in *2021 International Conference on Electronics, Communications and Information Technology (ICECIT)* (2021), pp. 1–4
12. E. Petrucci, V. Balian, G. Filippini, L. Mainardi, Atrial fibrillation detection algorithms for very long term ECG monitoring. Comput. Cardiol. **2005**, 623–626 (2005)
13. X. Ruan, C. Liu, C. Liu, X. Wang, P. Li, Automatic detection of atrial fibrillation using R-R interval signal, in *2011 4th International Conference on Biomedical Engineering and Informatics (BMEI)*, vol. 2 (2011), pp. 644–647
14. R. Couceiro, P. Carvalho, J. Henriques, M. Antunes, M. Harris, J. Habetha, Detection of atrial fibrillation using model-based ECG analysis, in *2008 19th International Conference on Pattern Recognition* (2008), pp. 1–5
15. H.-G. Kim, U. Erdenebayar, C.-H. Kang, D.-W. Kang, K.-J. Lee, Estimation of atrial fibrillation using arbitrary normal ECG segments based on convolutional neural networks, in *2018 International Conference on Electronics, Information, and Communication (ICEIC)* (2018), pp. 1–2
16. Y. Liu, B. Fang, Y. Zhao, J. Chen, Ensemble learning for atrial fibrillation screening from a single lead ECG wave of wearable devices, in *2021 IEEE 3rd International Conference on Frontiers Technology of Information and Computer (ICFTIC)* (2021), pp. 590–594
17. X. Fan, Q. Yao, Y. Cai, F. Miao, F. Sun, Y. Li, Multiscaled fusion of deep convolutional neural networks for screening atrial fibrillation from single lead short ECG recordings. IEEE J. Biomed. Health Inf. **22**(6), 1744–1753 (2018)
18. S. Sadasivuni, R. Chowdhury, V.E.G. Karnam, I. Banerjee, A. Sanyal, Recurrent neural network circuit for automated detection of atrial fibrillation from raw ECG, in *2021 IEEE International Symposium on Circuits and Systems (ISCAS)* (2021), pp. 1–5
19. J. Lee, B.A. Reyes, D.D. McManus, O. Mathias, K.H. Chon, Atrial fibrillation detection using a smart phone, in *2012 Annual International Conference of the IEEE Engineering in Medicine and Biology Society* (2012), pp. 1177–1180
20. S.-M. Shan, S.-C. Tang, P.-W. Huang, Y.-M. Lin, W.-H. Huang, D.-M. Lai, A.-Y.A. Wu, Reliable PPG-based algorithm in atrial fibrillation detection, in *2016 IEEE Biomedical Circuits and Systems Conference (BioCAS)* (2016), pp. 340–343
21. P. Cheng, Z. Chen, Q. Li, Q. Gong, J. Zhu, Y. Liang, Atrial fibrillation identification with PPG signals using a combination of time-frequency analysis and deep learning. IEEE Access **8**, 172692–172706 (2020)
22. A.G. Bonomi, F. Schipper, L.M. Eerikäinen, J. Margarito, R.M. Aarts, S. Babaeizadeh, H.M. de Morree, L. Dekker, Atrial fibrillation detection using photo-plethysmography and acceleration data at the wrist, in *2016 Computing in Cardiology Conference (CinC)* (2016), pp. 277–280
23. L.M. Eerikäinen, A.G. Bonomi, F. Schipper, L.R.C. Dekker, H.M. de Morree, R. Vullings, R.M. Aarts, Detecting atrial fibrillation and atrial flutter in daily life using photoplethysmography data. IEEE J. Biomed. Health Inf. **24**(6), 1610–1618 (2020)
24. S.P. Shashikumar, A.J. Shah, Q. Li, G.D. Clifford, S. Nemati, A deep learning approach to monitoring and detecting atrial fibrillation using wearable technology, in *2017 IEEE EMBS International Conference on Biomedical Health Informatics (BHI)* (2017), pp. 141–144
25. M.R.G. Moody, A new method for detecting atrial fibrillation using R-R intervals. Comput. Cardiol. **10**, 4 (1983)
26. I. Moody, B. Moody G. Villarroel M. Clifford G. Silva, MIMIC-III waveform database matched subset (2020)
27. A.E. Johnson, T.J. Pollard, L. Shen, L. Wei, H. Lehman, M. Feng, M. Ghassemi, B. Moody, P. Szolovits, L.A. Celi, R.G. Mark, MIMIC-III, a freely accessible critical care database. Sci. Data **3**(1), 160035 (2016)
28. S.K. Bashar, M.B. Hossain, E. Ding, A.J. Walkey, D.D. McManus, K.H. Chon, Atrial fibrillation detection during sepsis: study on MIMIC III ICU data. IEEE J. Biomed. Health Inf. **24**(11), 3124–3135 (2020)

29. U. Satija, B. Ramkumar, M. Sabarimalai Manikandan, Real-time signal quality-aware ECG telemetry system for IoT-based health care monitoring. IEEE Internet Things J. **4**(3), 815–823 (2017)
30. L. Smital, C.R. Haider, M. Vitek, P. Leinveber, P. Jurak, A. Nemcova, R. Smisek, L. Marsanova, I. Provaznik, C.L. Felton, B.K. Gilbert, D.R. Holmes III., Real-time quality assessment of long-term ECG signals recorded by wearables in free-living conditions. IEEE Trans. Biomed. Eng. **67**(10), 2721–2734 (2020)
31. C. Orphanidou, *Signal Quality Assessment in Physiological Monitoring: State of the Art and Practical Considerations* (Springer International Publishing, 2018)
32. W. Karlen, K. Kobayashi, J.M. Ansermino, G.A. Dumont, Photoplethysmogram signal quality estimation using repeated gaussian filters and cross-correlation. Physiol. Meas. **33**(10), 1617–1629 (2012)
33. D. Blalock, J.J.G. Ortiz, J. Frankle, J. Guttag, What is the state of neural network pruning? (2020)
34. M. Courbariaux, I. Hubara, D. Soudry, R. El-Yaniv, Y. Bengio, Binarized neural networks: training deep neural networks with weights and activations constrained to +1 or –1 (2016)
35. L. Zhu, V. Nathan, J. Kuang, J. Kim, R. Avram, J.E. Olgin, J. Gao, Atrial fibrillation detection and atrial fibrillation burden estimation via wearables. IEEE J. Biomed. Health Inf. 1–1 (2021)

Chapter 8
Discussions and Conclusions

Keywords Data fusion · Generalized fusion architecture · Health monitoring · Wearables · Explainable AI · Federated learning · Algorithmic fairness · Responsibility in clinical decision-making

8.1 Summary

This book presented the potential of data fusion techniques applied to wearable sensor signal processing for health monitoring and addressed gaps in data fusion literature. The classes of fusion frameworks developed in this book were briefly discussed in Chap. 1.

Chapter 2 discussed existing literature on data fusion, summarized commonly used fusion frameworks, and identified gaps in the literature.

Chapter 3 presented a study on using various machine learning algorithms in conjunction with previously studied signal quality indicators for diagnostic quality assessment of ECG signal segments. The **bagged ensemble of neural networks** was the best-performing model and computationally less expensive compared to other well-performing classifiers and was found to be most suited for deployment on an energy-efficient IoT device. The method discussed was found to be suitable for improving the quality of ECG signals obtained from wearable IoT devices based on performance accuracy as well as computational complexity.

Chapter 4 explored fusion algorithms based on three different characterizations: selection-based, combination-based, and elimination- and combination-based fusion. The algorithms studied were for the fusion of signals obtained from multi-sensor sources acquiring data of similar modalities. Various methods that belong to these categories were studied, and the best fusion strategy based on the quality of the input signals was determined. The performance of the fusion algorithms on signals corrupted by muscle artifacts, electrode motion, and baseline wander indicates

A. John et al., *Deep Learning and Signal-Processing Methods for Multisensor Data Fusion*, https://doi.org/10.1007/978-3-031-96724-5_8

the suitability of the fusion algorithm for deployment on a wearable IoT device. The developed methods were applied to two-channel ECG records for heart rate estimation. It was observed that the **Kalman fusion algorithm** exhibits the best performance when both channels have signals with low SNRs, and the **α-trim mean filtering algorithm** is suitable for all the other noise scenarios. Consequently, the feasibility of using SQIs as a surrogate for signal SNR for algorithm selection is proposed, and it was observed that the strategy of using Kalman fusion when SQIs of both channels are lower than a threshold and α-trim mean filtering elsewhere in the input signal space is effective in obtaining the best estimate of the RR interval.

Chapter 5 presented a wavelet-based fusion framework for the fusion of signals of different modalities. The wavelet transform domain was used to capture signal information in a common representational format for fusion. The proposed method used an SQI-weighted technique for the fusion of the wavelet transform features. A generalized **signal quality index-based weighted fusion for multimodal signals in the wavelet transform domain** was proposed to ensure that only signals of reliable signal quality can contribute to decision-making. The proposed fusion was applied to the heartbeat detection problem through the fusion of ECG and PPG signals. The performance of the proposed fusion algorithm was tested at different SNRs on the addition of various kinds of noise from −50 to 50 dB. The fusion algorithm yielded a sensitivity of 99.69 %, a positive predictive value of 99.64 %, E_{rel} of 0.01, and E_{90} of 0.02 from −30 to 50 dB SNRs for all noise scenarios considered. On comparing the performance of the proposed fusion algorithm to some state-of-the-art single-lead ECG beat detection algorithms and a single-lead PPG beat detection algorithm, the fusion algorithm yielded better results in almost all noise scenarios considered.

Chapter 6 developed a **1D-CNN-based fusion algorithm for multiresolution signals**, i.e., signals sampled at different sampling frequencies for temporal inferences, without having to resort to padding/resampling (which leads to additional computational costs). An experimental study in which selective dropout of the features obtained from the sensor with a larger sampling rate is used to prevent the fusion model from overfitting to features from the sensor with the larger sampling rate was carried out. The proposed model is applied to the sleep apnea detection problem using ECG, SpO2, and abdominal movement signals with the signals sampled at different sampling rates. The performance of the three signal fusion models and various combinations of two signal fusion models were analyzed to exhibit the merits of the fusion algorithm. Pruning methods were also studied to compensate for the increase in computational complexity arising from additional sensor sources. It was observed that although an increase in model sparsity may lead to a drop in performance of the fused model, the pruned fusion model performance can outperform individual model performances at lower computational complexities—indicating that pruned fusion models are a promising direction for fusion in wearable devices.

Chapter 7 presented a **multi-level fusion framework using 1D-CNNs**, where the network learns the best information abstraction level at which fusion needs to be carried out while also using the signal quality indicators of the individual signals to aid in the training and inference stage. This method allows for more flexibility as relevant information in the early stages is fused and propagated to the later stages if required when fusion at later stages also occurs. This fusion model is applied to the atrial fibrillation detection problem using ECG and PPG signals. The performance of the fusion model was evaluated by comparing two training approaches: one where only the fusion network loss was considered and another where the training was based on the average of the losses from both the fusion network and the individual networks. It was found that using the average loss showed a performance improvement of 0.13% compared to using just the central loss, indicating that individual well-performing models improve the performance of the fusion network. The performance of the fusion models when the signals were corrupted by noise was observed, and it was observed that the fusion model that used the average loss outperformed the ECG signal model by 5.49% and the PPG signal model by 5.99% in terms of accuracy. The benefits of providing the signal quality indicators of the input signals as input streams to the fusion model were demonstrated by comparing the performance of the SQI-based multi-level fusion model to a generic multi-level fusion model. The performances of the fusion models were compared against other fusion models in the literature, and the proposed models outperformed the state-of-the-art fusion models discussed in the literature.

8.2 Discussions

The findings from the various studies encapsulated in this book demonstrate the efficacy of multi-sensor and multimodal data fusion techniques for improving the accuracy and reliability of health monitoring systems. Through the implementation of various fusion methodologies, including rule-based, SQI-driven, and deep learning-based approaches, this research has provided a comprehensive analysis of fusion architectures applicable to wearable health monitoring systems.

The research, unsurprisingly, illustrates that the inclusion of SQI inputs significantly improves performance, especially in noisy environments where individual sensor readings might be unreliable. By dynamically weighting sensor contributions based on their respective SQI values, the proposed models ensure that decision-making is based on the most reliable data at any given time. This addresses a common problem in wearable health monitoring, where sensors may produce faulty or inconsistent readings due to external factors. The ability to automatically assess and prioritize high-quality signals represents an important step toward the deployment of more effective and autonomous wearable health monitoring solutions. The proposed sample-wise SQI-based fusion frameworks also introduce a novel approach to assessing and selecting reliable signals dynamically, reducing the impact of low-quality sensor data on overall model performance. Compared to state-of-the-art techniques, this framework exhibits superior robustness and adaptability, making it a viable candidate for real-world applications in continuous health monitoring, if low complexity sample-wise SQI assessment can be carried out. This innovation is particularly useful in real-time applications where instant decisions regarding a patient's health status are necessary. The system's ability to detect unreliable signals and compensate by leveraging other sources of data ensures that health monitoring remains consistent and accurate, even under challenging conditions.

A key contribution of this research is the development of a data-driven multi-level fusion architecture that integrates 1D time-series signals, capturing temporal information while leveraging SQI-based decision-making. This multi-level approach addresses a significant challenge in physiological signal processing: the fusion information abstraction level and the variation in signal quality due to motion artifacts or environmental interference. The experimental results confirm that the proposed fusion models outperform individual sensor models, showcasing the advantage of integrating complementary signals. The proposed multi-level fusion approach introduces a generalized paradigm for SQI-aware fusion for time-series detection tasks from one-dimensional time-series signals. The data-driven approach automatically determines the most relevant features for classification, reducing the need for manual feature engineering. This supports the hypothesis that combining diverse physiological signals leads to enhanced inferencing capabilities and better resilience against noise artifacts.

8.3 Limitations and Critical Assessment

Although the book aimed to develop a generalizable fusion framework for biomedical applications, the analysis presented in this book leans toward testing the feasibility of concepts and validating certain applications rather than developing an optimized implementation for a specific task deployable on hardware. Despite the promising results, certain limitations must be acknowledged. The dependency on high-quality labeled datasets for training deep learning models poses a challenge, as real-world data often includes inconsistencies and artifacts that are difficult to simulate. The proposed fusion frameworks can be extended to analogous classes of fusion problems; however, for different classes of problems, the fusion frameworks discussed in this book have to be modified accordingly.

Although fusion improves accuracy, the fusion methods need more computational power and sensors compared to using a single information channel for inference. The computational complexity is factored in a few of the methods introduced in this book, but the computational complexity study was carried out purely on the basis of the computations required for the operations. This method is only suitable to provide an initial estimate on the suitability of deployment in wearable devices. Implementation on IoT devices needs to be studied to ensure that the gains brought about by fusion are not overshadowed by the power consumption and area requirements. Future research should not only consider the broad class of problems discussed in this book but also focus on specific applications where fusion is the only viable solution for enabling deployment in wearable devices. This approach will ensure that the value proposition for its adoption is clear and compelling.

The performance of the fusion algorithms is dependent on the fact that at least one channel of information is clean. In a scenario where all the input signals are noisy, the fusion process may provide worse results than when any one of the noisy channels is used for inference.

The signal quality indicators developed in this work rely on prior knowledge of the signals used. This is feasible due to the quasi-periodic nature of signals like ECG and PPG, which are analyzed in this book. Statistical features such as kurtosis and skewness are utilized to assess the diagnostic quality of these signals. Moreover, the instantaneous signal quality indicator developed in this book makes use of the signal cycle template to estimate the signal quality. Therefore, the developed signal quality indicators are not suitable for all biomedical signals. Thus, depending on the type of signal used for fusion, signal quality indicators suitable for those signals need to be developed, else the fusion process needs to forego the use of signal quality indicators.

The computational complexity of deep learning-based fusion models may hinder their deployment on low-power wearable devices, necessitating further optimization techniques such as pruning, quantization, and edge computing strategies to enhance efficiency. In terms of computational complexity, the tradeoff between gains due to fusion and computational complexity needs to be evaluated through the implementation of the fusion frameworks in an IoT wearable device.

8.4 Recommendations for Future Works

A guideline for developing signal quality indicators of non-quasi-periodic signals needs to be developed to extend the SQI-based fusion framework to non-quasi-periodic signals such as electroencephalogram signals.

In the case of the 1D-CNN-based fusion models, an analysis of the frequency responses of the filters learned by the network can be carried out to map human understanding of the requisite features maps for a task to the features required for the task from a machine perspective. This would aid in the development of deterministic models that may be more suited in scenarios where deep learning-based methods fail.

Optimizing the computational efficiency of deep learning-based fusion techniques will be crucial for facilitating real-time deployment on resource-constrained wearable devices. Techniques such as model compression, edge computing, and federated learning should be explored to balance computational efficiency with predictive performance.

The development of a multi-level and multiresolution fusion model for automatic fusion level detection for multiresolution data can be developed, deriving from the results of Chaps. 6 and 7.

Exploring hybrid approaches that combine heuristic and data-driven fusion methodologies may further enhance adaptability and performance across different healthcare applications. A promising avenue for future research is the integration of explainable AI (XAI) techniques to improve the explainability and interpretability of fusion-based decision-making, ensuring that clinicians and users can understand and trust the system's outputs.

Finally, efforts toward the development of a generalizable fusion framework that can be extended to any class of applications needs to be expended. For example, the multi-level fusion framework developed in Chap. 7 can be considered to be a generic case of the fusion algorithms developed in Chaps. 4 and 5, where the heart rate estimation problem can be posed as a regression problem, while the heartbeat detection problem can be posed as a classification problem. Further analysis in this direction may point us toward a generalizable fusion framework that can be extended to different classes of problems. The development of open-access databases with multi-sensor physiological recordings would greatly benefit the field by enabling researchers to train and evaluate fusion models under a wider range of conditions. Therefore, the future of wearable health monitoring powered through data fusion appears to hold significant potential as projected by the results of the methods proposed in this book.

8.5 Considerations for Implementation

To implement data fusion-based inference in wearables and remote patient monitoring, the explainability of AI-based decision-making needs to be prioritizied. Developments in explainable AI techniques are crucial to bringing multi-sensor data fusion in remote patient monitoring to support clinical decision-making [1]. Developing uncertainty modeling approaches, such as Bayesian methods, can also improve the reliability of predictions by managing ambiguous or incomplete data.

Data quality remains a fundamental priority for accurate health monitoring. Improvements in data pre-processing, such as adaptive filtering [2], can enhance data quality prior to fusion. Incorporating methods to handle missing data through the development of robust imputation techniques are crucial for reliable remote patient monitoring. Additionally, due to the lack of large reliable datasets for training fusion models, especially with the restrictions placed on data privacy, federated learning, which allows model training across multiple devices without centralizing data, can also play a transformative role [3]. By processing data locally, federated learning strengthens privacy and security while enabling more personalized and adaptive health monitoring systems without compromising patient confidentiality. However, relying on localized datasets for model development can introduce biases influenced by demographic, geographic, and socioeconomic factors, as the training data may not adequately represent diverse populations. This can lead to AI models that unintentionally reinforce existing healthcare disparities, resulting in less accurate predictions or recommendations for underrepresented groups. To mitigate these biases, it is crucial to implement fairness-aware algorithms. Techniques such as fair representation learning and disentanglement aim to ensure that AI models provide equitable outcomes across different demographic groups [4]. Additionally, adopting the FAIR (Findable, Accessible, Interoperable, and Reusable) principles for data management can enhance the quality and representativeness of datasets used in AI model training. By ensuring that research data is stored and managed in a FAIR manner, we can promote more equitable and unbiased decision-making in AI-driven healthcare systems.

Personalized health monitoring represents a valuable future direction for multi-sensor remote patient monitoring systems. Adaptive models that account for individual patient characteristics can improve monitoring accuracy, as they adjust data fusion techniques based on each patient's unique context and health history. Additionally, advances in the field of human-in-the-loop AI can support the personalization of these algorithms through input from the patient and their healthcare practitioner, further reducing responsibility concerns [5] associated with remote patient monitoring solutions.

To conclude, this book underscores the transformative potential of data fusion in wearable health monitoring, leading to the development of more reliable, accurate, and intelligent healthcare solutions. By addressing the existing challenges and leveraging emerging technologies, future advancements in this field could significantly improve personalized healthcare and early disease detection, ultimately contributing to better health outcomes globally. The advancements presented in this book provide a foundation for further innovation in multi-sensor fusion, with the goal of creating fully autonomous, highly reliable health monitoring systems that operate seamlessly in real-world environments.

References

1. A. Gerdes, The role of explainability in AI-supported medical decision-making. Discover Artif. Intell. **4** (2024)
2. A. John, J. Sadasivan, C.S. Seelamantula, Adaptive savitzky-golay filtering in non-gaussian noise. IEEE Trans. Signal Process. **69**, 5021–5036 (2021)
3. T. Wang, Y. Du, Y. Gong, K.-K.R. Choo, Y. Guo, Applications of federated learning in mobile health: scoping review. J. Med. Internet Res. **25**, e43006 (2023)
4. R.J. Chen, J.J. Wang, D.F. Williamson, T.Y. Chen, J. Lipkova, M.Y. Lu, S. Sahai, F. Mahmood, Algorithmic fairness in artificial intelligence for medicine and healthcare. Nat. Biomed. Eng. **7**, 719–742 (2023)
5. A.V. Silven, P.G. van Peet, S.N. Boers, M. Tabak, A. de Groot, D. Hendriks, H.J.A. van Os, T.N. Bonten, D.E. Atsma, T.J. de Graaf, M.P. Sombroek, N.H. Chavannes, M. Villalobos-Quesada, Clarifying responsibility: professional digital health in the doctor-patient relationship, recommendations for physicians based on a multi-stakeholder dialogue in the Netherlands. BMC Health Serv. Res. (2022)

Appendix A
Low Complexity Signal Quality Indicators

A.1 Implementation of the k-NN-Based Quality Assessment Using Approximate Multipliers

The feature vectors derived in Chap. 3 are used by the k-NN algorithm for classifying records into acceptable (0) and unacceptable (1) categories. k-NN is used for this classification task because it is well-suited for case-based decision-making [1]. The implementation of the k-NN-based classifier requires several multiplication operations. Multipliers are usually costly in terms of power and complexity. Since k-NNs are distance-based metrics that merely use the calculated distances for comparisons within the training set, accurate distance computation is not important. The possibility of using an approximate 16×16 multiplier for feature extraction and classification for the implementation of the data integrity assessment in an IoT device is explored.

The k-NN classifier can be modified with new training data easily in an edge wearable device if required during the update stages. In this experiment, the k-NN feature space is 6 dimensional and populated with 512 objects (number of training examples) during the training phase. Euclidean distance was chosen as the distance measure. The selection of the hyperparameter k for this experiment was chosen based on the performance of various k values for voting as discussed in Chap. 3. An area-optimized low-latency multiplier is used to reduce computational cost and power consumption [2]. This approximate multiplier offers 25–31.5% area reduction, 8.6–53.2% reduction in latency, and 8.86–67% gains in Energy-Delay-product (EDP) when compared to the accurate multiplier implementation offered by Xilinx Vivado [3]. The multiplier method utilizes 6-input Look-Up Tables (LUTs) for the generation of approximate partial products. The approximate multiplier works based on the approximate multiplication of a 4-bit number $A(A_3A_2A_1A_0)$ to a 2-bit number $B(B_1B_0)$, which generates a 6-bit output. Truncation of the last bit in the product, P_0, limits the output error to the least significant product bit and the final output accuracy to 75% with a maximum error magnitude of “1” for all input combinations. The approximation of any other product bit results in a higher magnitude of error

A. John et al., *Deep Learning and Signal-Processing Methods for Multisensor Data Fusion*, https://doi.org/10.1007/978-3-031-96724-5

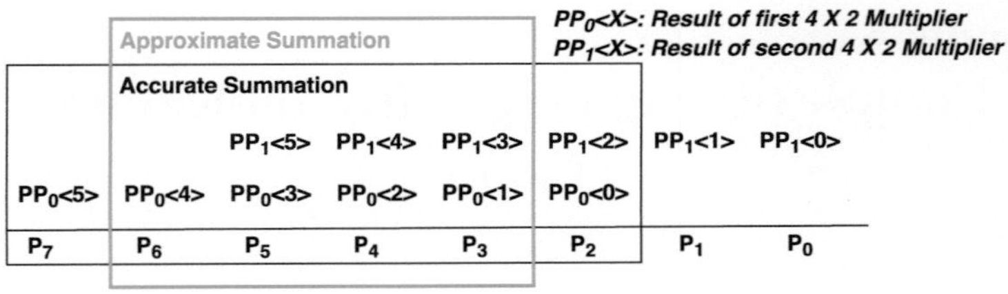

Fig. A.1 4 × 4 multiplier using two 4 × 2 multipliers [2]

in the final output. This method of approximation uses four 6-input LUTs for its implementation by truncating "P_0".

The implementation of a 4 × 4 multiplier requires two 4 × 2 multipliers, consuming 8 LUTs for partial products generation. As shown in Fig. A.1, the accurate summation of the approximate partial products generated by the two 4 × 2 multipliers requires the use of two carry chains. Therefore, the approximate 4 × 4 multiplier, with an accurate summation of partial products, requires 16 LUTs (2 LUTs wasted by the second carry chain). However, the 4 × 4 multiplier design utilizes approximate addition of the partial products and employs only 4 LUTs for the computation of the final product.

For the design of higher-order approximate multipliers, the approximate 4 × 2 and 4 × 4 multipliers are utilized recursively. For example, an approximate 8 × 8 multiplier can be implemented by accurately adding the approximate partial products generated by four instances of approximate 4 × 4 multipliers. Although in [2], the authors discuss unsigned number multiplications, here a signed-unsigned converter is used to carry out signed number multiplication using the approximate unsigned multiplier. The approximate 16 × 16 multiplier discussed has an average dynamic power of 2535.1323 μW with a total delay of 7.328 ns, yielding an EDP of 0.5445×10^{-18} Js.

A.1.1 Performance Analysis

The results and performance comparison of the *k*-NN classifier with the 10-fold cross-validation method utilizing the features generated using both accurate and approximate multipliers are shown in Table A.1. During prediction, the *k*-NN classifier presented an accuracy of 97.18% and a high detection sensitivity of 95.42%. Energy consumption drops significantly when using a 16 × 16 approximate multiplier, instead of accurate multipliers, and these results are discussed in the next section. However, as a tradeoff, the detection sensitivity drops to an acceptable 94.01%. Utilizing a large and diverse dataset can further improve the performance of the *k*-NN classifier, but this leads to an increase in the number of objects stored in the feature space and thereby increases the search cost.

Table A.1 Performance of k-NN in identifying noisy records with approximate multipliers

	Acc (%)	Se (%)	Sp (%)
Accurate multiplier	97.18	95.42	98.94
Approximate multiplier	96.48	94.01	98.94

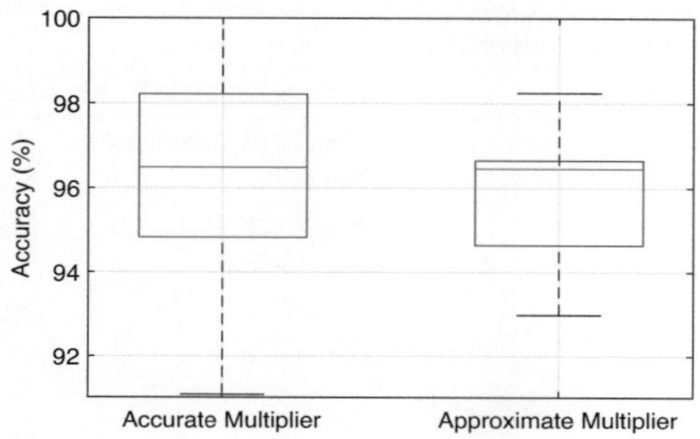

Fig. A.2 Boxplot of accuracy values during testing over the 10 folds for signal quality assessment

The box plot of accuracy values over 10-fold cross-validation with the accurate multiplier and approximate multiplier is shown in Fig. A.2 for performance comparison over all folds of testing.

A.1.2 Complexity Calculation

The computational complexity of the proposed algorithm is evaluated in terms of the number of multiplications required for the quality assessment of a 30-second long record. Feature extraction and normalization stages require a total of 6591 multiplications. Computations of kSQI and sSQI can share a few multipliers. The shared number of multipliers required was 4386. Computation of kSQI alone using the shared values requires 2187 multiplications while sSQI specific calculation requires 12 multiplications. Six multipliers are required for the normalizing stage.

For estimating the computational complexity of the k-NN classifier, the complexity of Euclidean distance calculation to each object needs to be evaluated. k-NN- with 512 training examples- requires approximately 8704 multiplications for classifying a record based on its features. Table A.2 summarizes these values as well as the improvement in EDP when using an approximate multiplier compared to accurate multipliers. For this comparison, Xilinx's area/speed optimized accurate 16×16 signed multiplier is used. All the values of power and total delay are obtained using Vivado 17.4 for Xilinx Virtex-7 xc7v585tffg1157-3 FPGA for verification purposes and the results could be extended to a standard cell-based or custom integrated circuit implementation. The EDP of the area optimized 16×16 accurate multiplier (AOAM) with a total delay of 5.035 ns and dynamic power of 8203.109 μW was

Table A.2 Multipliers required at each stage for quality assessment of one record

Multiplications				Total EDP (f Js)		
Feature generation	Normalization	k-NN classification	Total	AOAM	SOAM	Approximate
6585	6	8704	15301	12.728	8.974	8.332

Table A.3 Comparison of the approximate classifier with state-of-the-art ECG record quality assessment algorithms

Article	Clifford et al. [4]	Li et al. [5]	This work
Dataset	CinC 2011 challenge: Set B	CinC 2011 challenge: Set B	Subset of Cinc'17 challenge: training
Method	F and NF SQIs; ML and fusion	F and NF SQIs, SVM	NF SQIs; ML
Accuracy	**92.6%**	**80.26%**	**96.48%**

found to be 0.8318×10^{-18} Js, while the EDP of the speed optimized 16×16 accurate multiplier (SOAM) with a total delay of 4.269 ns and dynamic power of 8045.487 μW was found to be 0.5864×10^{-18} Js. From Table A.2, it can be observed that the proposed use of an approximate multiplier in place of an accurate multiplier exhibits good improvements in EDP.

The performance of the discussed SQI-based k-NN classifier using approximate multiplications is compared against the performance of state-of-the-art ECG record quality assessment algorithms in Table A.3, and it can be observed that the proposed algorithm exhibits the best performance.

In this Section, we explored the possibility of data integrity assessment of ECG signal segments on wearable devices using approximate multiplier-based computing. To improve the quality of the signal acquired, records of poor signal quality need to be discarded and re-recorded, calling for algorithms that are highly sensitive to poor signal quality. The method discussed used kurtosis SQI and skewness SQI as features for data integrity assessment using a k-NN classifier. The experiment carried out proves that approximate multipliers can be used for data integrity assessment, in place of accurate multipliers, with just a slight drop in sensitivity but with a gain in EDP, which is an important consideration in wearable devices. This would be a helpful add-on to IoT physiological monitoring devices that are currently available. In the next section, we discuss how the complexity of the sample-wise signal quality assessment (cSQI), introduced in Chap. 5, is determined. Additionally, we explain how the fast correlation method reduces computational complexity compared to conventional correlation calculations.

A.2 Complexity Evaluation of the cSQI Algorithm

The cSQI algorithm proposed in Chap. 5 with the fast correlation calculation method was found to have a computational complexity as expressed in Tables A.4, A.5, and A.6 in terms of the number of computations required. From the computational complexity analysis done, it is observable that the fast correlation calculation algorithm cuts down the number of required computations significantly, and therefore the algorithm is suitable for implementation on a wearable device in real-time.

As can be seen from Tables A.5 and A.6, a large number of computations are required to calculate the correlation-based cSQI for the first sample, but for every subsequent sample, the fast correlation calculator reduces the number of computations required by utilizing the correlation and Toeplitz matrices calculated for the previous samples. Now that we have shown that the cSQI is computationally efficient, we now assess the performance of cSQI method as a standalone signal quality indicator for ECG quality assessment on the Physionet/CinC 2011 challenge dataset and the results are discussed in the next section.

Table A.4 Complexity of template generation algorithm (Algorithm 1) for template of length 2M+1 with N cycles

Operation	Count	Count for N=10, M=35 for F_s=125
Multiplication	N ($8M^2$+6M+1)	100110
Addition	N ($8M^2$+4M+1)	99410

Table A.5 Complexity of cSQI calculation during initialization or when calculating for the first sample, c[0]

Operation	Count	Count for M=35 for F_s=125
Multiplication	$8M^2$+8M+3	10083
Addition	$8M^2$+6M+1	10011

Table A.6 Complexity of cSQI calculation for every subsequent sample after initialization

Operation	Count	Count for M=35 for F_s=125
Multiplication	10M+3	353
Addition	12M+3	423

A.3 Performance of the cSQI Algorithm on Physionet/CinC 2011 Challenge Dataset

The performance of the cSQI algorithm proposed in Chap. 5 on the records in the Physionet/CinC 2011 challenge dataset [6] was also analyzed. This was done by using the averages of the cSQI obtained per channel for a record as features for training a decision tree ensemble. This model used the adaptive boosting ensemble creation method with 100 learners for binary classification. For training and validation, Set A of the dataset was used, and the score obtained on validation was 0.938. This model was then tested on the records in Set B of the dataset, and a score of 0.904 was obtained in Event 1 of the challenge. These results were found to be only comparable and did not outperform the scores obtained by the top competitors in the Physionet/CinC 2011 challenge with the top competitor exhibiting a score of 0.932 in Event 1 [6, 7]. But as mentioned in [6], the top competitors also employed a variety of techniques, including machine learning with a large number of fiducial and non-fiducial features to ascertain the acceptability of the records. In this work, a single non-fiducial feature-based signal quality indicator calculated from each of the 12 ECG leads is used to classify records to aid implementation in an IoT device.

Appendix B
Low-Complexity Sleep-Apnea Detection Methods

B.1 Pruning and Binarization of the ECG-Based Sleep Apnea Detection Model

The feasibility of using pruning and binarization techniques on the ECG-based sleep apnea detection model discussed in Chap. 6 is discussed here. The ECG-based 1D-CNN model for sleep apnea detection has 1408 (128 samples per second) nodes in the input layer. The samples are standard normalized prior to the input layer and the input layer is followed by the batch-normalization stage. The 1D-CNN uses 3 convolution layers with 3, 50, and 30 filters of size 100, 10, and 30, respectively. The stride of the first convolution layer of 3 filters was set to 2, while the stride was maintained as 1 for the other two convolution layers. The model considered has a total of 50,909 parameters. This model was referred to as $M_1 1$ in Chap. 6.

B.1.1 Pruning

A popular approach for reducing resource requirements at test time is pruning, which entails systematically removing parameters from an existing network [8]. Here, we carry out weight-pruning, i.e., magnitude-based weight pruning to gradually zero out model weights during the training process to achieve model sparsity. This brings about improvements in model compression and is therefore suitable in resource-constrained environments. We use the model $M_1 1$ and attempt to sparsify the model with sparsity starting at 50% and slowly increase it up to the desired final sparsity level of 80% over 50 epochs. We refer to the pruned model as $M_{1,\,\text{pruned}} 1$.

A. John et al., *Deep Learning and Signal-Processing Methods for Multisensor Data Fusion*, https://doi.org/10.1007/978-3-031-96724-5

Table B.1 Average performance of the models $M_1 1$, $M_{1,\text{pruned}} 1$, and $M_{1,\text{binarized}} 1$ for sleep apnea detection, in terms of accuracy, sensitivity, and specificity

Model	Accuracy (%)	Sensitivity (%)	Specificity (%)
$M_1 1$	99.56	96.05	99.66
$M_{1,\text{pruned}} 1$ at 80%	97.34	86.48	97.64
$M_{1,\text{pruned}} 1$ at 70%	98.49	92.10	98.66
$M_{1,\text{pruned}} 1$ at 60%	99.30	92.54	99.48
$M_{1,\text{pruned}} 1$ at 50%	99.24	94.44	99.37
$M_{1,\text{binarized}} 1$	75.59	63.23	75.92

B.1.2 Binarized CNN

A method to develop neural networks with binary weights and activations at run-time was proposed in [9]. Binarized kernel elements and weights reduce memory size and accesses and substantially improve power-efficiency. In this work, we binarize the kernel parameters in the convolution layers and the weights in the fully connected layers of the model. The activations and batch-normalization layers are not binarized. This model has 50,824 parameters as we do not use bias terms in this model. We refer to the binarized model as $M_{1,\text{binarized}} 1$ (Table B.1).

B.1.3 Performance Analysis

For the pruned model, we study the performance when the final model sparsity is at 50, 60, 70, and 80%. The pruning method reduces the model complexity by increasing the sparsity of the convolutional layers and the fully connected layer to the desired level of sparsity. The performance of the pruned model in terms of accuracy, specificity, and sensitivity over varying sparsity levels is shown in Fig. B.1. As expected,

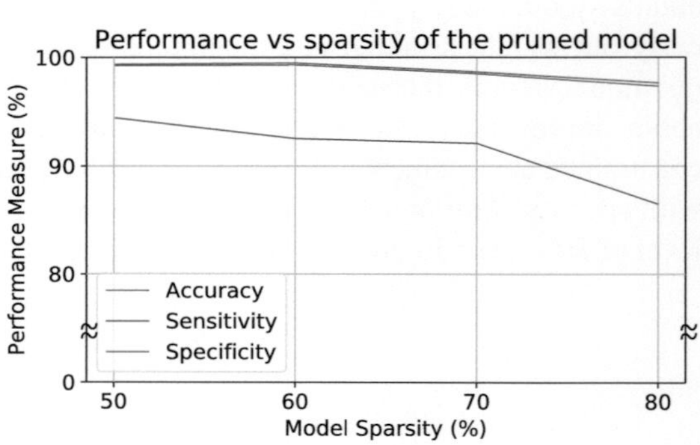

Fig. B.1 Performance of the pruned ECG model on the test set when sparsity is increased from 50 to 80%, for sleep apnea detection

the performance drops with an increase in sparsity levels, and therefore a tradeoff between power/resource consumption and performance is required to determine the optimal sparsity level. The pruned model M2 exhibited an accuracy of 97.34%, a specificity of 97.64%, and a sensitivity of 86.48% on the test set consisting of data from all the patients at the highest sparsity level of 80%.

The binarized model $M_{1,\text{binarized}}1$ was trained over the full training set and simultaneously validated on the validation set for each epoch. The model $M_{1,\text{binarized}}1$ exhibited an accuracy of 75.59%, a specificity of 75.92%, and a sensitivity of 63.23% on the test set consisting of data from all the patients. When compared to the models $M_1 1$ and $M_{1,\text{pruned}}1$, the performance of the binarized model is poor. Further investigations into various combinations of binarized layers and normal layers need to be carried out to achieve the desired performance levels with improvements in power-efficiency.

B.1.4 Complexity Calculation

The model complexity was calculated in terms of the number of multiplications and additions required for sleep apnea detection per second. Prediction with the model $M_1 1$ requires 6534116 multiplications and 6546647 additions. Prediction with the pruned model $M_{1,\text{pruned}}1$ requires the same number of multiplications and additions as $M_1 1$. However, we can approximate the gains due to pruning by calculating the number of multiplications and additions with non-zero numbers. When $M_{1,\text{pruned}}1$ is 80% sparse compared to $M_1 1$ in the convolutional layers and dense layers, we can approximately calculate the number of multiplications required as 1312232 and additions as 1325763. When the sparsity is 70%, the number of multiplications required is 1964936, and additions are 1977467. When the model sparsity is 60%, the model effectively requires 2617676 multiplications and 2630207 additions for prediction. 3270416 multiplications and 3289663 additions are required for prediction when the sparsity is 50%. Prediction with the binarized model $M_{1,\text{binarized}}1$ requires a series of addition operations at the convolution layers and fully connected layer since the weights are either +1 or –1. Therefore, prediction with model $M_{1,\text{binarized}}1$ requires 4766 multiplications and 6301530 additions. The total energy consumption during prediction using $M_1 1$ is found to be 2.55 μJ, $M_{1,\text{pruned}}1$ at 80% sparsity to be 0.52 μJ, $M_{1,\text{pruned}}1$ at 70% sparsity to be 0.77 μJ, $M_{1,\text{pruned}}1$ at 60% sparsity to be 1.03 μJ, $M_{1,\text{pruned}}1$ at 50% sparsity to be 1.28 μJ, and $M_{1,\text{binarized}}1$ to be 0.13 μJ. This is calculated by assuming that the energy required is 0.39 pJ for a 16-bit by 16-bit multiplication accumulation (MAC) operation [10, 11], and approximately 20 fJ for a 16-bit by 16-bit addition operation [12] in 28nm FD-SOI technology. The computational complexity and energy consumption are discussed in Table B.2.

Table B.2 Computational complexity and energy consumption of the sleep apnea detection models during prediction

Model	Net parameters	Multiplications	Additions	Energy (μJ)
$M_1 1$	50909	6534116	6546647	2.55
$M_{1,\,\text{pruned}} 1$ at 80%	10349	1312232	1324763	0.52
$M_{1,\,\text{pruned}} 1$ at 70%	14247	1964936	1977467	0.77
$M_{1,\,\text{pruned}} 1$ at 60%	20489	2617676	2630207	1.03
$M_{1,\,\text{pruned}} 1$ at 50%	25559	3270416	3289663	1.28
$M_{1,\,\text{binarized}} 1$	50824	4766	6301530	0.13

B.2 Pruning and Binarization of the SpO2-Based Sleep Apnea Detection Model

The feasibility of using pruning and binarization techniques on the SpO2-based sleep apnea detection model discussed in Chap. 6 is discussed here. The SpO2-based 1D-CNN model for sleep apnea detection has 88 (8 samples per second) nodes in the input layer and the input layer is followed by a batch-normalization stage. The 1D-CNN uses three convolution layers with 6, 50, and 30 filters of size 25, 10, and 30, respectively. The strides of all the convolutional layers were maintained as 1, and zero padding was used. The model considered has a total of 27,182 parameters. This model was referred to as $M_1 2$ in Chap. 6.

B.2.1 Pruning

Here, we carry out weight-pruning, i.e., magnitude-based weight pruning to gradually zero out model weights during the training process, the same as the pruning method discussed in Sect. B.1.1. We use $M_1 2$ and attempt to sparsify the model with sparsity starting at 50% and slowly increasing it up to the desired final sparsity level over 50 epochs. We refer to the pruned model as $M_{1,\,\text{pruned}} 2$.

B.2.2 Binarized CNN

Binarized kernel elements and weights reduce memory size and accesses and substantially improve power-efficiency [9]. Here, the binarized model has 27,094 parameters

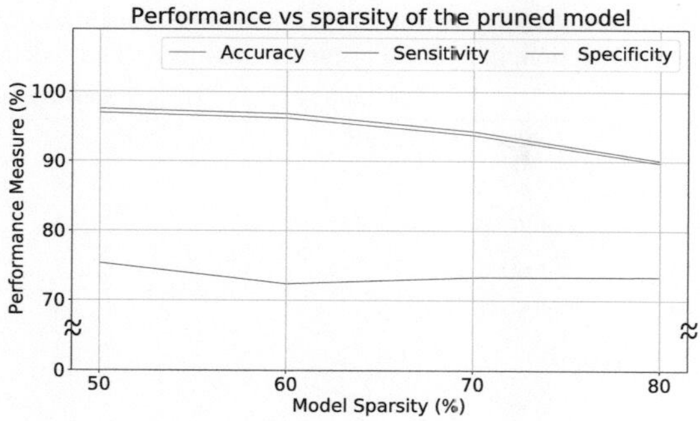

Fig. B.2 Performance of the pruned SpO2 model on the test set when sparsity is increased from 50 to 80%, for sleep apnea detection

as we do not use the bias terms in this model. We refer to the binarized model as $M_{1,\ \text{binarized}}2$.

B.2.3 Performance Analysis

$M_1 2$ was trained over the full training set and simultaneously validated on the validation set for each epoch. Validation callback was used to save the best weights during training based on which set of weights exhibited the highest validation accuracy. $M_1 2$ exhibited an accuracy of 97.08%, a specificity of 97.42%, and a sensitivity of 84.65% on the test set.

For the pruned model, we study the performance when the final model sparsity is at 50, 60, 70, and 80%. The pruning method reduces the model complexity by increasing the sparsity of the convolutional layers and the fully connected layer to the desired level of sparsity. The performance of the pruned model in terms of accuracy, specificity, and sensitivity over varying sparsity levels is shown in Fig. B.2. As expected, the performance drops with an increase in sparsity levels, and therefore a tradeoff between power/resource consumption and performance is required to fix the optimal sparsity level. However, it can be observed that sensitivity drops when model sparsity is increased from 50 to 60%, but sensitivity increases when sparsity is increased to 70 and 80%. It may appear that even though overall performance drops with a drop in accuracy, the weights that make the model highly specific are pruned away with an increase in sparsity, making the model more sensitive. The pruned model $M_{1,\ \text{pruned}}2$ exhibited an accuracy of 89.75%, a specificity of 90.19%, and a sensitivity of 73.39% on the test set consisting of data from all the patients at the highest sparsity level of 80%.

The binarized model $M_{1,\ \text{binarized}}2$ was trained over the full training set and simultaneously validated on the validation set for each epoch. This model exhibited an accuracy of 68.22%, a specificity of 67.94%, and a sensitivity of 78.44% on the test set consisting of data from all the patients. When compared to $M_1 2$ and $M_{1,\ \text{pruned}}2$,

Table B.3 Average Performance of the three models M_12, $M_{1,\,pruned}2$, and $M_{1,\,binarized}2$ for sleep apnea detection, in terms of accuracy, sensitivity, and specificity

Model	Accuracy (%)	Sensitivity (%)	Specificity (%)
M_12	97.08	84.65	97.42
$M_{1,\,pruned}2$ at 80%	89.75	73.39	90.19
$M_{1,\,pruned}2$ at 70%	93.77	73.32	94.32
$M_{1,\,pruned}2$ at 60%	96.21	72.37	96.86
$M_{1,\,pruned}2$ at 50%	96.97	75.36	97.56
$M_{1,\,binarized}2$	68.22	78.44	67.94

the performance of the binarized model is poor. When compared to $M_{1,\,pruned}2$ in terms of sensitivity alone, $M_{1,\,binarized}2$ performs better at the cost of accuracy.

The performance parameters of all the models are detailed in Table B.3. From the results, it can be concluded that M_12 and $M_{1,\,pruned}2$ are suitable for second-by-second sleep apnea detection, and $M_{1,\,binarized}2$ requires further investigation.

B.2.4 Complexity Calculation

The model complexity was calculated in terms of the number of multiplications and additions required for sleep apnea detection per second. Prediction with M_12 requires 2260016 multiplications and 2262876 additions. Prediction with the pruned model $M_{1,\,pruned}2$ requires the same number of multiplications and additions as M_12. However, we can approximate the gains due to pruning by calculating the number of multiplications and additions with non-zero numbers. When $M_{1,\,pruned}2$ is 80% sparse compared to M_12 in the convolutional layers and dense layers, we can approximately calculate the number of multiplications required as 453200 and additions as 456060. When the sparsity is 70%, the number of multiplications required is 679316, and additions are 682176. When the model sparsity is 60%, the model effectively requires 904904 multiplications and 907764 additions for prediction. 1131020 multiplications and 1133880 additions are required for prediction when the sparsity is 50%. Prediction with the binarized model $M_{1,\,binarized}2$ requires a series of addition operations at the convolution layers and fully connected layer since the weights are either +1 or −1. Therefore, prediction with $M_{1,\,binarized}2$ requires 1496 multiplications and 2169946 additions. The total energy consumption during prediction using M_12 is found to be 0.8825 μJ, $M_{1,\,pruned}2$ at 80% sparsity to be 0.1179 μJ, $M_{1,\,pruned}2$ at 70% sparsity to be 0.2660 μJ, $M_{1,\,pruned}2$ at 60% sparsity to be 0.3540 μJ, $M_{1,\,pruned}2$ at 50% sparsity to be 0.0043 μJ, and $M_{1,\,binarized}2$ to be 0.13 μJ. This is calculated by assuming that the energy required is 0.39 pJ for a 16-bit by 16-bit multiplication accumulation (MAC) operation [10, 11], and approximately 20 fJ for a 16-bit

Table B.4 Computational Complexity and Energy consumption of the sleep apnea detection models during prediction

Model	Parameters	Multiplications	Additions	Energy (μJ)
$M_1 2$	27182	2260016	2262876	0.8825
$M_{1,\text{ pruned}} 2$ at 80%	5537	453200	456060	0.1779
$M_{1,\text{ pruned}} 2$ at 70%	8242	679316	682176	0.2660
$M_{1,\text{ pruned}} 2$ at 60%	10948	904904	907764	0.3540
$M_{1,\text{ pruned}} 2$ at 50%	13653	1131020	1133880	0.4422
$M_{1,\text{ binarized}} 2$	27094	1496	2169946	0.0043

by 16-bit addition operation [12] in 28nm FD-SOI technology. The computational complexity and energy consumption are discussed in Table B.4.

Appendix C
Generalized Multi-level Fusion Architecture

A generalized version of the multi-level fusion architecture introduced in Chap. 7 is discussed here. To reiterate, consider the fusion of $N + 1$ multi-sensor inputs $S_k \forall k \in [0, \cdots, N]$, with their corresponding signal quality indicators $Q_k \forall k \in [0, \cdots, N]$. Let us consider models for the signal inputs $M_k \forall k \in [0, \cdots, N]$ and for the corresponding SQIs as $Ms_k \forall k \in [0, \cdots, N]$. The initial input to the central network is calculated as

$$h^0_{C_{init}} = \sum_{k=0}^{N} \lambda_k h^{-1}_{C_k} \tag{C.1}$$

where

$$h^{-1}_{C_k} = tanh[conv[S_k]] \tag{C.2}$$

and λ_k are trainable weights. The first hidden feature set is computed as

$$h^0_C = Block_{C_0}[h^0_{C_{init}}] \tag{C.3}$$

The first set of fused features is calculated as

$$h^1_{C_{init}} = \alpha_{C_0} h^0_C + (1 - \alpha_{C_0}) \sum_{k=0}^{N} g^0_{k,j} h^0_{k,j} \tag{C.4}$$

where α_{C^0} is a trainable parameter, j is used to indicate that the multiplications are carried out element wise, $h^0_{M_{k,j}}$ corresponds to the hidden features obtained from the first computational block of network M_k (the neural network model for signal S_k), as shown here:

$$h^0_k = Block_{M_{0,k}}[S_k]. \tag{C.5}$$

The term $g^0_{k,j}$ is obtained after a normalization stage as shown:

A. John et al., *Deep Learning and Signal-Processing Methods for Multisensor Data Fusion*, https://doi.org/10.1007/978-3-031-96724-5

$$g^0_{k,j} = \frac{g^0_{k,j,init}}{\sum_{k=0}^{N} g^0_{k,j,init}} \tag{C.6}$$

and $g^0_{k,j,init}$ corresponds to the hidden features obtained from the first computational block of network Ms_k (the neural network model for the SQI signal Q_k), as shown here:

$$g^0_{k,j,init} = Block_{Ms_{0,k}}[Q_k]. \tag{C.7}$$

Here, j indicates that the multiplications are carried out element-wise. After Eq. (7.5), the final hidden features of the central network are then calculated as

$$h^1_C = Block_{C_1}[h^1_{C,init}]. \tag{C.8}$$

It should be noted that for Eq. 7.5 to perform as expected, the outputs from $Block_{M_{0,k}} \forall k \in [0, \cdots, N]$, $Block_{Ms0,k} \forall k \in [0, \cdots, N]$, and $Block_{C_0}$ must produce outputs of the same dimensions for sample-wise fusion. Further to this, the following

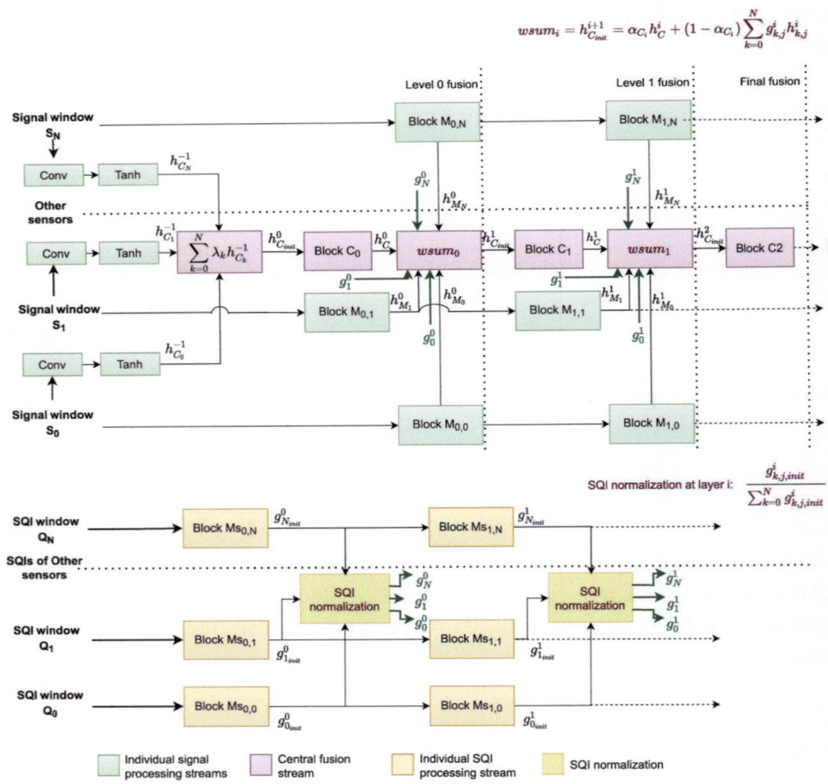

Fig. C.1 Multi-level fusion architecture with signal quality indicators for fusion multi-sensor inputs

computations are carried out for every subsequent layer denoted by i, and T is the number of the final computational block/layer. The following equations are generalized versions of the network equations discussed above, where i indicates the layer number.

$$g^{i}_{k,j,init} = Block_{Ms_{i,k}}[g^{i-1}_{k,j,init}] \quad \forall i \in [1, \cdots, T]. \tag{C.9}$$

$$g^{i}_{k,j} = \frac{g^{i}_{k,j,init}}{\sum_{k=0}^{N} g^{i}_{k,j,init}} \quad \forall i \in [1, \cdots, T]. \tag{C.10}$$

$$h^{i}_{k} = Block_{M_{i,k}}[h^{i-1}_{k}] \quad \forall i \in [1, \cdots, T]. \tag{C.11}$$

$$h^{i}_{C_{init}} = \alpha_{C_{i-1}} h^{i-1}_{C} + (1 - \alpha_{C_{i-1}}) \sum_{k=0}^{N} g^{i-1}_{k,j} h^{i-1}_{k,j} \quad \forall i \in [1, \cdots, T]. \tag{C.12}$$

$$h^{i}_{C} = Block_{C_i}[h^{i}_{C,init}] \quad \forall i \in [0, \cdots, T]. \tag{C.13}$$

The generalized flow diagram of the fusion architecture with $N + 1$ number of sensors is as shown in Fig. C.1.

References

1. Y.A. Gerhana et al., The implementation of K-nearest neighbor algorithm in case-based reasoning model for forming automatic answer identity and searching answer similarity of algorithm case, in *2017 5th International Conference on Cyber and IT Service Management (CITSM)* (2017), pp. 1–5
2. S. Ullah, S. Rehman, B.S. Prabakaran, F. Kriebel, M.A. Hanif, M. Shafique, A. Kumar, Area-optimized low-latency approximate multipliers for FPGA-based hardware accelerators, in *2018 55th ACM/ESDA/IEEE Design Automation Conference (DAC)* (June 2018), pp. 1–6
3. I. Xilinx, Xilinx Logicore IP Multiplier v11.2. (2011)
4. G. Clifford et al., Signal quality indices and data fusion for determining clinical acceptability of electrocardiograms. Physiol. Meas. **33**, 1419–33 (2012)
5. Q. Li, C. Rajagopalan, G.D. Clifford, A machine learning approach to multi-level ECG signal quality classification. Comput. Methods Programs Biomed. **117**(3), 435–447 (2014)
6. I. Silva, G.B. Moody, L. Celi, Improving the quality of ECGs collected using mobile phones: the PhysioNet/computing in cardiology challenge 2011, in *2011 Computing in Cardiology* (Sept. 2011), pp. 273–276
7. H. Xia, G.A. Garcia, J.C. McBride, A. Sullivan, T. De Bock, J. Bains, D.C. Wortham, X. Zhao, Computer algorithms for evaluating the quality of ECGs in real time, in *2011 Computing in Cardiology* (2011), pp. 369–372
8. D. Blalock, J.J.G. Ortiz, J. Frankle, J. Guttag, What is the state of neural network pruning? (2020)
9. M. Courbariaux, I. Hubara, D. Soudry, R. El-Yaniv, Y. Bengio, Binarized neural networks: training deep neural networks with weights and activations constrained to +1 or –1 (2016)

10. H. Reyserhove, N. Reynders, W. Dehaene, Ultra-low voltage datapath blocks in 28nm UTBB FD-SOI, in *2014 IEEE Asian Solid-State Circuits Conference (A-SSCC)* (Nov. 2014), pp. 49–52
11. R. Taco, I. Levi, M. Lanuzza, A. Fish, An 88-fJ/40-MHz [0.4 V]– 0.61-pJ/1-GHz [0.9 V] dual-mode logic 8 × 8 bit multiplier accumulator with a self-adjustment mechanism in 28-nm FD-SOI. IEEE J. Solid-State Circuits **54**, 560–568 (2019). (Feb)
12. R. Taco, I. Levi, M. Lanuzza, A. Fish, Evaluation of dual mode logic in 28nm FD-SOI technology, in *2017 IEEE International Symposium on Circuits and Systems (ISCAS)* (2017), pp. 1–4

Index

A
Abdominal movement, 5, 36, 170
Accelerometer, 16, 18, 19
Accuracy, 1, 29, 70
Adam, 163, 166, 168, 170, 203
Adaptive boosting, 226
Adaptive filtering, 218
Adaptive linear filter, 19
Adaptive nonharmonic model, 36
Adder, 75, 177
Alarm fatigue, 2, 119, 161
Amplitude modulation, 19, 33
Apparent noise variance, 95
Approximate coefficients, 123
Approximate multiplier, 221
Approximate multipliers, 77, 221, 224
Approximate partial products, 221
Apriori probability distribution, 92
Area under the precision-recall curve, 37
Arrhythmia, 10, 11, 38, 40, 191
Arrhythmia detection, 26
Arterial blood pressure, 18
Arterial oxygen saturation, 37
Artificial intelligence, 2
Artificial neural networks, 14, 15
Associated beliefs, 14
Asystole, 40
Atrial fibrillation, 3, 5, 7, 10, 11, 42, 191
Atrial fibrillation detection, 26
Atrial flutter, 44, 192
Autocorrelation, 34
Autoencoder, 24
Autonomous driving, 10, 22
Autoregression, 34
Average loss, 204

B
Backward shortcut connections, 37
Bagging, 68, 69
Ballistocardiogram, 25, 31
Baseline modulation, 19, 33
Baseline wander, 84, 120, 122, 126, 129, 130, 133, 140
Batch-normalization, 163, 166, 227
Bayes-eliminated fusion, 97
Bayes-eliminated Kalman fusion, 107
Bayesian elimination, 85, 94, 97, 106, 115
Bayesian fusion, 20, 31
Bayesian inference, 15
Bayesian probability, 66
Bayes probabilities, 92, 94, 107, 108, 115
Bayes theorem, 65
Beats-per-minute, 20, 146, 148, 149
Behavioral knowledge framework, 11
Behavioral rules, 11
Best Bayes, 94, 106
Best SQI, 94, 106, 107
Bidirectional gated recurrent units, 41
Bidirectional long short-term memory, 35, 41
Binarization, 165, 167, 227, 230
Binary classification, 198
Binary cross-entropy, 163, 166, 170, 203, 204
Biomedical signal processing, 16

A. John et al., *Deep Learning and Signal-Processing Methods for Multisensor Data Fusion*, https://doi.org/10.1007/978-3-031-96724-5

Biorthogonal 6.8 wavelet, 123
Bland Altman plot, 108–110, 112, 148, 149
Blind deconvolution, 24, 26
Blood pressure, 16, 120
Boosting, 68, 69
Bootstrap aggregation, 69
BP-derived respiration, 34
Bradycardia, 40
Breaths-per-minute, 34

C
Capacitive ECG, 20, 28, 31
Cardiac vibration signals, 31
Case-based reasoning, 40
Catastrophic fusion, 25–27, 158
Central limit theorem, 134, 169
Central loss, 204
CentralNet, 156, 190, 193
Cepstrum, 31
Channel-wise attention, 41
Chi-squared distribution, 92
Chromatograph, 192
Class covariance inverse, 77
Cochleogram, 18
Common representational format, 4, 116, 120, 190, 214
Competitive fusion, 16
Complementary fusion, 16, 74, 138, 149
Computed tomography, 23
Condorcet Jury Theorem, 68
Constant Q-transform, 18
Continuous-valued annotation, 31
Contourlet transform, 23
Convolutional neural networks, 5, 28, 37, 191
Cooperative fusion, 16, 74
Correlation coefficient, 32
Cross-validation, 60, 222
Curve SQI, 124, 199
Cutoff frequency, 86

D
DAI-DAO, 19
DAI-FEO, 19
Dasarathy model, 19, 46
Data-driven, 5, 158, 185, 190, 208, 209
Data-driven fusion, 157, 160
Data fusion, 1, 9, 156
Data in-data out, 13, 19
Data in-feature out, 13, 19
Decision in-decision out, 13, 19
Decision-level fusion, 5, 13, 18, 27, 29, 59, 70, 84, 189
Decision tree, 40, 63, 226
Deconvolutional single-shot detector, 22
Deep learning, 5, 190
Defense, 10
DEI-DEO, 20
Dempster-Shafer theory, 15, 23
Detail coefficients, 123
Discrete stationary wavelet transform, 120, 122, 148
Discrete wavelet transform, 123
Dropout, 159, 163, 166
Durrant-Whyte model, 16, 46
Dynamics-level fusion, 13

E
Early-stage fusion, 190
ECG-derived respiration, 34, 161
Ego-localization and mapping, 22
Elastic net logic, 44, 192
Electrocardiogram, 4, 84, 190, 191, 227
Electrode motion, 2, 84, 113, 119, 122, 125, 130, 132, 133, 140, 161, 169
Electroencephalogram, 18, 162
Electromyogram, 18
Electrooculogram, 18, 120
Emotion recognition, 23
Encoder-decoder, 22
Energy-delay-product, 221
Ensemble, 36, 68, 77, 79, 170, 192
Ensemble empirical mode decomposition, 34
Environmental monitoring, 10
Error rate, 35
Euclidean distance, 64, 76, 221
Event detection, 5
Evidence-level fusion, 13
Exhaustive search, 71
Explainability, 209, 217
Explainable AI, 217, 218
Extended Kalman filter, 22
EXtreme gradient boosting, 41, 69

F
F_β-score, 164
F6-score, 165, 169, 172, 173, 176, 185
False alarms, 2, 3, 84, 119
False alarm suppression rate, 42

Faster R-CNN, 22
Fast R-CNN, 22
Feature extraction, 2
Feature fusion, 31
Feature in-decision out, 13, 19
Feature in-feature out, 13, 19
Feature-level fusion, 4, 5, 13, 18, 29, 36, 37, 189
Federated learning, 217, 218
FEI-DEO, 20, 28, 36, 37
FEI-FEO, 20
Force sensors, 31
Fourier transform, 18, 34
Frequency modulation, 19, 33
Frequency SQI, 88
F1-score, 42, 70, 142
F SQI, 59, 89
Fully depleted silicon on insulator, 77, 177
Fusion framework, 4, 10, 213
Fuzzified rule-based method, 20
Fuzzy logic, 14, 58
Fuzzy probabilities, 20
Fuzzy probability distribution, 20
Fuzzy structural algorithm, 36, 161

G
Gaussian radial basis kernel, 65
Generic specialists, 13
Gini's diversity index, 64
GQRS, 87, 94, 121, 138, 148, 149
Gradient boosting, 69

H
Healthcare, 10
Heartbeat, 3, 10, 11, 19, 119, 151
Heartbeat detection, 6, 26, 121, 150
Heart rate, 3, 5, 11, 83, 84, 120, 125
Heart rate estimation, 26, 120
Heart rate variability, 34, 120, 161
Heuristics, 14
Hidden representation, 193
Hierarchical framework, 14
High-frequency noise, 122, 140
Hilbert transform, 38
Hjorth descriptor, 34
HR, 18
Human activity recognition, 23
Human-in-the-loop AI, 219
Human-robot interaction, 22
Hyperbolic tangent activation, 68
Hyperparameters, 71, 158, 198
Hyperplane, 65

I
Image-based PPG, 28
Image classification, 156
Image fusion, 13, 14
Image stitching, 17
Impulse response, 123
Information abstraction level, 5, 6, 13, 190, 193, 209, 216
Information needs, 14
Instantaneous signal quality, 85
Instantaneous signal quality assessment, 200
Instantaneous signal quality indicator, 121, 133
Internet of things, 2, 57

J
Joint Directors of Laboratories, 11

K
Kalman filter, 15, 22, 24, 34, 38, 46, 95
Kalman filter residuals, 31
Kalman fusion, 95, 107, 112, 115, 116, 214
k-dimensional tree search, 71
k-nearest neighbor, 40, 63, 221
Knowledge matrix, 14
KSQI, 61, 62
Kurtosis, 34, 61, 88, 224

L
Late fusion, 29
Latent ensemble layer, 156
Leaky ReLU, 198, 203
Learning rate scheduler, 163
Left ventricular failure, 20
Lempel-Ziv, 58
Light amplification by stimulated emission of radiation, 22
Light detection and ranging, 21
Linear discriminant, 40, 63
Linear programming, 14
Linear regression, 34
Local weighted linear prediction, 26
Long short-term memory, 22, 37, 161, 170, 192
Look-up tables, 62, 221, 222

Low-frequency noise, 122, 140
Low-level specialists, 13
Luo and Kay classification, 16, 46

M
Machine learning, 5, 15, 20, 63, 192, 213
Magnetic resonance imaging, 23
Markov model, 192
Matched network structures, 159
Mean absolute error, 32
Mean error, 32
Medical imaging, 23
MIMIC III waveform database, 122, 126, 130, 138, 197
MIT-BIH arrhythmia database, 84, 86, 98, 115
MIT-BIH atrial fibrillation database, 197
MIT-BIH noise stress test database, 86, 122, 140
MLBF-Net, 41
Modalities, 4, 5, 213
Mother wavelet, 123
Motion artifacts, 2, 3, 10, 88, 119, 120, 125, 140, 161, 169, 216
Moving sample mean, 92
Moving sample variance, 92
Multi-intelligence, 14
Multi-level fusion, 5, 192, 193, 205, 208, 214, 235
Multimodal, 6, 116, 120, 121
Multimodal data association, 19
Multimodal fusion, 149, 151, 156
Multi-model fusion, 79
Multiplication accumulation, 177
Multiplier accumulator, 77
Multipliers, 75, 221
Multiresolution fusion, 157, 182, 217
Multi-sensor fusion, 1, 156, 158
Muscle artifacts, 84, 108, 122, 126, 130, 133, 140, 148

N
Naïve Bayes, 63
Neural networks, 23, 34, 35, 63, 68, 79, 193

NF SQI, 59, 60, 78, 79
Noise variance, 96
Non-parametric classifier, 64
Normal sinus rhythm, 38, 44

O
Object detection, 156
Objective criterion, 15
Omnibus model, 14
1-dimensional convolutional neural network, 5, 156, 227
1-dimensional time-series, 11, 16, 216
Overfitting, 73, 156, 159, 173, 177, 185

P
Padding, 5, 156
Pan-Tompkins algorithm, 26, 137, 150
Parametric models, 78
Particle filtering, 15, 18, 22, 31, 84
Pattern recognition, 5
Perception maps, 22
Peripheral oxygen saturation, 5, 10, 36, 161, 230
Personalized medicine, 7
Photoplethysmogram, 4, 120, 191
Physiological signals, 1
Physionet/ CinC 2011 challenge, 57, 58, 85, 225, 226
Physionet/ CinC 2014 challenge, 26, 84, 120, 122
Physionet/ CinC 2015 challenge, 38
Physionet/CinC 2017 challenge, 59
Pixel-level fusion, 13
Poincare plot, 192
Polysomnography, 36, 160, 162
Positive predictive value, 29, 70, 142
Possibility theory, 13
Power-line interference, 120, 122, 126, 129, 130, 133, 140, 161
Power spectral density, 34
PPG-derived respiration, 34
Precision, 29, 70
Premature ventricular contraction, 41
Prior probabilities, 20, 66
Probabilistic model, 31, 84
Probabilistic network-based knowledge, 14
Pruning, 157, 165, 167, 179, 185, 214, 227, 230
Pulmonary oedema, 20
Pulse rate variability, 34
Pulse transit time, 34, 122

Q
QRS, 18, 26, 31, 58, 59, 79, 120
Quasi-periodic, 121, 124, 125, 133

R
Radio detection and ranging, 22
Random forest, 40

Recall, 29
Recurrent neural networks, 22, 37, 170
Redundant fusion, 16
Region-based CNN, 22
Regularization, 159, 176
Reinforced learning, 13
Relative error, 32
ReLU activation, 163, 166
Resampling, 5, 156
Resource management process, 15
Respiration quality index, 33
Respiration rate, 10, 11, 19, 32
Respiration rate estimation, 26, 120
Respiratory airflow, 161
Respiratory modulation signals, 32
Respiratory sinus arrhythmia, 34
Responsibility, 219
Right ventricular failure, 20
RMSprop, 163
Robotics, 10
Root mean square error, 29, 113
Rule-based association models, 19

S
Sample variance, 92
Sampling frequencies, 5
Seismocardiogram, 35
Self-improving multi-sensor fusion, 13
Self-learn, 190, 191, 196, 209
Semantic object specialists, 13
Sensitivity, 29, 70, 121, 142, 143, 169, 173, 175
SensorNet, 156
Sensor supervisor, 13
Sequential Monte Carlo, 15
Sigmoid activation, 203
Signal-based selective dropout, 156, 159, 160, 173, 177, 183, 185, 214
Signal-level fusion, 4, 5, 13, 29, 121, 189
Signal purity index, 34
Signal quality indicators, 3, 4, 10, 25, 28, 58, 62, 85, 88, 123, 124, 150, 184, 194, 213, 235
Signal-to-noise ratio, 98, 112, 113, 123, 214
Simultaneous localization and mapping, 22
Single-input multi-output system, 26
Single-shot multibox detector, 22
Skewness, 61, 224
Sleep apnea, 3, 5, 6, 10, 11, 36, 157, 160–162, 170, 227
Sleep apnea detection, 26
Smartwatches, 120
Snowballing, 11
Soft computing, 14
Softmax activation, 167, 198
Sound navigation and ranging, 22
Sparsity, 227
Spatial pyramid pooling net, 22
Specificity, 70, 169, 173, 176
Spectrogram, 18
SQRS, 26, 87, 94, 121, 138, 148, 149
SSQI, 61, 62
Stable classifiers, 68
Standard normalization, 63, 163
State estimation, 24
State vector, 95
Stationary wavelet transform, 23
Stochastic gradient descent, 163
Stroke volume, 26, 120
Structural health monitoring, 10
Supervised learning, 63, 65
Support vector machine, 20, 23, 40, 58, 63
Support vectors, 65, 77
Supraventricular ectopic beat, 20
Symmlet 8, 123
System-on-a-chip, 58, 71

T
Tachycardia, 40
Taylor series approximations, 75
Teager-Kaiser energy operator, 27, 84
Temporal fusion, 18, 19, 148
Thoracic volume displacement, 19
Time-frequency images, 17
Time-series images, 37
Transfer learning, 158
Transform-domain fusion, 121
Transform layer, 198
Translation-invariant, 123
Tucker-decomposed images, 41
2-dimensional convolutional neural networks, 156

U
Uncertainty modeling, 218
U-Net, 34
Unified Modeling Language, 14
Unimodal, 4, 87, 116
Unstable classifiers, 68
Unsupervised Bayesian framework, 84

V
Validation callback, 164, 166, 168, 170, 203, 231
Ventricular ectopic beat, 20
Ventricular flutter fibrillation, 41
VGG net, 192
Vital signs, 83

W
Waterfall model, 14
Wavelet, 18, 24, 121, 138, 170, 214
Weak learners, 77
Wearable devices, 1
WFDB toolbox, 87, 121, 138
Wireless body sensor networks, 23
World model, 11
WQRS, 26, 121, 138, 148, 149

Y
You only look once, 22

MIX
Papier aus verantwortungsvollen Quellen
Paper from responsible sources
FSC® C105338

If you have any concerns about our products,
you can contact us on
ProductSafety@springernature.com

In case Publisher is established outside the EU,
the EU authorized representative is:
Springer Nature Customer Service Center GmbH
Europaplatz 3, 69115 Heidelberg, Germany

Printed by Libri Plureos GmbH
in Hamburg, Germany